The Ecology of Malaria Vectors

The Ecology of Malaria Vectors

Jacques Derek Charlwood

CRC Press is an imprint of the
Taylor & Francis Group, an **informa** business

CRC Press
Taylor & Francis Group
6000 Broken Sound Parkway NW, Suite 300
Boca Raton, FL 33487-2742

Printed in Canada on acid-free paper

International Standard Book Number-13: 978-0-367-24848-2 (Paperback)
International Standard Book Number-13: 978-0-367-24860-4 (Hardback)

Visit the Taylor & Francis Web site at
http://www.taylorandfrancis.com

and the CRC Press Web site at
http://www.crcpress.com

'We aim for simplicity and hope for truth'

Nelson Goodman

Dedication

This day relenting God hath placed within my hand a wondrous thing;
and God be praised.
At his command seeking his secret deeds
With tears and toiling breath,
I find thy cunning seeds,
O million-murdering Death.
I know this little thing a myriad men will save.
O Death where is thy sting?
Thy victory, O Grave?

In addition to being a scientist, Ronald Ross considered himself a poet. He wrote this the night he first saw oocysts of the malaria parasite on the stomach wall of two 'dapple-winged' mosquitoes. After the amount of work that he'd done trying to find them, I think I too might have been driven to poetry! Ross, like his mentor Manson, believed that the parasites were liberated and the mosquito died while laying eggs in water and that it was through drinking contaminated water that the disease was transmitted. How wrong he was…

For my friend, inspiration and my own mentor

Tony Wilkes

Contents

Foreword

Whilst the author traces his firm intention to finish his book on *The Ecology of Malaria Vectors* to a moment lazing on an island in the Mekong in Cambodia, it was something his friends and colleagues had already witnessed coming into existence for the past 30 years. Working with Derek, we had the pleasure and privilege to enjoy, live and experience the ecology of vectors and their respective complexities every day in the field, at the bench, and in the laboratory. The book we enjoy today in print was not only written, but truly *lived* by a wonderfully curious scientist and an exceptional personality. Humbly, he shows how 'life is a journey and not a destination' and clearly it is not about creating great things, but mainly about how one sees and does the ordinary things with conviction, appreciation and joy of, and for, their basic intrinsic values.

In 1988, when Derek Charlwood had just joined the Kilombero Malaria Programme in Ifakara, Tanzania—long after his initiation as entomologist at the University of Sussex and after his fieldwork in The Gambia, Papua New Guinea and Brazil—we celebrated the 10-year jubilee of the Tanzanian National Institute of Medical Research in Dar es Salaam. We all gave talks. Wearing his usual *kikoi* (a Tanzanian sarong) from the local weaving school in Ifakara and his dogtooth necklace (from Madang, Papua New Guinea), sporting shoulder-length hair, Derek cut an eccentric figure (no matter what shirt he wore!). It was like this that he gave his talk. He presented just a single overhead (a hastily drawn mosquito). Not everyone was impressed but for those who listened he gave a tour through the ecology of the vectors in the Kilombero Valley, one of the most malarious areas in the world. His entomological insight has been enriched since then. Derek has worked in São Tomé and Príncipe, Mozambique, Ghana, Cambodia and elsewhere in Tanzania since then. Not forgetting Eritrea, the trigger for writing this book being the preparation of short courses there.

Necessity dictated that he used many of his own works there (indeed, it became a challenge to include them all), because the Internet was unavailable. This, however, means that it is more than a textbook—it is Derek's journey as an entomologist and as a committed, environmentally and community-sensitive, modern, vibrant 'naturalist'. It is also a never-ending source of scientific inspiration as well as a call to action towards effective public health actions for malaria control and elimination. We are harmonically carried from inventions and innovations, over most careful validation in different settings, towards the application of scientific research into public health practice. You are not just reading a book: you have the privilege to live and work with Derek Charlwood across different systems and cultures, and to realize how much we do not live in the first, second or third world, but only in *one world* where no part can live and survive without the remainder.

Marcel Tanner
Professor, PhD, Epidemiologist, MPH

Preface

I was lucky to start my career as an accidental entomologist at the University of Sussex with the legendary Mick Gillies (still the doyen of medical entomologists) and his sidekick Tony Wilkes at the Mosquito Behaviour Unit (MBU – Gillies's joke because *mbu* in Kiswahili means mosquito). I have spent the large part of my career trying, unsuccessfully, to disprove their work. During my PhD studies, I was also lucky to undertake a limited amount of fieldwork in the Gambia. On arrival in Banjul I had an epiphany. I realised that studying the mosquitoes in the field was what I wanted to do. I have been fortunate in that I have managed to do this for much of my career. The life of a medical entomologist is more perspiration than inspiration. I have managed to work in many places where 'the hand of man rarely sets foot' (Figure 1) and despite being 'a bear of little brain', because not much had previously been done in these places it was easy to develop a reputation of sorts, if not a career.

The original motivation for the book was preparation for a short course on malaria vectors that I gave at the College of Health Sciences in Asmara, Eritrea. There was no internet available and so I had to rely on the information available on my computer. The book was amplified while lazing on an island in the Mekong, Cambodia, at my friend Fred's house, and further additions have been made during a more recent stint in Asmara.

My profound thanks go to my friends and former students, especially João Pinto, and Carla Sousa, and Corey LeClair for their input and help during the preparation of the book. Also, Nick White, Janet Hemingway, Hilary Ranson, Duane Gubler, Greg Lanzaro and George McGavin for accepting that I liberally use their own work in this offering. I would also like to express my thanks to my sister Dominique and brother-in-law, Jonathan, for their support and to Alan Bullion and Rija Tariq of Informa UK for seeing the initial effort through to completion and Alice Oven for her guidance in producing the final product. Also, of course, my other friends and colleagues, especially Tom Smith, Marcel Tanner, Pete Billingsley, Olivier Briet Virgilio do Rosario, Erling Pedersen and Jason Pitts, who have helped and encouraged me in my work over the years. And, of course, Elsa (entomological assistant extraordinaire) and Nelma for their support and love over the years. Mistakes and errors are mine alone. … if I don't know I don't know, I think I know (Figure 1).

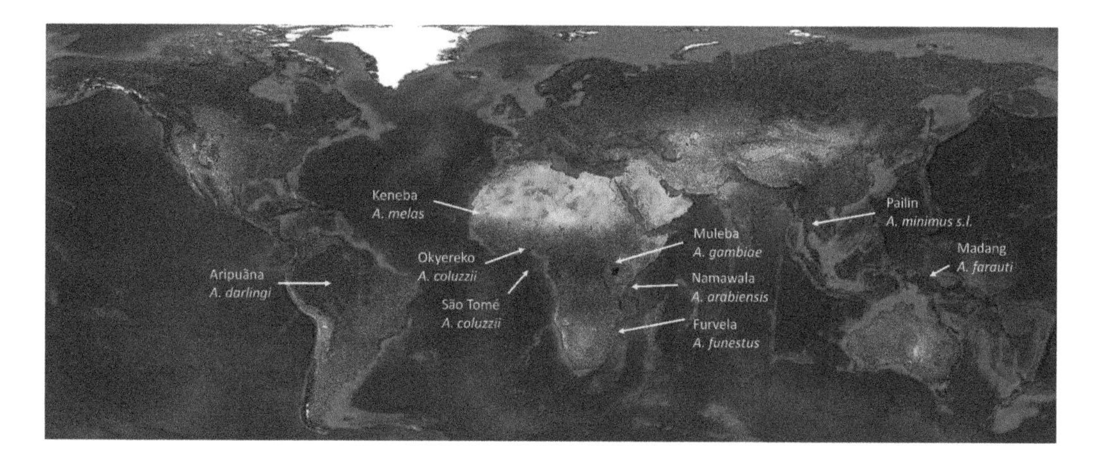

Figure 1 Map showing some of the locations where the author has worked, with the principal vectors species, mentioned in the text.

Author

Jacques Derek Charlwood is an honorary fellow of Global Health and Tropical Medicine, IHMT, Lisbon, Portugal, and an honorary fellow of the Liverpool School of Tropical Medicine, UK. He has 43 years of experience working in the field. He has lived with the Yanomama in Brazil, the Kukuku and Amele in Papua New Guinea, the Maasai and Wandamba in Tanzania, the Phnong in Cambodia, Monco and Forro in São Tomé and Príncipe, among others. He has authored more than 100 scientific papers. His scientific friends think he is a good photographer.

CHAPTER 1

Classification and Systematics

The following comes mainly from the 'Introduction to Insects' by George C. McGavin in *The Insects: Structure and Function, Fifth Edition:*

> It is, I think, both informative and entertaining. If you haven't thought much about the importance of insects before, you may do so after reading this.
>
> **Stephen J. Simpson and Angela E. Douglas**

The ancestor of the *Arthropoda* was in all probability a segmented worm-like marine creature that lived in oceans during the late Precambrian era. By the early to mid-Cambrian (540–520 million years ago) the early arthropods had already evolved into a range of clearly recognisable groups with distinct body plans. Arthropods are characterised by a number of features: the possession of a periodically moulted, chitinous cuticle that acts as a rigid exoskeleton for the internal attachment of striated muscles; segmental paired legs; and the aggregation and/or fusing of body segments into discrete functional units, of which the most universal is the head. The exoskeleton is the key to their successes and limitations. It has several functions – to support the body, protect the animal from the external environment, reduce water loss, store energy and (through the development of limbs and wings) assist in locomotion.

Flexible cuticle between sections of the limbs and body segments forms joints and allows movement by muscles attached to the cuticle. The exoskeleton also extends into the gut (foregut and hindgut) and lines the tracheae, the tubes used in respiration by the myriapods and insects. The combination of hardened plates with soft membranes between them gives both strength and rigidity to the body as well as flexibility. Thus, the construction of an arthropod as a series of jointed tubes gives it considerable mechanical advantage; the skeleton has greater power of resistance to bending than the endoskeleton of vertebrates. For the same cross-sectional area of muscle and skeleton, a solid endoskeleton would be nearly three times weaker than a hollow exoskeleton. Similarly, to have the same strength as an exoskeleton, an endoskeleton would need to be considerably thicker, leaving little additional space for musculature. Because of its mechanical efficiency, and being composed of remarkably flexible material, the exoskeleton of arthropods has been expressed in an astonishing range of body forms and structures unrivalled anywhere in the animal kingdom. However, the presence of an exoskeleton has a number of limitations: it puts a maximum limit on the physical size of an organism; it limits growth, thereby necessitating the shedding of the outer covering for any increase in size; and the skeleton needs to be perforated with sensilla to monitor the outside world. Growth can only be achieved in organisms with a hardened outer covering by the process of moulting. The inner layers of the cuticle are digested by a series of enzymes to separate the old 'skin' from the newly formed cuticle. After moulting, the arthropod swallows water or air to inflate its flexible cuticle until this hardens. The process of moulting is controlled by a complicated interplay between special hormones. The process of respiration has been solved in different ways by the various groups of arthropods. Insects and myriapods have small tubes, tracheae, through which oxygen diffuses to all parts of the body. The physical constraints governing the diffusion process are such that an increase in size of an insect or myriapod is not accompanied by a proportional increase in the rate at which oxygen is delivered to the tissues; this is one of the principal factors governing the maximum size of insects. However, respiration via tracheae can be very efficient and is able to operate with a very small difference in the partial pressure of gases between the tissue end of the tracheae and the outside atmosphere. Tracheal respiration delivers oxygen to insect muscle, which is the most active tissue in the animal kingdom.

A little over 1.5 million species of living organism have been scientifically described to date. The vast majority (66%) are arthropods such as crustaceans, arachnids, myriapods and insects. Insects represent 75% of all animals, and one insect order – the beetles (Coleoptera) – is famously species-rich (J. B. S. Haldane once famously said 'If God loved life he loved insects and if he loved insects he loved beetles'). One thing is clear, however – the full extent of Earth's biodiversity remains a mystery. From attempts more than 30 years ago to estimate the number of extant species to the present day, we still only have a rough idea of how many species live alongside us. Estimates range from as few as 5 million to perhaps as many as 10–12 million species. The task of enumerating them may become substantially easier as the loss and degradation of natural habitats, especially the forests of the humid tropics, continues unabated. It is certain that the majority of insect species will become extinct before they are known to science.

Insects are the dominant multicellular life form on the planet, ranging in size from minute parasitic wasps at around 0.2 mm to stick insects measuring 35 cm in length. Insects have evolved diverse lifestyles and although they are mainly terrestrial, there are a significant number of aquatic species. Insects have a versatile, lightweight and waterproof cuticle, are generally small in size and have a complex nervous system surrounded by an effective blood–brain barrier. Insects were the first creatures to take to the air and have prodigious reproductive rates. These factors, together with the complex interactions they have with other organisms, have led to their great success both in terms of species richness and abundance. The very high diversity of insects today is the result of a combination of high rates of speciation and the fact that many insect taxa are persistent – that is, they show relatively low rates of extinction. In comparison to insects, vertebrate species make up less than 3% of all species. As herbivores, vertebrates are altogether out-munched by the myriad herbivorous insects. In tropical forests, for example, 12%–15% of the total leaf area is eaten by insects as compared with only 2%–3% lost to vertebrate herbivores. Termites remove more plant material from the African savannahs than all the teeming herds of wildebeest and other ungulates put together. Vertebrates also fail to impress as predators. Ants are the major carnivores on the planet, devouring more animal tissue per annum than all the other carnivores. In many habitats ants make up one-quarter of the total animal biomass present. Indeed, it is estimated that the biomass of ants on the planet is the same as that of humans. Insects pollinate the vast majority of the world's 250,000 or so species of flowering plant. The origin of bees coincides with the main radiation of the angiosperms approximately 100 million years ago, and without them there would be no flowers, fruit or vegetables. At least 25% of all insect species are parasites or predators of other insect species. Insects are also important in nutrient recycling by disposing of carcasses and dung. Insects are the principal food source for many other animals. Virtually all birds and a large number of other vertebrates feed on them. An average brood of great tit chicks will consume around 120,000 caterpillars while they are in the nest and a single swallow chick may consume upwards of 200,000 bugs, flies and beetles before it fledges.

Insects are also nutritious. They are often a prized part of the diet in many cultures. In Papua New Guinea, the larvae of wood-boring longhorn beetles are barbecued or roasted. The larvae are about the size of a person's thumb and are very greasy (not to everyone's taste). More appetising are the various beetles and orthopterans eaten in Cambodia and other Southeast Asian countries (copious small light traps are scattered throughout the rice fields to catch them). In Tanzania, the grasshopper (locally called senene) (Figure 1.1) is a seasonal delicacy caught in huge light traps.

Figure 1.1 '*Senene*' are a delicacy in the Kagera region of Tanzania. Fried, they remain good to eat (and they are delicious) for a year – until the next short season of abundance.

Insects can also have a huge negative impact on humans. One-sixth of all crops grown worldwide are lost to herbivorous insects and the plant diseases they transmit. About one in six human beings alive today is affected by an insect-borne illness such as plague, sleeping sickness, river blindness, yellow fever, filariasis and leishmaniasis. About 40% of the world's population are at risk of malaria.

Malaria transmission varies regionally, and sometimes over very short distances, as a consequence of factors such as transmission intensity, which vector species are dominant, and characteristics of the human populations. At a global level, there are important differences between sub-Saharan Africa and the rest of the world. The first is that the African vectors *Anopheles gambiae*, *A. coluzzii* and *A. funestus* are the most efficient vectors of malaria and the ones with the strongest preferences for humans. This is perhaps because these species evolved with humans. Indeed, to paraphrase Voltaire 'If *Anopheles gambiae* did not exist man would have created it'. Africa has two other anopheline species, *A. arabiensis* and *A. nili* that are also very efficient vectors. All these species tend to bite indoors and at night, and because of these vectors, Africa overall has very intense transmission.

The other source for the following is from the Introduction to *Medical Insects and Arachnids* (Lane and Crosskey, eds., 1993).

TAXONOMY

Confucius said something along the lines that the first thing one must do is give something a name, for it is only then that one can proceed. Thus, biological classification aims to group and categorise biological entities that share some unifying characteristics. Identification is a precise term describing the allocation of an unknown specimen to a predefined group. In any system, it is important that it is easy to retrieve information, and this is best done if similar animals are grouped together. There is, however, no single 'right' classification; they are suggestions of how organisms are related.

Taxonomy is the arrangement of similar entities (objects) in a hierarchical series of nested classes, in which each, more inclusive, higher-level class is subdivided comprehensively into less inclusive classes at the next lower level. These classes (groups) are known as taxa (singular: taxon). The level of a taxon in a hierarchical classification is referred to as a taxonomic rank or category.

There are two basic stages in systematics: first, recognition of biologically operational units (species, subspecies, etc.), which are the basic units of any classification; and secondly, the ordering of these units into higher categories or levels (genera, families, etc.).

The familiar binomial system for naming animals and plants was devised by the Swedish naturalist Carl Linnaeus, and the tenth edition of his great work titled the *Systema Naturae* (published in 1758) marks the beginning of scientific biological nomenclature. The name of a species consists of two words, the first being the generic name (used for all member species of a genus) and the second word being the specific (species) name.

WHAT IS A SPECIES?

Taxa at all levels above the species are relatively arbitrary creations of taxonomists, which attempt to show how assemblies of species are related to each other. Identification is a precise term describing the allocation of an unknown specimen to a predefined group.

The basic unit of taxonomy is the species. In ascending order: species, genus (plura, genera), family, order, class and phylum (plural phyla) (Figure 1.2). A working definition of a species is a group of interbreeding individuals.

The main arthropods and insects that attack humans and other animals are shown in Figure 1.3.

Flies (Diptera) have a mobile head with large compound eyes and three simple eyes (ocelli). The mouthparts are adapted for lapping and sponging liquids or piercing and sucking. A characteristic feature of the order is the possession of a single pair of membranous front wings, although some ectoparasitic species are wingless. The hindwings in all species are reduced to form a pair of balancing organs called halteres. These insects were given the name Diptera (Di – two, pteron – wings) by the Greek philosopher Aristotle around 500 BC so that Linnaeus did not need to find a new name when he first produced his classification.

The order is divided into two suborders, the Nematocera and the Brachycera. Nematocera is the more primitive suborder and includes crane flies, mosquitoes, black flies, midges and fungus gnats, with delicate threadlike antennae. The Brachycera are more robust, with short, stout antennae of less than six segments, and include the orthorrhaphan groups, typified by horseflies and robber flies,

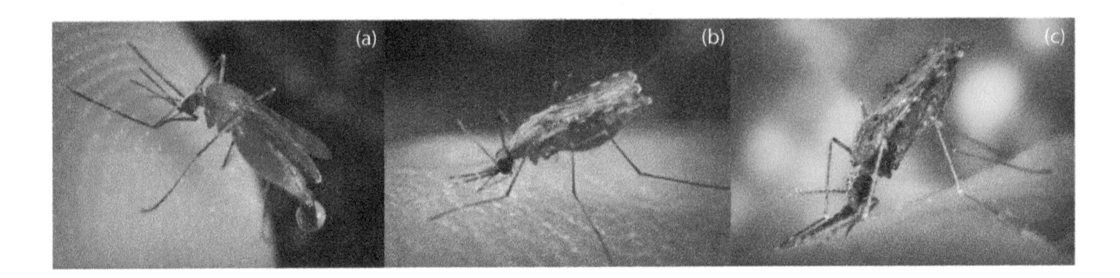

Figure 1.2 Hierarchical classification (from species to general) of (a) *Anopheles freeborni*, (b) *Anopheles minimus* and (c) *Anopheles albimanus*. (a) *Anopheles freeborni*, Freeborni Subgroup, Maculipennis Group, Anopheles Series, Angusticom Section, Subgenus *Anopheles*; (b) *Anopheles minimus*, Minimus Complex, Minimus Subgroup, Funestus Group, Myzomyia Series, Subgenus *Cellia*; and (c) *Anopheles albimanus*, Albimanus Series, Albimanus Section, Subgenus *Nyssorhynchus*. (From Harbach, R.E., The Phylogeny and classification of Anopheles. Chapter 1, in *Anopheles Mosquitoes—New Insights into Malaria Vectors*, Manguin, S. (Ed.), InTech, Rijeka, Croatia, 828 p., 2013.)

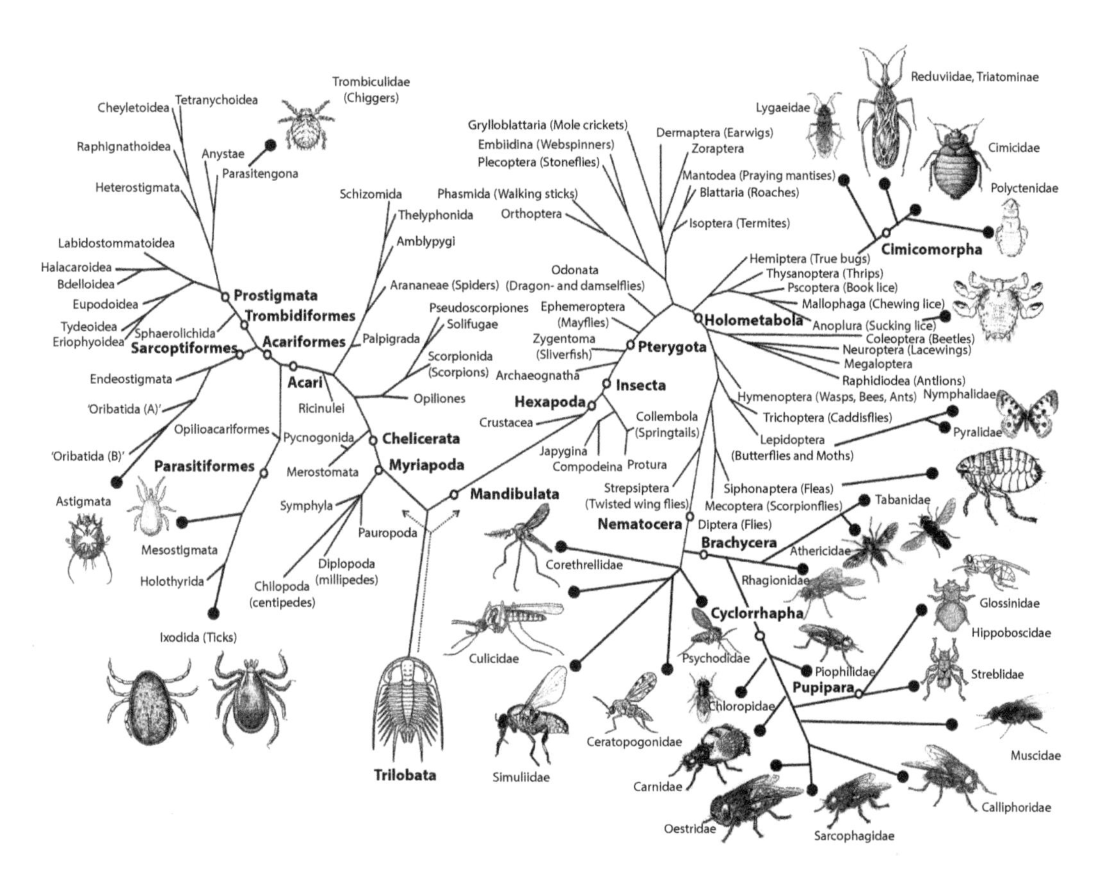

Figure 1.3 The evolutionary and systematic relationships between the different arthropod groups that are pests of man and animals. (Reprinted from *Biology of Disease Vectors*, 2nd ed., Black, W.C. IV and Kondratieff, B.C., Evolution of arthropod disease vectors, Copyright 2005, with permission from Elsevier.)

and the cyclorrhaphan species such as fruit flies, hoverflies, blowflies and flesh flies. Larval habits vary from fully aquatic to terrestrial and many larvae are serious plant pests.

MOSQUITOES (CULICIDAE)

Mosquitoes belong to the order Nematocera and are divided into three subfamilies: Anophelinae, Toxorhynchites and Culicinae. The important morphological characteristics of (female) mosquitoes are shown in Figure 1.4.

The subfamily Anophelinae comprise the genera *Chagasia*, *Bironella* and *Anopheles*. Adults of most *Anopheles*, which have a worldwide distribution, rest with the head, thorax and abdomen in a straight line and held at an angle of 30–45° to the surface. There are some 400+ species, of which 70 are malaria vectors and of these approximately 30 are important, or primary vectors. The other genera *Chagasia* (consisting of four species) and *Bironella* (containing nine species) are restricted to Neotropical and Australasian regions, respectively, and are 'missing links' between the Anophelinae and the Culicinae because they share characteristics of both subfamilies. The subfamily Toxorhynchites consists of 65 species. They are non-blood feeding and have predatory larvae that feed on the larvae of other mosquitoes. By far the largest subfamily is the Culicinae. This consists of 10 tribes, 30 genera, 109 subgenera and more than 3000 species. The relationship between the subfamilies and genera are shown in Figure 1.5.

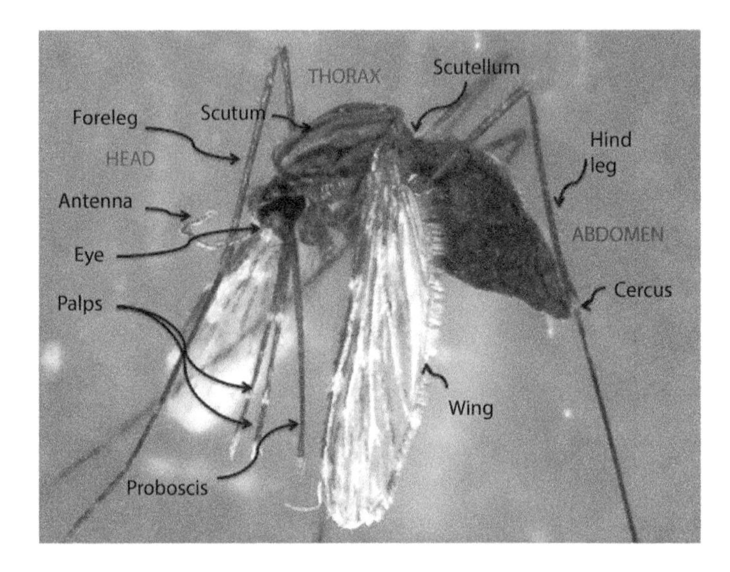

Figure 1.4 The main morphological features of an adult female *Anopheles* mosquito. Note the banding on the palps and the dark and light areas on the wing – these are often important for taxonomic purposes.

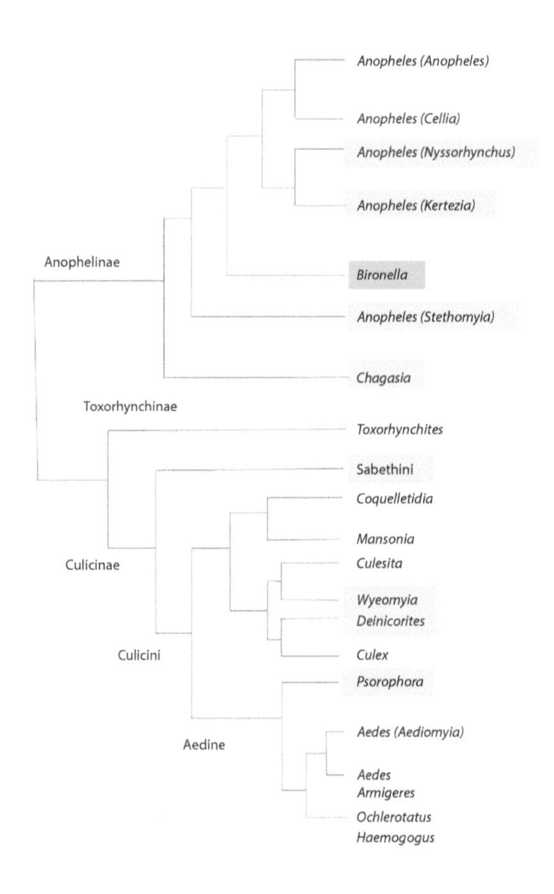

Figure 1.5 Classification of the Culicidae based on evolutionary relationships. Yellow highlights indicate species confined to the New World and Green to those restricted to the Australasian region.

Thus, the anophelines were the ancestral group from which the other genera were derived. The classification also implies that the non-blood feeding genera, the *Toxorhynchites*, lost the ability to feed on blood at some time.

Anopheles, *Culex* and *Aedes*

The main morphological characteristics that distinguish anophelines from culicines are shown in Figure 1.6.

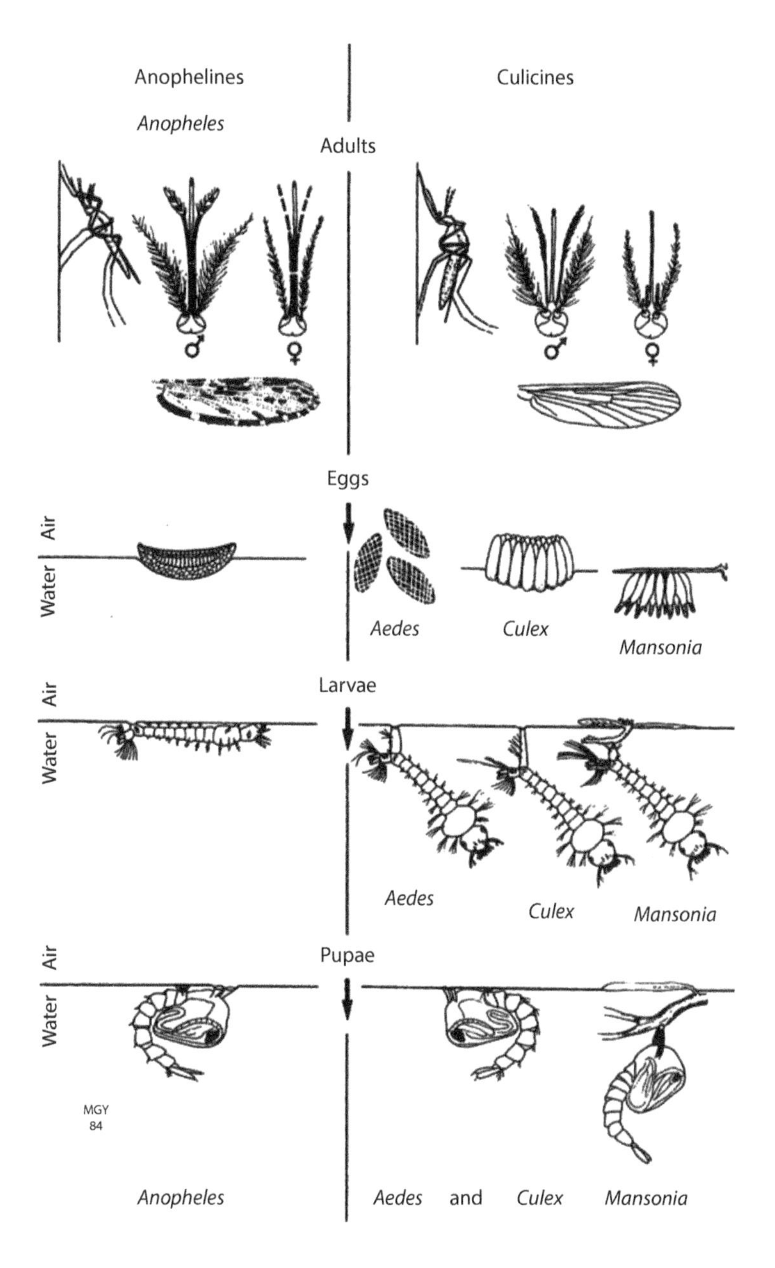

Figure 1.6 Distinguishing features of the different mosquito genera. (From Marshall, J.F., *The British Mosquitoes*, British Museum of Natural History, London, UK, 1938.)

- *Anopheles* typically rest with their bodies at a 45-degree angle to the surface (although *A. culicifacies*, as its name implies, rests in the manner of a culicine).
- The palps of female anophelines are almost the same length as the proboscis; those of culicines are shorter. The palps of anophelines are used for identification purposes. They often have bands of light and dark scales whose number and position help discriminate between species.
- The palps of male anophelines are club shaped at the ends whereas those of culicines are not.
- Anophelines have dark and light scales on their wings that are divided into 'blocks' (often used to identify them to species); the wings of culicines are uniform in colour. Differences, sometimes slight, in wing scaling are often used for the identification of different species (Figure 1.7).

Mosquitoes of the different genera have different kinds of eggs. Those of anophelines are characterised by having floats on the side. Those of culicines are laid on the water surface together and form a 'raft' visible to the naked eye. Eggs of *Aedes* are laid separately and for species like *Aedes aegypti* can withstand desiccation for long periods (although it is not the egg but rather the unhatched first instar larva that is inside the egg at these times). Eggs of *Mansonia* are attached to emergent vegetation just below the waterline. When they hatch, the larvae attach themselves by their siphon to the plant and obtain their oxygen through the plant itself. Eggs of anophelines do not survive desiccation.

Culex and *Aedes* larvae have siphons, of various shapes and lengths, and hang from the water surface (so that they feed below the water surface). Larvae of *Mansonia* attach themselves to vegetation and so do not come to the surface of the water to breathe.

The larvae of anophelines lie parallel to the water surface. Examination under a microscope shows that the abdomen has small, brown, sclerotised plates, called tergal plates, on the dorsal surface of abdominal segments 1–8; there may also be 1–3 small accessory plates behind a main tergal plate. In addition, most or all of these segments have a pair of well-developed palmate hairs, sometimes called float hairs. Laterally on each side of segment 8 there is a sclerotised comblike structure with teeth called the pectin (Figure 1.8).

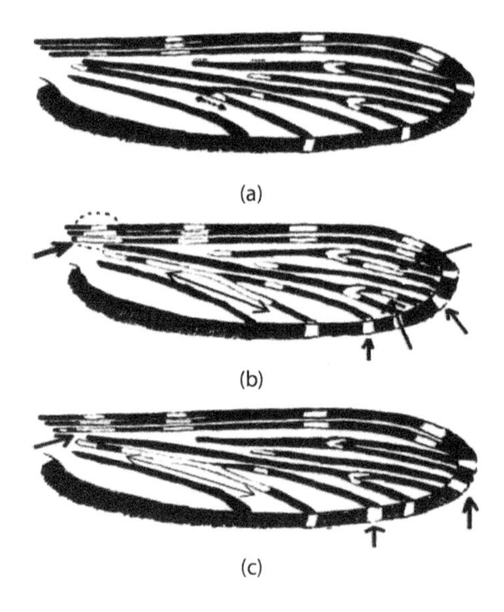

Figure 1.7 Morphological differences in the wings of the *A. nili* complex (a) *A. nili* and *A. somalicus* (b) *A. carnevale* (c) *A. ovengenisis*. (From Antonio-Nkondjio, C. and Simard, F., Highlights on *Anopheles nili* and *Anopheles moucheti*. Chapter 8, in *Anopheles Mosquitoes—New Insights Into Malaria Vectors*, Manguin, C. (Ed.), 828 p., InTech, 2013.)

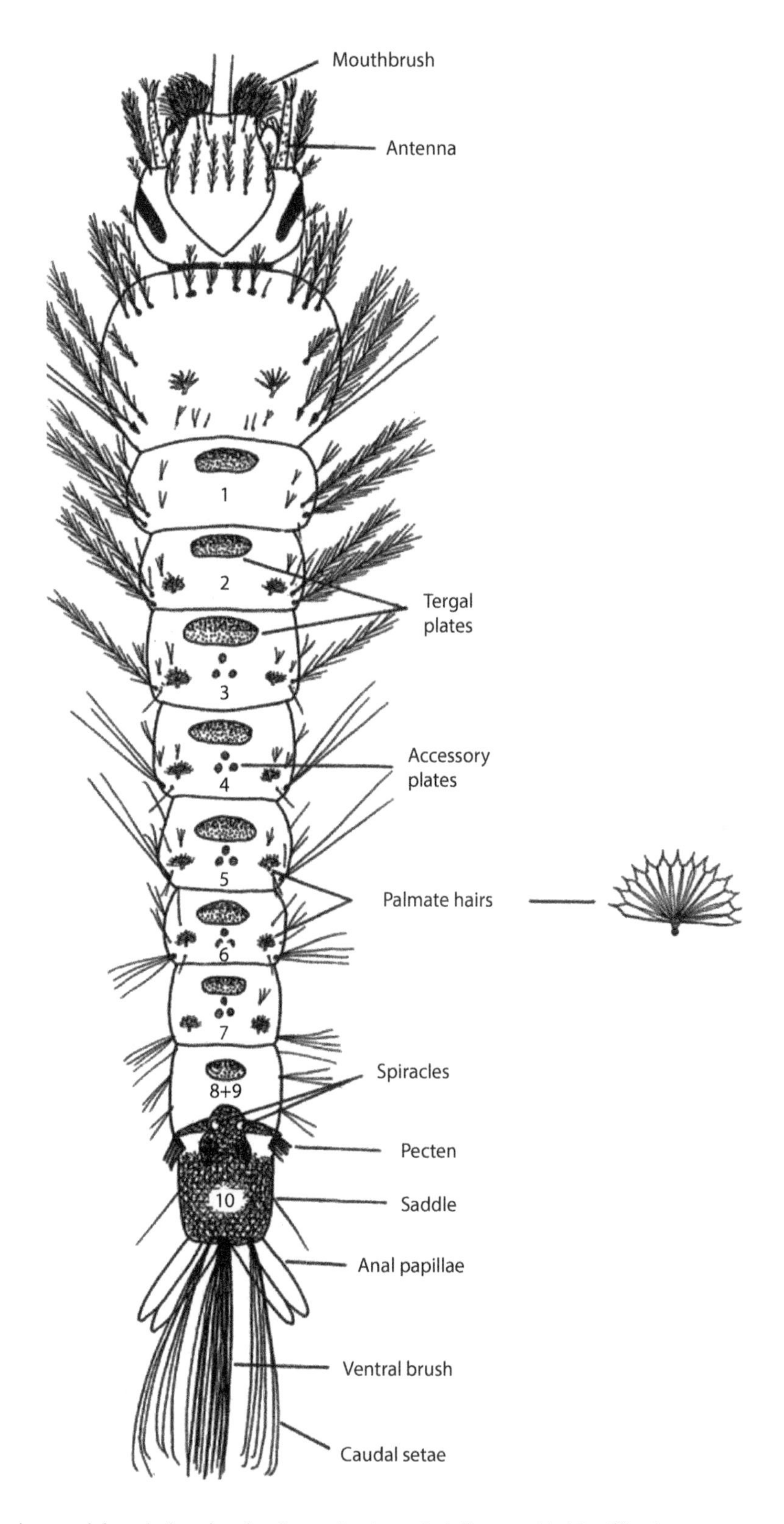

Figure 1.8 Larva of *Anopheles* showing the main characteristics used in identification.

Larvae of freshwater-adapted species have the problem of excess water uptake and loss of ions. To minimise these adverse effects, they maintain osmotic balance by restricting water intake, by producing dilute urine, and by uptake of ions from water through four (rarely two) large bladder-like structures on the tip of the abdomen called anal papillae (Figure 1.9). The anal papillae of freshwater species tend to be very large, and their size can vary according to water quality.

The comma-shaped pupae of the different genera are difficult to distinguish one from the other. The trumpets of *Anopheles* are generally broader than those of other culicines. Anopheline pupae have a series of short spines on the distal segments of the pupa that are not present in culicines.

Other distinguishing features in the adults include the presence of a single spermatheca in anophelines whilst culicines have three. Similarly, the scutellum (at the distal end of the thorax) is trilobed in culicines but presents a single curve in anophelines (Figure 1.10).

The Ecological Niche

The concept of the ecological niche is that each species occupies a unique 'hyperspace' that involves biological and environmental characteristics. Some of these characteristics may be similar, in which the species will be in competition for that particular feature. In general, natural selection will operate so that competition is avoided, perhaps in one life stage but may occur in another. So, for example, the larval habitat of the various genera differs: *Anophelines* occupy clean water; *Culicines* may occupy water with a high organic content, whilst many *Aedes* occupy small containers. Even within genera there are differences that tend to reduce competition. The typical breeding sites of *A. gambiae* are small sunlit

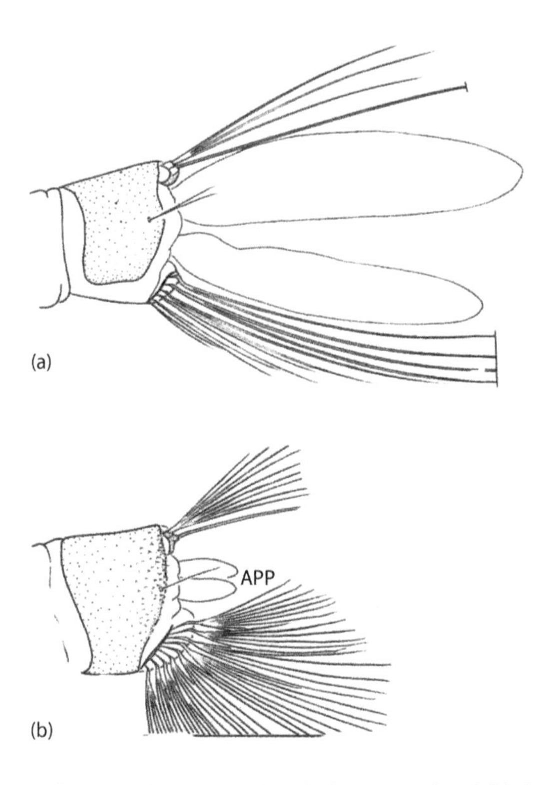

Figure 1.9 Anal papillae of (a) the freshwater species *Aedes aegypti* and (b) the saltwater species *Aedes taeniorhynchus* showing the differences in size of the anal papillae (APP). (From Darsie, R.F. Jr & Ward, R.A., Identification and geographical distribution of the mosquitoes of North America north of Mexico, *Mosquito Syst.*, 1–313, 1981.)

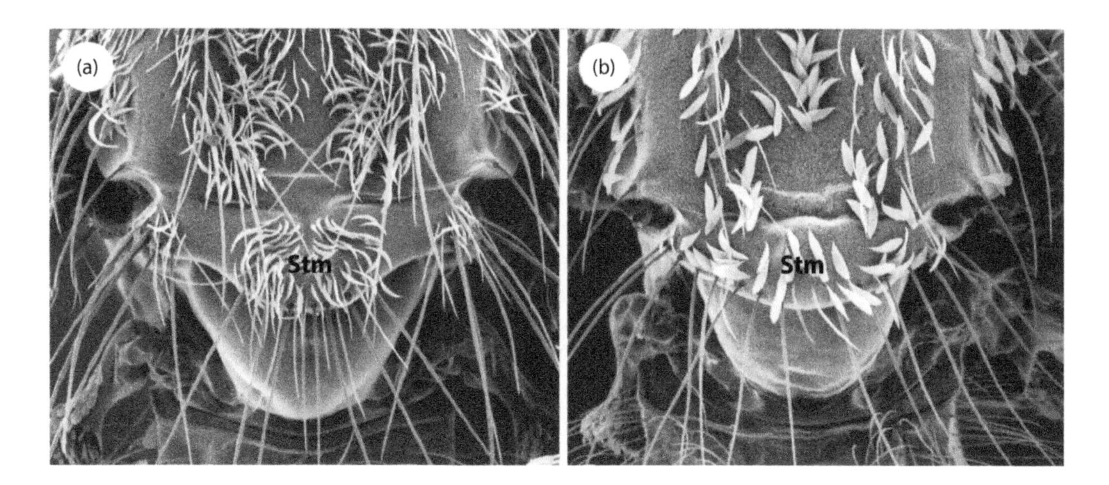

Figure 1.10 The shape of a typical *Culex* (a) and *Anopheles* (b) scutellum (Stm) as seen with a scanning electron microscope. (From Harbach, R.E., The Phylogeny and classification of Anopheles. Chapter 1, in *Anopheles Mosquitoes—New Insights into Malaria Vectors*, Manguin, S. (Ed.), p. 828, InTech, Rijeka, Croatia, 2013.)

pools whilst the sympatric *A. funestus* occupies more permanent water bodies that have emergent vegetation. Within the *A. gambiae* complex the different members of the complex occupy different habitats: *A. merus* and *A. melas* occupy brackish water habitats but on different sides of Africa; *A. coluzzii* appears to occupy more temporary habitats, especially rice fields, than *A. gambiae*. On the other hand, similar niches may be occupied by different species in different places. Thus, *Anopheles punctulatus* breeds in sunlit pools in Papua New Guinea similar to those occupied by *A. gambiae* in Africa. Other members of the *A. punctualtus* complex occupy different larval habitats and so also avoid competition (Charlwood and Galgal, 1985). Forest-dwelling *Anopheles dirus* in Southeast Asia and *A. darlingi* in South America face similar challenges to each other and so they too have similar ecologies.

Should the species occupy the same larval habitat (as can occur with *A. gambiae* and *A. arabiensis*), then competition between the adults may be reduced – in this case by the insects tending to feed on different hosts; *A. gambiae* largely feeds on humans whilst *A. arabiensis* has more catholic feeding habits, as does *A. quadriannulatus*, which largely feeds on animals (and so is not involved in malaria transmission).

Niches can arise *de novo*. There was not a niche for dog fleas until there were dogs. There was not a niche for mosquito larvae in rice fields until there were rice fields.

Species, groups of interbreeding natural populations that are reproductively isolated from other such groups, have a *fundamental* niche (i.e., the entire set of conditions under which an animal [population, species] can survive and reproduce itself) and a *realised* niche, the actual set of conditions used by the species, after interactions with other species (predation and especially competition) have been considered.

Although his book was called *The Origin of Species*, Darwin did not actually describe how they might arise. Coluzzi et al. (1985) described the ecological theory of species origin. Basically, in this model speciation occurs when some habitats at the edge of a species distribution (likely to be marginal habitats) are occupied, often as a group of populations that are separated by space (in other words as a metapopulation). These spatially separated populations interact as individual members move from one population to another but are liable to extinction and recolonisation. Eventually, the populations occupying these habitats adapt to them so that they become the new optimal ‘habitat’. Should the adaptation include modifications to the animals mating behaviour, then they will become separate species.

SPECIES COMPLEXES

In the Culicidae in general, and the Anophelinae in particular, many morphological species form species complexes or cryptic species. These are species that generally cannot be separated from one another by morphological features and which are often sympatric (meaning that they occur in the same place).

The possibility that certain morphologically similar species might actually be different was postulated because in Europe there were areas where the malaria vector *Anopheles maculipennis* occurred but where there was no malaria transmission. In some regions, the mosquito preferred to feed on animals, whilst in others it was anthropophilic. This led to the concept of *anophelism sans malaria*. Differences in mating behaviour were also observed. In 1926, Falleroni described distinct morphological differences in the eggs of certain populations. His observations were rediscovered five years later. Today, eight members of the complex are recognised. They can still be distinguished according to the morphology of their eggs (Figure 1.11).

Species complexes are not simply phenomena of interest to taxonomists: they have fundamental importance in the biology of disease and must be considered by medical entomologists and parasitologists. It is often observed in the field that a morphologically defined species is not biologically uniform but shows, in different parts of its range or at different seasons, puzzling inconsistencies in its behaviour or ecology, including susceptibility to parasites and host feeding patterns.

The *Anopheles gambiae* Complex

The following description of the present state of the art concerning the A. gambiae *complex comes largely from Lanzaro & Lee, 2013.*

Among the global vectors of human malaria, arguably the most important species belong to the *Anopheles gambiae* complex. The *Anopheles gambiae* species complex includes eight recognized sibling species: *A. gambiae* s.s., *A. arabiensis*, *A. bwambae*, *A. melas*, *A. merus*, *A. quadriannulatus*, *A. amharicus*, and most recently *A. coluzzii*.

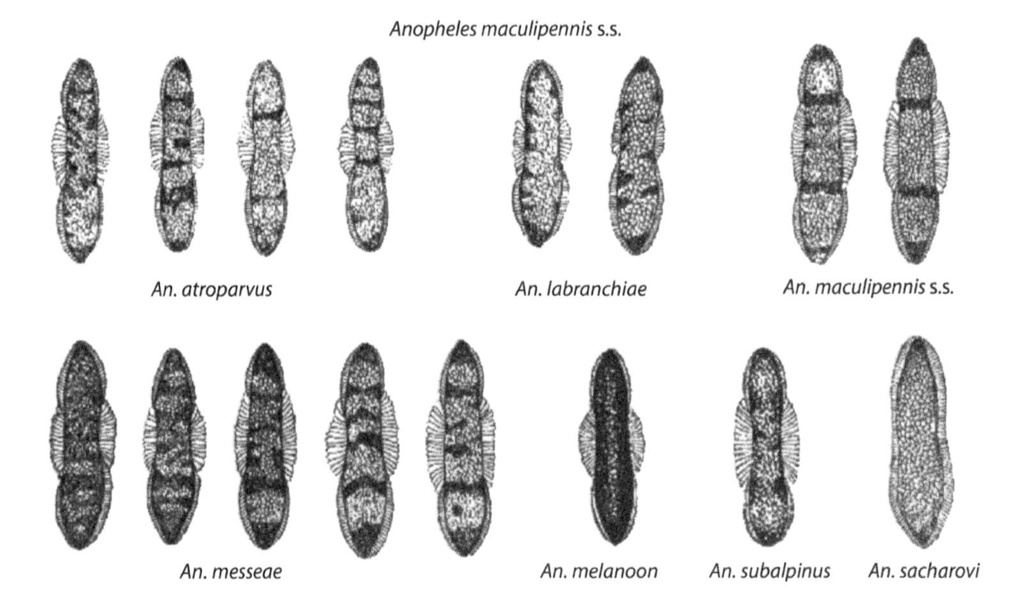

Figure 1.11 Eggs of the *Anopheles maculipennis* complex. (From Gutsevich, A. et al., *Fauna. SSSR. Diptera.*, 100, 1–408, 1974.)

Although the species cannot be reliably distinguished morphologically, they do differ in terms of their ecology and geographic distributions. Distribution of the main members of the *A. gambiae* complex is shown in Figure 1.12. The distribution of *A. gambiae* and *A. coluzzii* are considered together in this figure (from Lanzaro & Lee, 2013).

Two species, *A. merus* and *A. melas*, are associated with brackish water larval habitats and are restricted in distribution to sites along the east and west coasts of Subsaharan Africa, respectively. *Anopheles bwambae* is only known to occur in association with hot springs in Semliki Forest National Park, eastern Uganda. The species, *A. quadriannulatus* and *A. amharicus* are primarily zoophilic and are not considered to be involved in the transmission of malaria. *Anopheles quadriannulatus* occurs in Southeastern Africa and *A. amharicus* in Ethiopia. There is also a population on the island of Grande Comore in the Indian Ocean that was proposed as a distinct species of the complex, *A. comorensis*, on the basis of morphological characters of a single specimen but genetic confirmation is still missing (Brunhes et al., 1997; Coetzee et al., 2013).

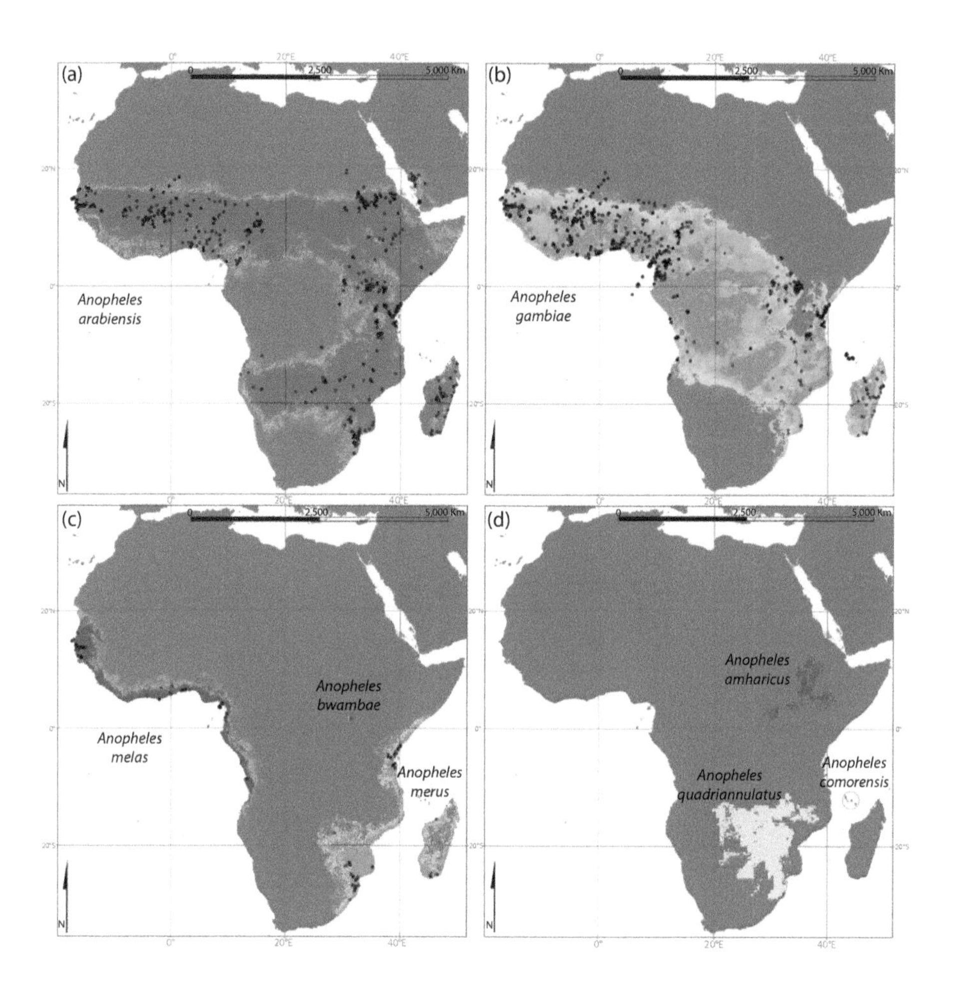

Figure 1.12 Distribution of the members of the *Anopheles gambiae* complex in Africa. (a) *A. arabiensis* (b) *A. gambiae* and *A. coluzzii* combined (c) *A. melas*, *A. merus* and *A. bwambae* (d) *A. amharicus*, *A. quadriannulatus* and, the as yet to be confirmed member of the complex, *A. comorensis*. (From Lanzaro, G.C. and Lee, Y., Speciation in *Anopheles gambiae*—The distribution of malaria vectors in Africa genetic polymorphism and patterns of reproductive isolation among natural populations, Chapter 6, in *Anopheles Mosquitoes—New Insights into Malaria Vectors,* Manguin, S. (Ed.), InTech, 828 p., 2013.)

Two of the three remaining freshwater species, *A. gambiae* (formerly known as the S Form) and *A. arabiensis*, have the broadest geographic distribution whilst the third freshwater member *A. coluzzii* (formerly known as the M form of *A. gambiae*) is restricted to West and Central Africa. These three species are the most important vectors of human malaria. *A. gambiae* and *A. coluzzii* have been the most studied with respect to molecular and population genetics.

Diptera like *Anopheles gambiae* have, in some of their cells (e.g., ovarian nurse cells, salivary glands of larvae), massive chromosomes formed by repeated lengthwise multiplication of DNA. These so-called polytene chromosomes contain light and dark banding patterns that are of cytotaxonomic value (Figure 1.13). The *A. gambiae* genome is organised on three chromosomes: two autosomes and one X/Y sex chromosome. For descriptive purposes, the autosomes are divided into two 'arms' at the centromere. The longer arm is referred to as the right arm and the shorter the left arm. Inversions are rearrangements of sections of chromosome turned around between break points so that the linear order of the genes is reversed on that part of the chromosome. The nature of the inversions often makes it possible to decide if specimens belong to the same or different species and to detect natural hybrids between them. The presence of fixed species-specific fixed inversions, particularly in chromosome X, allows the identification of most of the *Anopheles gambiae* complex members (Coluzzi et al., 1985; Coetzee et al., 2013).

In addition to species-specific fixed inversions, a high degree of chromosomal polymorphism, in the form of paracentric inversions (i.e., excluding the centromere-inversions that include the centromere are called pericentric inversions), has been described in natural populations of *A. gambiae*. They result from a double break in a segment of the chromosome with end-for-end rotation of the fracture between the fracture lines and re-fusion of the fragments. Inversions may not cause abnormalities in the carriers (as long as there is no genetic loss or extra genetic information), but heterozygotes (in the case of *A. gambiae* species these may be hybrids)

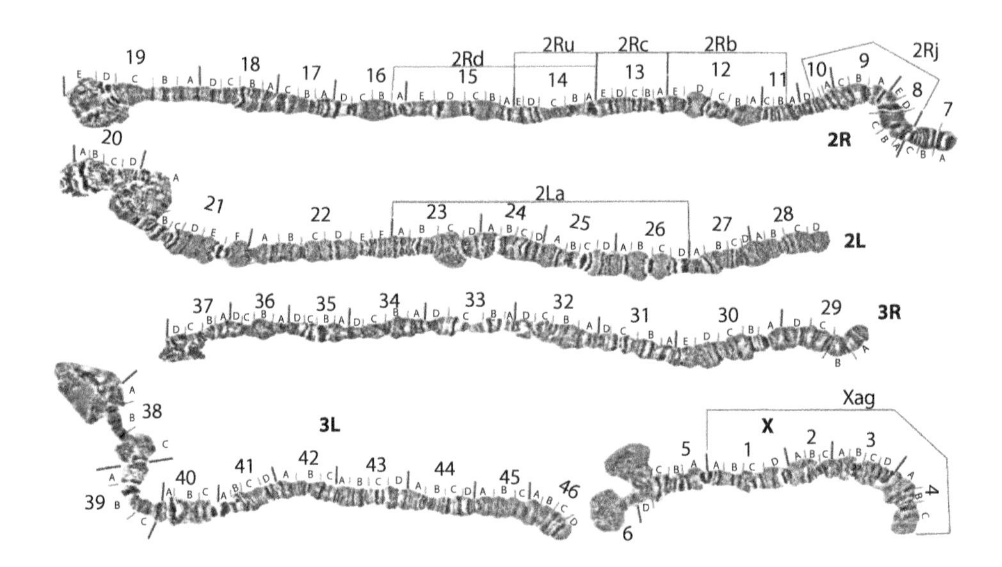

Figure 1.13 Ovarian polytene chromosomes of *A. gambiae* showing the light and dark bands used to distinguish chromosomal forms and the major paracentric inversions found in the members of this complex. Numbers and small capital letters refer to chromosomal divisions and subdivisions. The approximate location of each inversion is delimited by square brackets with the name of the inversion above (e.g., 2La: inversion a at left arm of chromosome 2). Inversion Xag is specific to *Anopheles gambiae* s.s. (From Lanzaro, G.C. and Lee, Y., Speciation in *Anopheles gambiae*—The distribution of malaria vectors in Africa genetic polymorphism and patterns of reproductive isolation among natural populations. Chapter 6, in *Anopheles Mosquitoes—New Insights Into Malaria Vectors,* Manguin, S. (Ed.), InTech, 828 p., 2013.)

may be at a disadvantage. The ecotypic speciation model is founded on the observation that certain paracentric inversions that are polymorphic in *A. gambiae* are nonrandomly distributed in nature. They occur most often on the right arm of chromosome 2 (2R). These are thought to contain multilocus genotypes that are adaptive to specific ecological niches. Under this model, populations carrying alternate gene arrangements would inhabit different, spatially isolated, habitats. Genetic divergence, enhanced by reduced recombination associated with the inversions, would then evolve. Ultimately, divergence would include genes resulting in different mating behaviour, in particular different swarming behaviour, and so reproductive isolation would ensue. Hybrid disadvantage may be overcome if one species transfers genes that may be under considerable selective pressure – for example, genes that confer resistance to insecticides. Should hybrids mate with the susceptible species then the resistance genes will spread into the second species, as appears to be the case in Ghana (Clarkson et al., 2014). Other genes that might 'hitchhike' with the adaptive genes tend to be actively selected against and are eventually removed from the population (Hanemaaijer et al., 2018).

Interbreeding between populations that have ecologically diverged, by either niche specialisation or invasion of a new niche (such as the water bodies created during rice cultivation), produces hybrid individuals of lower fitness in each of the parental habitats. Ecological speciation theory predicts that reproductive isolation is environment-dependent; in other words, it is driven by ecological selective forces such as resource competition or predation. The strength of reproductive isolation is correlated with the degree of ecological divergence, rather than time since lineage splitting. Thus, the separation of, say, optimal breeding sites may result in ecological speciation. Although the incipient species may for many aspects occupy the same geographical space and the same niche (i.e., they are sympatric), at certain points they must be separated (i.e., they are allopatric). Not all available habitats may be occupied at any one time. Populations, particularly at the margins of the mosquitoes' distribution, may become locally extinct leaving the habitat unoccupied until it is recolonised. In other words, they are likely to form a metapopulation.

There is general agreement that inversions represent co-adapted gene complexes that may enable individuals carrying them to occupy different ecological niches. The best example is the strong association of inversions 2La and 2Rb with aridity. Five chromosomal forms of *A. gambiae* have been described and named Mopti, Bamako, Bissau, Forest and Savanna according to different chromosomal arrangements and the geographic regions from which they were first collected, indicating an association of each with a particular type of habitat. The Savanna form has the broadest distribution, occurring throughout sub-Saharan Africa; the Mopti form predominates in drier habitats in West Africa; the Forest form occurs in wetter habitats in Africa; the Bamako form occurs in habitats along the Niger River in West Africa and the Bissau form is restricted to West Africa (Figure 1.14).

In an extensive study by Costantini et al. (2009) from Burkina Faso, the distribution of all three freshwater members of the *A. gambiae* complex differed. *Anopheles coluzzii* and *A. arabiensis* occupied the drier northern regions whilst *A. gambiae* predominated in the wetter, southern, part of the country (Figure 1.15). This means that *A. coluzzii* and *A. arabiensis* are more likely to be in direct competition with each other than are the two more closely related species. In the more southern part of its distribution the species occupies a much wetter habitat and has a different chromosomal arrangement, the standard 'Forest' form. No doubt, given time and the development of a distinct mate recognition system (maybe involving the use of different markers for swarms), the populations would become distinct species.

The molecular form concept has, however, now largely replaced chromosomal forms for defining discrete subpopulations of *A. gambiae*, which are to some extent reproductively isolated. Two molecular forms within *A. gambiae s.s.*, denoted M and S, have been described based on genetic polymorphisms found in the spacers of the ribosomal DNA. While the S-form (now designated *A. gambiae*) is the only one present in East Africa, both the S-form and M-form (now designated *A. coluzzii*) largely coexist in West Africa (Figure 1.16). However, in spite of large areas of sympatry, M/S hybrids are normally rare (<1%).

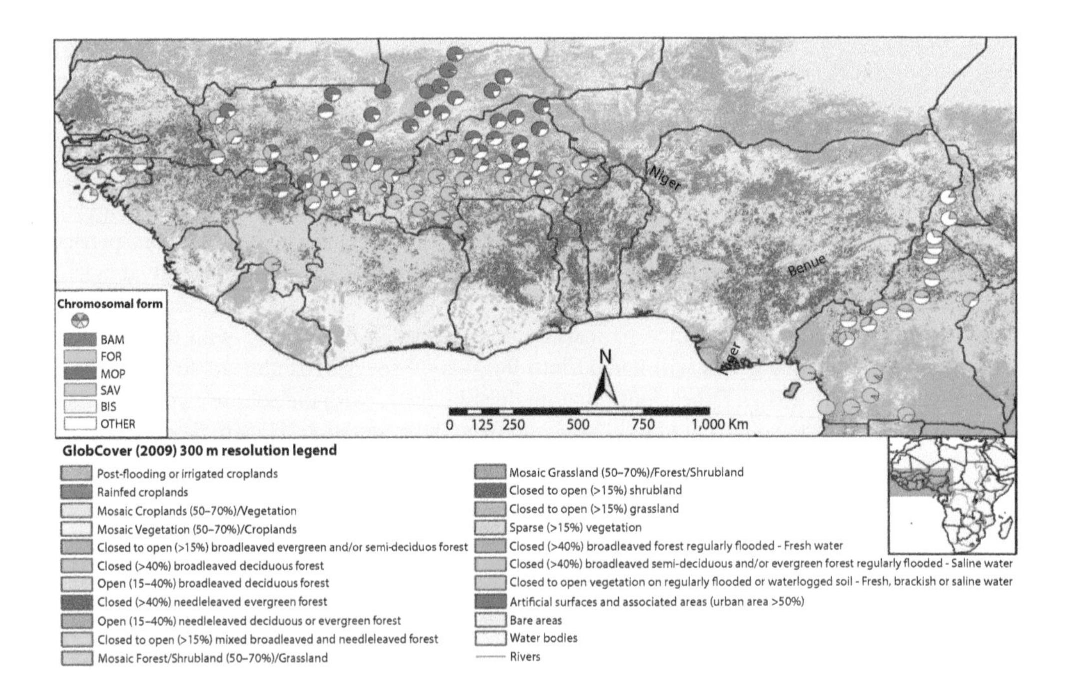

Figure 1.14 Distribution of the chromosomal forms of *A. gambiae* in West Africa and habitat ecotype. (From Lanzaro, G.C. and Lee, Y., Speciation in *Anopheles gambiae*—The distribution of malaria vectors in Africa genetic polymorphism and patterns of reproductive isolation among natural populations. Chapter 6, in *Anopheles Mosquitoes—New Insights into Malaria Vectors,* Manguin, S. (Ed.), 828 p., InTech, 2013; Coluzzi, M. et al., *Parassitologia*, 16, 107–109, 1974.)

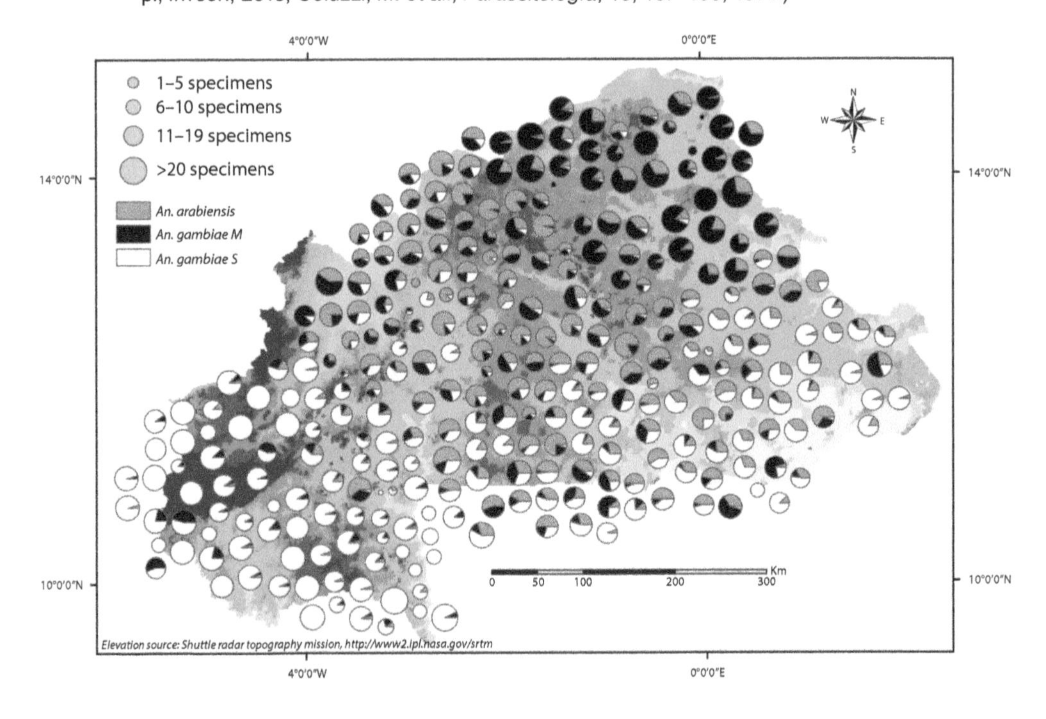

Figure 1.15 Relative abundance of *Anopheles gambiae* complex mosquitoes in Burkina Faso showing the results of molecular identification expressed as relative frequency and sample size. (From Costantini, C. et al., *BMC Ecol.*, 9, 16, 2009.)

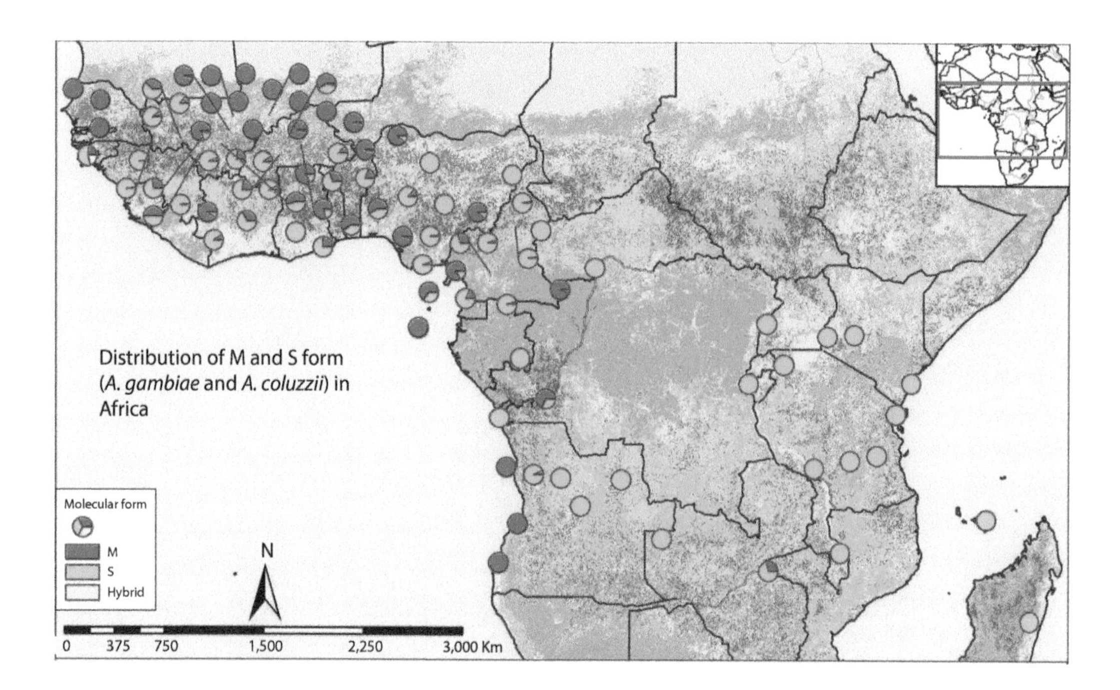

Figure 1.16 Distribution of *A. coluzzii* (M form) and *A. gambiae* (S form) in Africa. (From Lanzaro, G.C. and Lee, Y., Speciation in *Anopheles gambiae*—The distribution of malaria vectors in Africa genetic polymorphism and patterns of reproductive isolation among natural populations. Chapter 6, in *Anopheles Mosquitoes—New Insights into Malaria Vectors,* Manguin, S. (Ed.), InTech, 828 p., 2013.)

Both *A. gambiae* and *A. coluzzii* show considerable intraspecific variation that is associated with different larval habitats or niches, many of them man-made (such as rice fields). In many places, *A. coluzzii* is more likely to be found in such habitats while *A. gambiae* may be found in the 'classical' temporary habitats (which all field workers in Africa are familiar with), but this is not universal. In the village of Okyereko in Ghana, the *A. coluzzii* and *A. gambiae* larvae occurred in rice fields in the same proportion as they did in adult collections whilst puddles were colonised exclusively by *A. coluzzii* (Charlwood et al., 2012).

Along the Gambia River, different microsatellite polymorphisms are found that are associated with different levels of salinity. Populations of *A. gambiae* were differentiated even at alternate ends of an extensive region of rice cultivation that itself was only colonised by *A. coluzzii* (Figure 1.17). Should these differences result in different mate recognition systems (Paterson, 1985) then the process of speciation will follow. The region of the genome that is considered most likely to contain genes for reproductive isolation consists of highly diverged pericentromeric regions of chromosome X and (to a lesser extent) 2L, termed 'speciation islands'. In many cases these 'islands' may be considered 'atolls' because there may be a relative 'tsunami' of introgression from one species to the other if there is sufficient selective force (such as insecticide pressure) to outweigh the disadvantages of hybridisation.

From a collection at 15 locations across 8 African countries, including areas of rainforest, inland savannah and coastal biomes (Figure 1.18), it was found that populations are locally adapted, and migration between populations is limited by both geographical distance and major ecological discontinuities, notably the Congo Basin tropical rainforest and the East African rift system.

These data revealed complex population structure and patterns of gene flow, with evidence of ancient expansions, recent bottlenecks and local variation in effective population size. Within *A. gambiae* populations fall into two well-defined clades, based on analysis of microsatellite DNA.

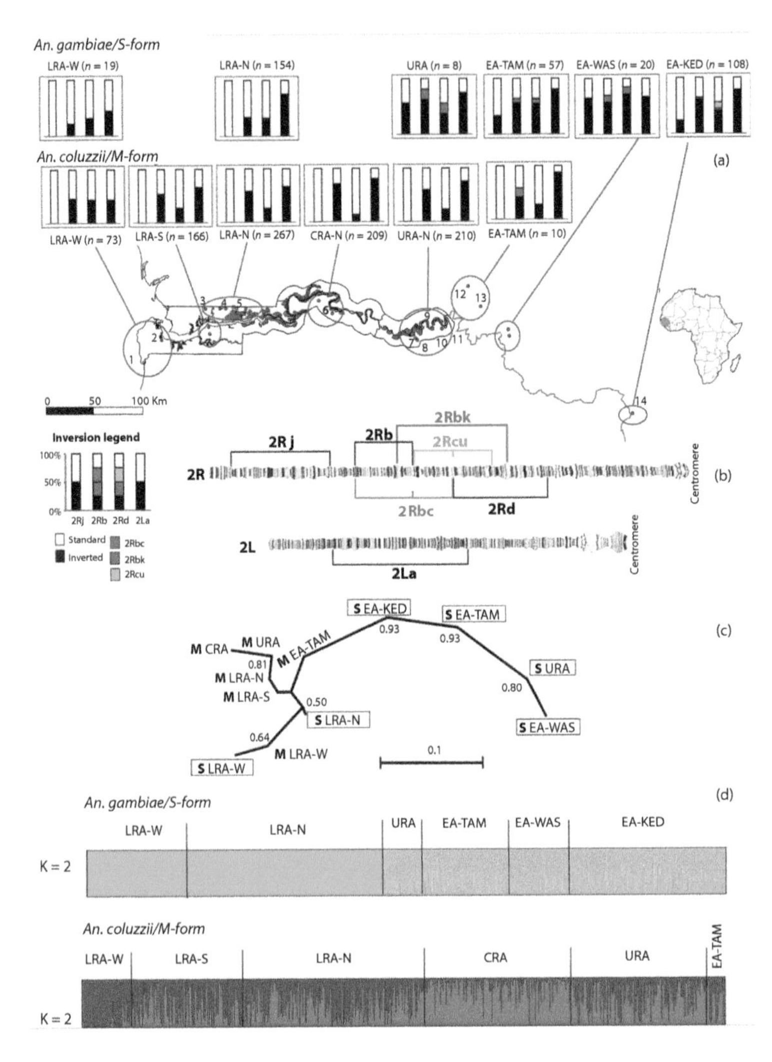

Figure 1.17 Microsatellite polymorphisms in *Anopheles coluzzii* and *A. gambiae* along the Gambia River. (a) Histograms represent frequencies of alternative chromosome-2 inversion arrangements. (b) Diagrammatic representation of inversion polymorphisms on right (2R) and left (2L) of standard chromosome. (c) Neighbour-joining dendogram from a matrix of F_{ST} among samples of each species assembled in ecological subareas. (d) Proportion of assignment of each specimen to clusters. LRA-W = Western Lower River Area; LRA-N = North Bank Lower River Area; CRA = Central River Area; URA = Upper River Area; EA-TAM = Tambacounda Eastern Area; EA-KED = Kedougou Eastern Area. (From Caputo, B. et al., *Mol. Ecol.*, 23, 4574–4589, 2014.)

These are referred to as the Northwest (Nigeria, Gabon, Democratic Republic of Congo, Northwest Kenya) and Southeast (Southwest Kenya, Tanzania, Malawi) divisions.

Although in most places the number of hybrids is low, there exists a relatively stable hybrid zone in Guinea-Bissau at the western limit (the 'far west') of the species distribution. As in The Gambia, this appears to be associated with differences in salinity tolerance among the larvae. In this case there is considerable introgression from *A. coluzzii* to a coastal population of *A. gambiae*, whilst inland the *A. gambiae* is similar to populations common across West Africa.

Individuals from coastal areas were largely *A. gambiae* but with a high proportion of introgressed genes from *A. coluzzii* and with a chromosomal form characterised as BISSAU; those from a central region were largely *A. coluzzii*, whilst inland a more typical *A. gambiae* with a chromosomal form characterised by SAVANNAH predominated (Figure 1.19). Coastal

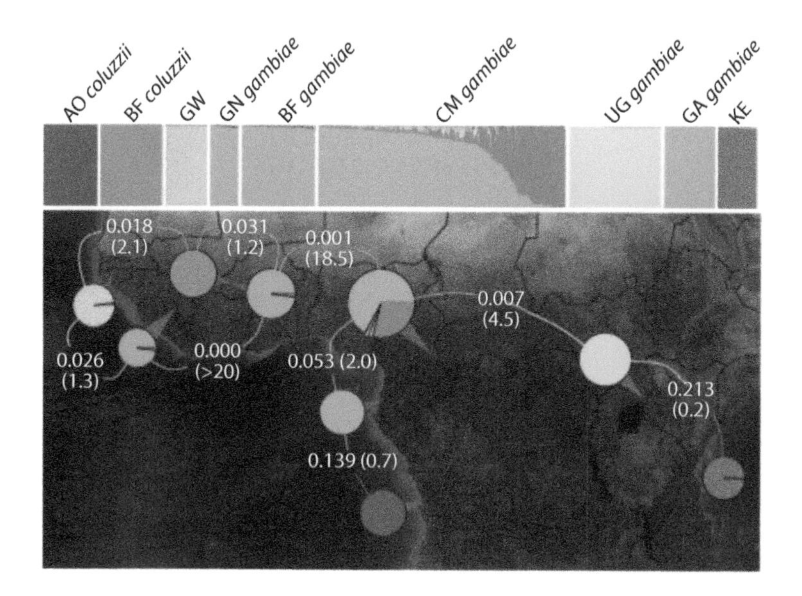

Figure 1.18 Geographical population structure and migration. Top of each mosquito is depicted as a vertical bar painted by the proportion of the genome inherited from each of $K = 8$ inferred ancestral populations. Pie charts on the map depict the same ancestry proportions summed over all individuals for each population. Text in white shows average fixation index (F_{ST}) followed in parentheses by estimates of the population migration index (2 *Nm*). (From The *Anopheles gambiae* 1000 Genomes Consortium, *Nature*, 552, 96–100, 2017.)

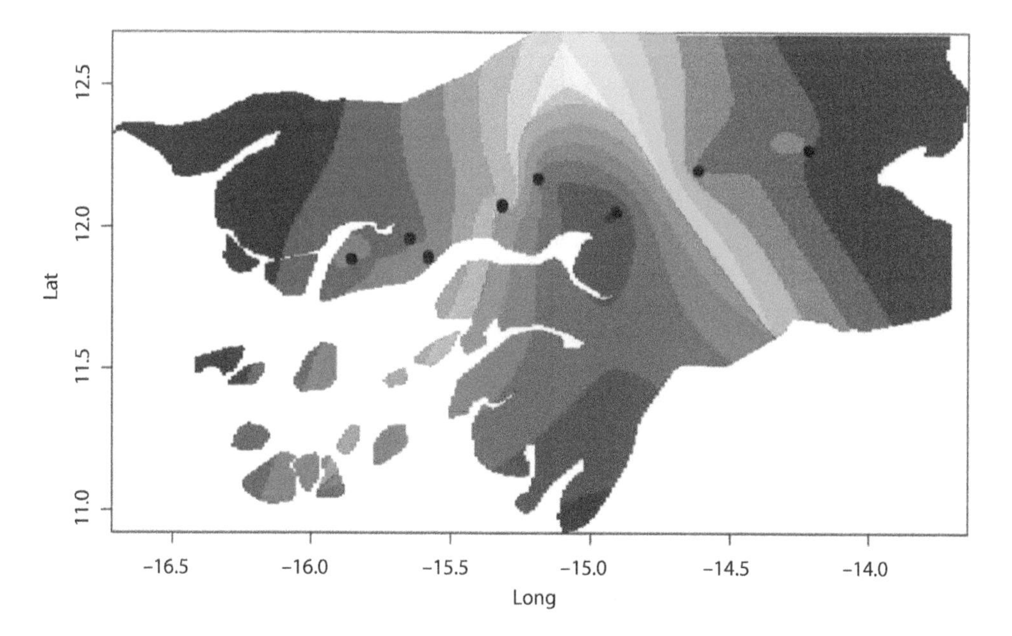

Figure 1.19 Map of Guinea-Bissau showing probability densities of the three clusters of the freshwater members of the *A. gambiae* complex derived from microsatellite data. Red cluster *A. coluzzii*; Green cluster *A. gambiae* inland and Blue cluster *A. gambiae* coastal. The black dots represent the locations from where samples were obtained. (From Vicente, J.L. et al., *Nat. Sci. Rep.*, 7, 46451, 2017.)

A. gambiae, with many introgressed *A. coluzzii* genes, may have larvae that are relatively tolerant to saline or brackish water and may eventually become a distinct species.

THE FUNESTUS GROUP

The other major vector of malaria in Africa is perhaps the most anthropophilic and synanthropic mosquito of all. Both male and female *Anopheles funestus* spend much of their adult life resting in houses. This previously made it the species most amenable to control by indoor residual spray (IRS). Such was the success that according to Mick Gillies (personal communication) earlier entomologists thought that the mosquito might in the future only be present in the drawers of museum cabinets. Presently, however, it presently is the major vector in many East African locations. Although some other members of the group (such as *A. rivulorum*) have been found with sporozoites, only *A. funestus* is a vector of any importance. Members of the group include *A. minimus* and its siblings.

The mosquito is more difficult to colonise than *A. gambiae* and its polytene chromosomes are more difficult to prepare and score. Studies of its genetics have therefore lagged behind those of the *A. gambiae* complex.

The Funestus group (rather than complex) was first designated by Gillies and de Meillon (1968) to describe a set of closely related species that showed small morphological differences (such as the presence or absence of a fringe spot of light scales at the distal end of vein 5.2 on the wing). Presently, there are 13 species in the group, which have a broad distribution throughout Africa (Figure 1.20).

The larval niche differs from that of *A. gambiae* (so although they may compete as adults, they avoid competition as larvae). Larvae are found in permanent or semipermanent sites with emergent or floating vegetation. Densities tend to increase with increasing height of the water table. In Tanzania, the mosquito was found during the dry season, at locally high densities, breeding in the 'lakes' left behind by non-flowing rivers. These became unsuitable habitat after the first rains (even though the streams only started flowing at a much later date) due to the muddy runoff from the nearby fields. Populations remained low until the end of the rainy season when rice in the fields was well developed, creating the habitat of emergent vegetation favoured by this species.

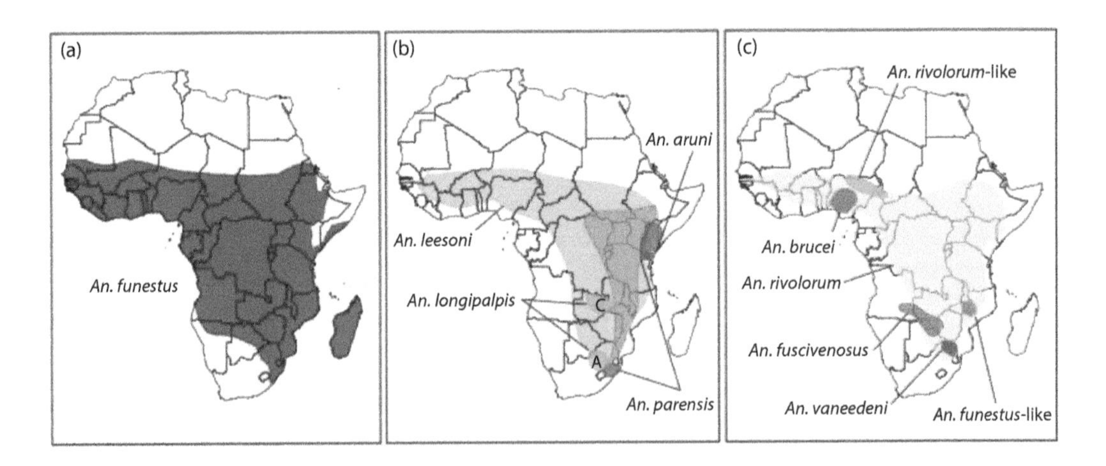

Figure 1.20 Distribution of the 13 species of the Funestus group in Africa. (a) *Anopheles funestus*, (b) *A. leesoni*, *A. longipalpis*, *A. aruni* and *A. parensis* (c) *A. rivulorum*, *A. rivulorum*-like, *A. funestus*-like, *A. vandeeni*, *A. fuscivenosus* and *A. brucei*. (From Dia, I. et al., Advances and perspectives in the study of the malaria mosquito *Anopheles funestus*. Chapter 7, in *Anopheles Mosquitoes—New Insights into Malaria Vectors*, Manguin, S. (Eds.), InTech, 828 p., 2013.)

Populations at this time were widespread and prolific. Elsewhere, where larval habitat is persistent, it is temperature that influences the population dynamics, numbers being reduced during the cooler months of the year. In other areas, it is rainfall that is the major driver of the population and in the absence of suitable larval habitat the mosquito may aestivate (Charlwood et al., 2013a).

There are abundant levels of genetic and chromosomal polymorphism in *A. funestus* across its range. Much of this variation is due to variation within rather than between populations. Thus, there is only a weak correlation between geographic distance between two populations and genetic distance between them (Michel et al., 2005).

As with *A. gambiae*, the Rift Valley may act as a geographical barrier to gene flow (Michel et al., 2005). On the basis of microsatellite allele frequencies, three clusters that correspond to (i) Africa east of the Rift Valley (coastal Tanzania, Malawi, Mozambique, Madagascar), (ii) western Africa (Mali, Burkina Faso, Nigeria) + Kenya west of the Rift Valley, and (iii) central Africa (Gabon, coastal Angola) have been described (Michel et al., 2005). Thus, in *A. funestus*, as in *A. gambiae*, samples from western Kenya are more closely related to those from West Africa than to samples from the coast of Kenya or Tanzania. In Burkina Faso, two chromosomally distinct incipient species of *A. funestus* designated Folonzo and Kiribina have been described (Costantini et al., 1999).

There are a number of anomalies concerning the mosquito. For example, following the cessation of earlier control campaigns, it took longer than anticipated for it to reappear in areas where it had previously been common. This may, in part, have been due to the limited flight range of the mosquito compared to *A. gambiae* (although during the dry season flights of more than 3 km were recorded in Tanzania) and the relatively isolated nature of the appropriate larval habitat. Following control with Indoor Residual Spraying, *A. funestus* is often replaced by other members of the group, notably *A. rivulorum* or *A. leesoni*. This implies that these species are in competition with *A. funestus* for a limited resource (but what that might be is not known) and that when *A. funestus* numbers are reduced they occupy the niche – again making it difficult for re-invasion to occur.

Despite the low re-invasion rates, the genetic information described above implies a considerable amount of gene flow between populations, as does the rapid spread of metabolic resistance from South Africa to Mali (including Mozambique) to pyrethroids in this species.

The reviews of the biology of both the *A. gambiae* complex and the *A. funestus* group in the books by Gillies and de Meillon (1968) and Gillies and Coetzee (1987) remain useful reading for people working on malaria vectors in Africa. The review by Antonio-Nkondjio and Simard (2013) is also most useful for information on the *A. nili* complex.

Species complexes are, if anything, more prevalent in other malaria endemic areas of the world. In the Asia-Pacific region among the 19 recognised dominant vectors, at least 10 belong to species complexes (Sinka et al., 2011). As an example, the Punctulatus Group is composed of at least 12 sibling species that collectively span most tropical areas of the Australasian region (Cooper et al., 2009).

In South America the picture is, if anything, more complicated (Conn et al., 2013). Thus, in a recent study examining Anophelinae species diversity from settler communities in the Amazonas region of Brazil, 23 species were collected, of which up to 13 appeared to be new. The contraction of the Amazon rainforest into isolated patches, so-called forest refugia (Haffer, 1974), during the Pleistocene (0.01–2.6 mya), with secondary intergradiation when rainfall increased is considered to be the primary driver for divergence in these mosquitoes (as well as butterflies, *Drosophila* and lizards). Isolation by distance is also likely to be important because most species do not have continuously distributed populations and gene flow between clumped demes is generally leptokurtic (because more individuals move smaller distances than they would if they moved at random). Hence, species with broad ranges are more likely to be divided by changes in geomorphology and climate than species with limited ranges. They are also more likely to have greater habitat diversity and so may develop clines and subspecies.

CHAPTER 2

Mosquito Life Histories

EGGS AND LARVAE

'A hen is only an egg's way of making another egg'. — Samuel Butler

Oviposition site selection by female mosquitoes is the primary determining factor in the habitat distribution of mosquito larvae, but the various clues used for oviposition remain only partially known. Colour, moisture and volatile chemical stimulants appear to play a role in most species.

Mosquito larvae pass through four instars. These are the only times when the mosquito can grow. Growth is continuous throughout each instar as seen in the progressive increase in size of the body regions that are covered with thin cuticle.

All four of the larval stages of culicine and aedine mosquitoes are easy to recognise because of the presence of an elongate siphon used for breathing. However, anopheline larvae lack a siphon and lie parallel to, and just beneath, the surface of the water to maintain contact with the air–water interface for breathing. Anophelines are able to maintain this orientation with the aid of pairs of palmate hairs that line the body and act as floats. Anophelines feed at the water surface by swiveling their head 180° so that the mouthparts face up toward the water surface.

The time required for development of the larval stages depends on several factors, the most important of which is water temperature (Figure 2.1).

The amount of reserves present at metamorphosis is largely dependent on the availability of nutrients during the fourth instar. *Anopheles* generally inhabit groundwater and exhibit rapid development and little ability to adjust their development time to accommodate variations in availability of nutrients. At higher temperatures, this means that the larvae may not obtain enough food to reach their optimum size. As a result, at higher temperatures the adults are smaller (Figure 2.2).

The numbers of adult mosquitoes also indicates the productivity of larval habitats with temperature (assuming temperature-independent survival in the adults). Productivity is greater at higher temperatures (Figure 2.3).

The effect of temperature on larval development is greater in *A. gambiae* complex mosquitoes than it is in *A. funestus*. This is because temperature extremes are greater in the temporary pools favoured by *A. gambiae* compared to the more permanent waters occupied by *A. funestus* and also because the fluctuations in temperature, which affect growth and feeding, are considerably greater in small sunlit pools. For example, in Mozambique, numbers of *A. funestus* increase approximately four times with an increase in temperature of 10 degrees (from 20°C to 30°C), but numbers of *A. gambiae* increased more than 10 times (Figure 2.4).

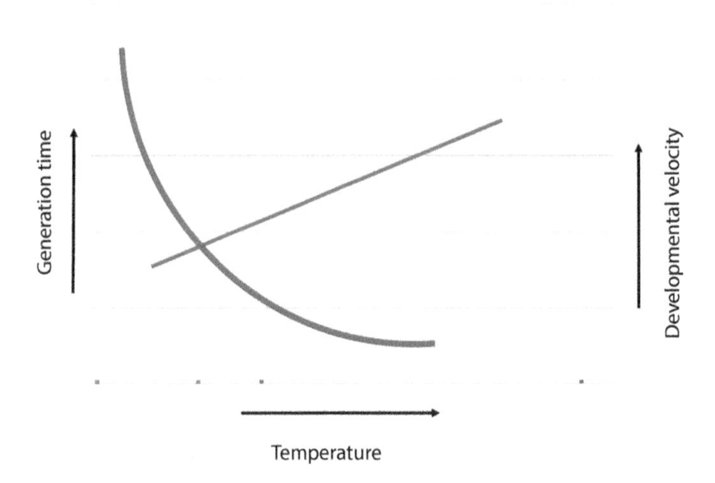

Figure 2.1 A schema showing the effect of increasing temperatures on developmental velocity (in red) and mean generation time (in blue) with air temperature.

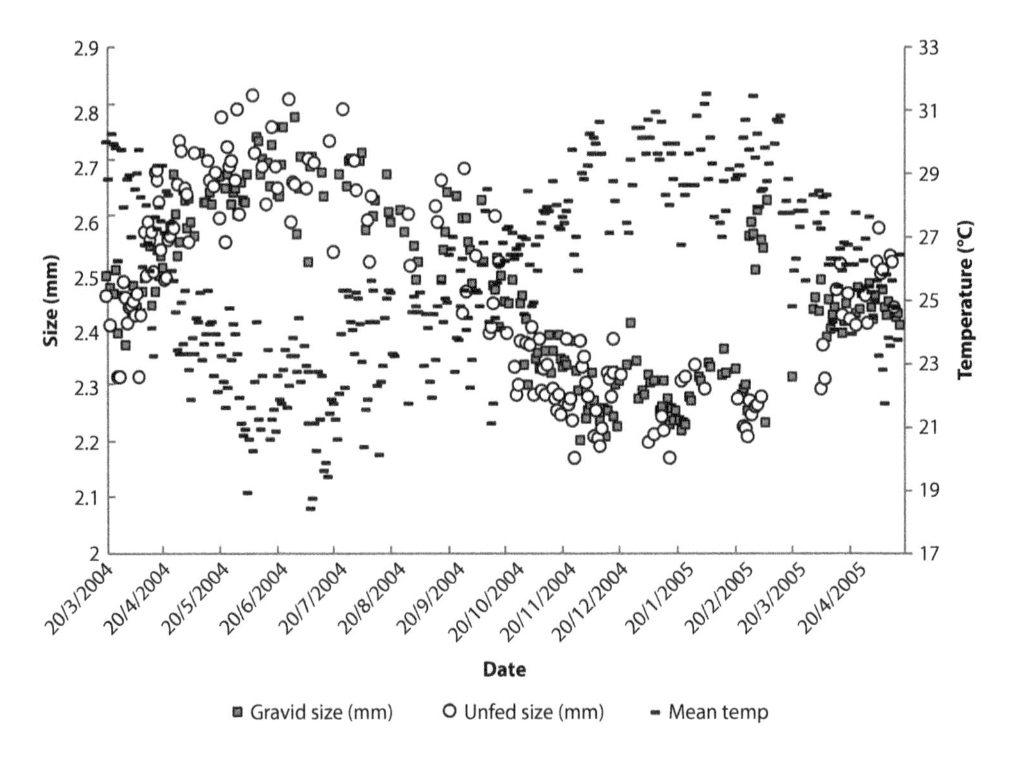

Figure 2.2 Wing length (a proxy for adult size and mass) of *A. funestus* and mean temperature from Furvela village in southern Mozambique. (From Charlwood, J.D. and Bragança, M., *J. Med. Entomol.*, 49, 1154–1158, 2012b.)

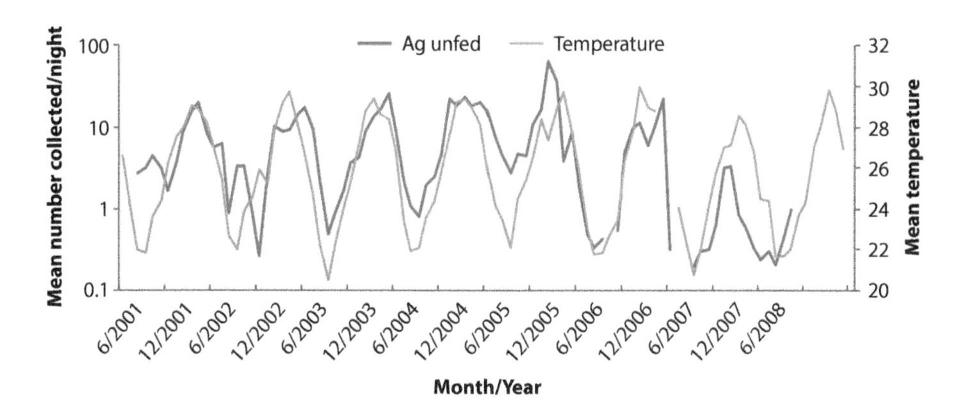

Figure 2.3 Numbers of *A. gambiae* s.l. collected in light traps in Furvela village, Mozambique, and temperature.

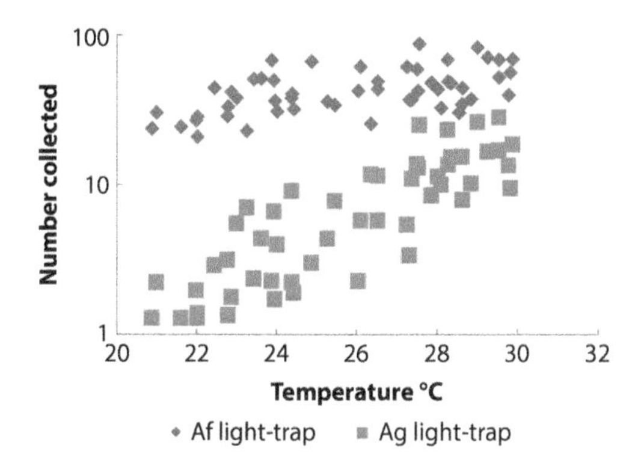

Figure 2.4 Numbers of *A. funestus* and *A. gambiae* s.l. females caught in light traps relative to ambient temperature, Furvela, Mozambique. (From Charlwood, J.D., *Peer. J.*, 5, 3099, 2017.)

WATER BITES

One commonly observed, but rarely commented on, phenomenon is the presence of parasitic water mites on mosquitoes. In multivoltine mosquitoes (that have several generations a year) the presence of such mites on a female mosquito is an indication that the female has not laid eggs. The mites are easily seen because of their red colour and relatively large size. They attach themselves to the emerging adult and hitch a ride (with the benefit of a free meal) until the mosquito lays its first batch of eggs; at which time the mite drops into the new water body. The more permanent a water body that is occupied by the larvae, the more water mites they have. *Mansonia africana* and *M. uniformis* commonly have such mites and, when they do, they usually have more than one attached (mainly to the thorax).

In univoltine mosquitoes, which only have a single generation a year, (such as *Ochlerotatus cantans*) the opposite is the case. These mosquitoes lay their eggs in dried out pools amongst leaf litter. The pools fill with rain water in the autumn, the eggs hatch and the mosquito overwinters as larvae. Adults emerge in the late spring and survive through the summer, but mites do not attach themselves at this time. It is only when the females go to oviposit that the mites attach; hence mites on these mosquitoes are an indication that the female has previously laid eggs.

EMERGENCE AND MATING

Mosquitoes are holometabolous; in other words, they undergo complete metamorphosis. The adult mosquito develops in the pupal stage. Some of the structures from the larvae continue, but in general there is a complete rearrangement of tissue. Eclosion, or emergence from the pupa, is a gated phenomenon – it only occurs at specific times even if the adult is physiologically ready to emerge earlier. In *Anopheles*, emergence takes place shortly after dusk. Newly emerged adults have a relatively soft and flexible cuticle. Appendages like the wings are blown up with air and then harden for the first flight. The insects will remain relatively inactive for the next 20 hours, although they may take a sugar meal that provides carbohydrate during this time. The male terminalia rotate through 180° in this period. Males are smaller (Figure 2.5) and, perhaps because of this, emerge a day or two before females.

Mating in *Anopheles* only takes place for a limited time of day. It takes place in swarms at dusk. Swarms are aggregations of males that generally form over distinct areas of contrast, called swarm markers, in the environment, be that a bush, a distinctly coloured patch of earth or some other area of contrast. Virgin females behave in many ways as surrogate males. They are active at dusk (Figure 2.6) and react to swarm markers or sites in much the same way as males.

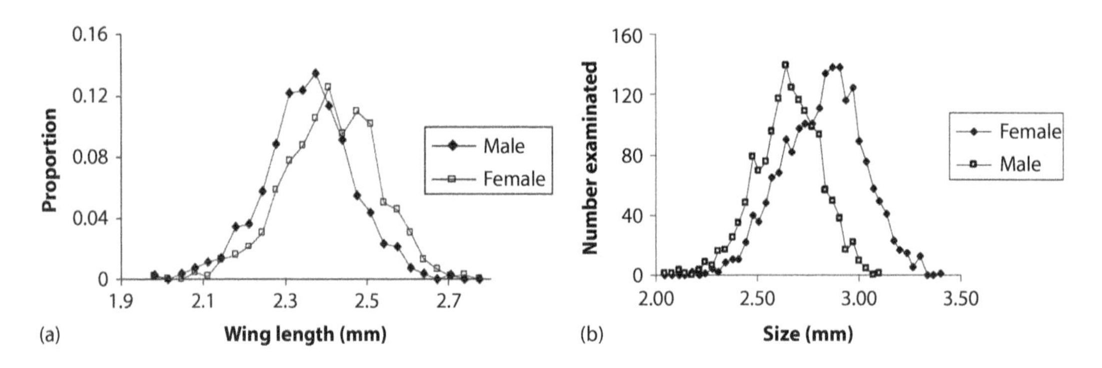

Figure 2.5 (a) Wing lengths of *A. funestus* (from Mozambique) and (b) *A. coluzzii* (from São Tomé). In both species males are smaller than females.

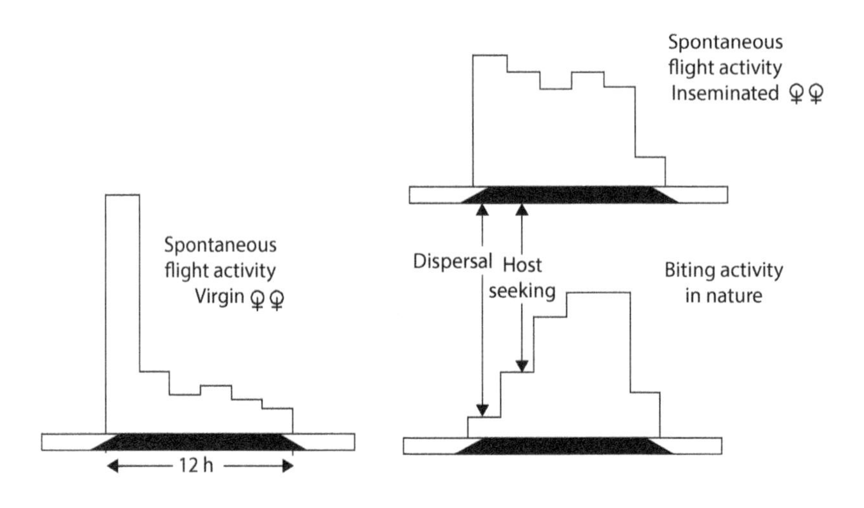

Figure 2.6 Flight activity of virgin and mated female *A. gambiae* in the laboratory compared to biting activity in nature. (From Jones, M.D.R., *J. Soc. Chem. Ind.*, 253–260, 1979.)

Smaller female *A. coluzzii* take a pre-gravid bloodmeal (which does not lead to the development of eggs) on the second night after emergence and mate on the third. Larger females mate on the second night after emergence after which they also take a pre-gravid meal (Figure 2.7). On the third night both large and small females feed again. This time with the blood-meal giving rise to egg development. According to the ecological theory of speciation, new variants, that might eventually develop into species with their own specific mate recognition system, occupy marginal habitats. These may become optimal habitats as the new species adapts. Whilst they remain marginal, the emerging adults may be nutritionally deprived compared to insects that developed in the usual, optimal, habitat. Under these circumstances, females may be more likely to require an initial, pre-gravid, meal compared to females that develop in the optimal habitat.

Environmental factors will also affect mating. In Ghana, the proportion of newly emerged female *A. coluzzii* that mated on a windy evening was lower than the mean, but the proportion that mated on the subsequent night were higher than normal (Figure 2.8).

Anopheles are eurygamic, meaning that they mate in large open spaces. Figure 2.9 shows the swarming site of *A. melas* in The Gambia whilst Figures 2.10 and 2.11 show typical swarming sites of *A. funestus* from Mozambique.

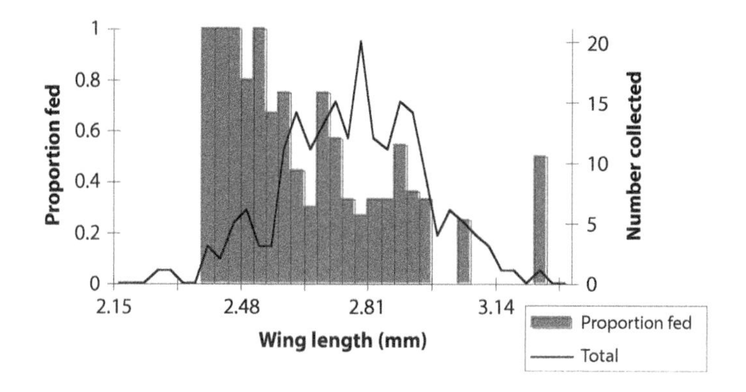

Figure 2.7 The proportion of *A. coluzzii* (formerly M form *A. gambiae*) females from São Tomé that took a blood meal before mating according to their wing length. (From Charlwood, J.D. et al., *Malar. J.*, 2, 7, 2003c.)

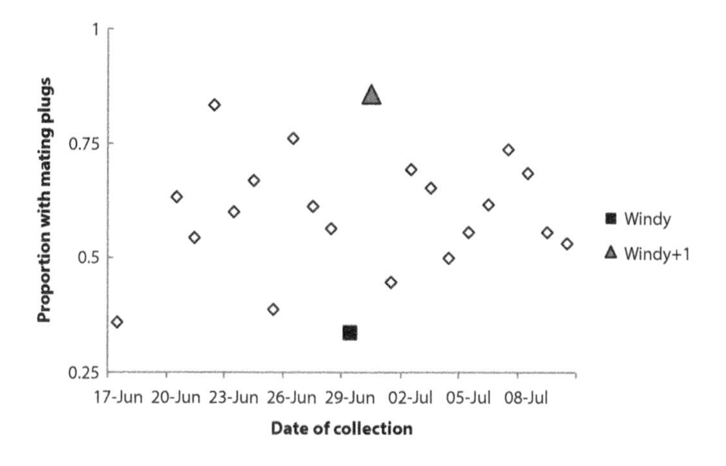

Figure 2.8 Proportion of female *A. coluzzii* with mating plugs (indicative of mating the previous night) by date of collection Okyereko, Ghana. Black square: windy night; filled triangle: following night. (From Charlwood, J.D. et al., *Med. Vet. Ent.*, 26, 263–270, 2012.)

Figure 2.9 Swarm site of *A. melas* in the salt flats close to the village of Keneba, The Gambia. (From Charlwood, J.D. and Jones, M.D.R., *Physiol. Entomol.*, 5, 315–320, 1980.)

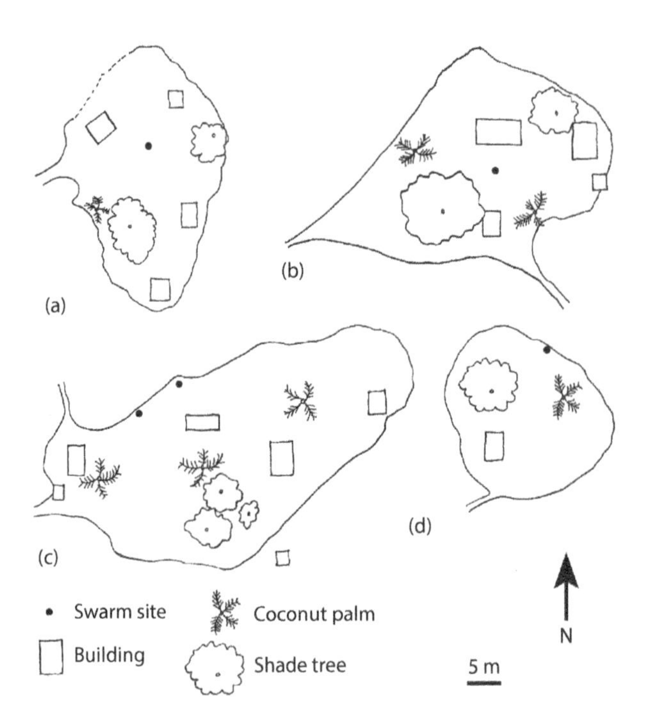

Figure 2.10 Swarming sites used by *A. funestus* in Furvela Village, Mozambique. The sites occur in sandy areas cleared around houses. (From Charlwood, J.D. et al., *Malar. J.*, 2, 3, 2003d.)

Figure 2.11 Swarm site at location (b) looking to the west. The arrow indicates the approximate position of the swarm. (From Charlwood, J.D. et al., *Malar. J.*, 2, 3, 2003d.)

Aedes, like *A. aegypti*, on the other hand are stenogamic, meaning that they can mate in small spaces. Individual male *Aedes* search out and mate with individual females whilst mating in *Anopheles* is a group phenomenon. This difference makes it more difficult to colonise anophelines compared to aedines. Male *Aedes* will encounter females of other species and previously mated females of their own species (who will not mate again). This means that *Aedes* males need to have a repertoire of techniques, such as contact chemical recognition, to recognise conspecific females that have already mated and members of other species that have the same flight tone (so as to avoid wasting their sperm), but male *Anopheles* are only likely to meet virgin females of their own species (who come to them to mate). Male *Anopheles*, therefore, do not need to have such refinements and in the laboratory will attempt to mate with anything that makes the right sound.

Swarms are characterised by a flight pattern, typically a figure of 8, that enables the male to stay over the marker (Figure 2.12). A single male can 'swarm'. Swarms may be considered stationary flocks. Flocking behaviour in simulated creatures ('boids') can be derived if they follow three simple rules. Each boid acts (1) to maintain a minimum distance from other objects in the environment, including other boids; (2) to match velocities with other boids in its neighbourhood; and (3) to move towards the perceived centre of mass of the boids in its neighbourhood. So, it is likely that mosquitoes follow similar rules to swarm.

The swarming behaviour described in Figure 2.12 was observed using a system that used an infrared sensitive video camera in the laboratory (Charlwood, 1974). More recently a much more sophisticated system using two cameras has been developed to monitor flights in wild swarming mosquitoes in three dimensions. This has enabled mating to be filmed directly in swarms (Figure 2.13).

Different species use different markers. For example, *A. coluzzii* swarms form over dark markers but these are avoided by the closely related *A. gambiae*, whilst *A. pharoensis* (Figure 2.14) swarm over light markers. In this way, the different species avoid each other.

Male mosquitoes have plumose, feathery antennae (Figure 2.15). In *Culex* and *Aedes* the fibrillae are always erect. In anophelines they are erect for only a short time each day, coincident with male swarming activity. Each hair on the antenna has an annulus at its base. A change in pH makes the annulus swell and as it does so the fibrillae become erect (Figure 2.16).

The erection of the fibrillae is under the control of a circadian clock and so continues when males are kept in constant dark (Figure 2.17).

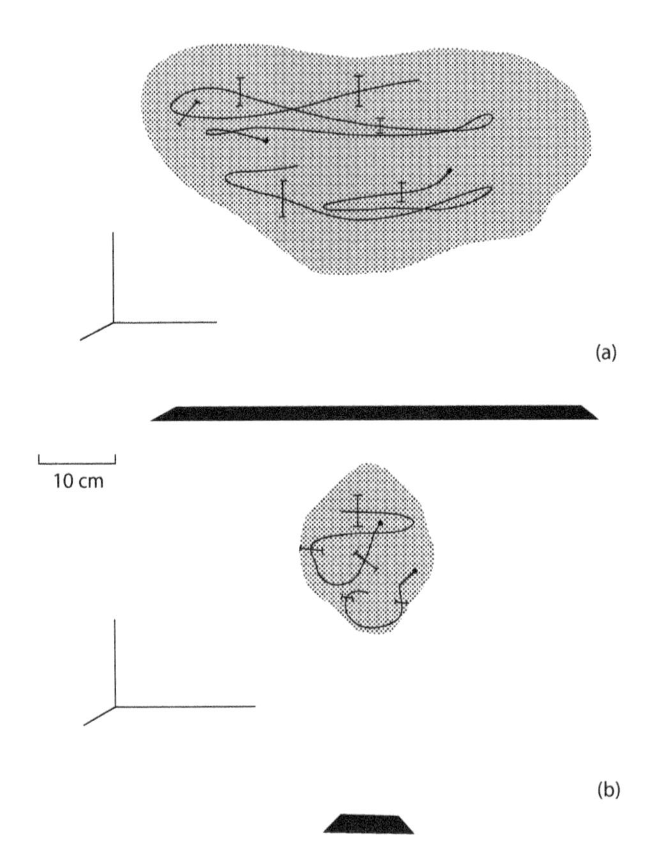

Figure 2.12 Two vertical elevations of representative 'swarming' flight paths (each c. 2 seconds duration) by male *A. coluzzii* over a 55 × 11 cm marker strip on the floor of a 1.2 m cube cage. Bars represent the range in accuracy of measurement of the same flight traced 10 times on a TV screen from an infrared video recording. (a) two flights seen in elevation parallel to axis of marker (drawn approximately in scale perspective view); (b) two other flights seen with the marker moved through 90° relative to the cage and TV camera. In each case the shaded area represents the limits within which four males flew in 1 min. Three-pronged lines indicate approximate perspective position of cage corner. (From Charlwood, J.D. and Jones, M.D.R., *Physiol. Entomol.*, 4, 111–120, 1979.)

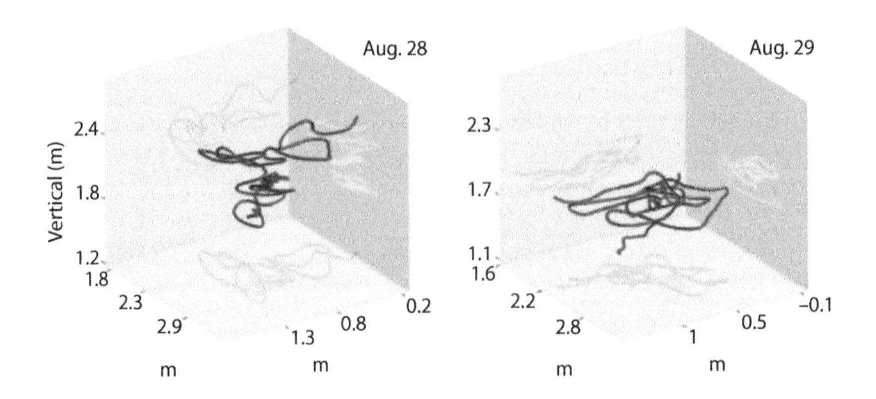

Figure 2.13 Three-dimensional reconstruction of *A. gambiae* mating events in the wild. The female mosquito track (red) and male track (blue) are shown. The couple is shown in purple. Pre-coupling tracks are projected onto two-dimensional planes on each side. (From Butail, S. et al., *J. Royal Soc. Interface,* 9, 2624–2638, 2012.)

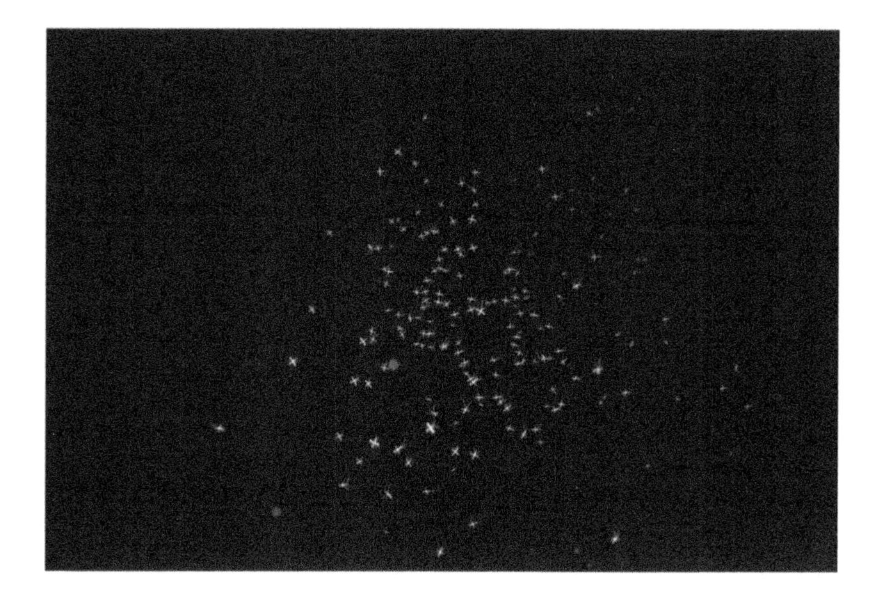

Figure 2.14 A swarm of *A. pharoensis* from Mozambique: note the erect antennae of the males.

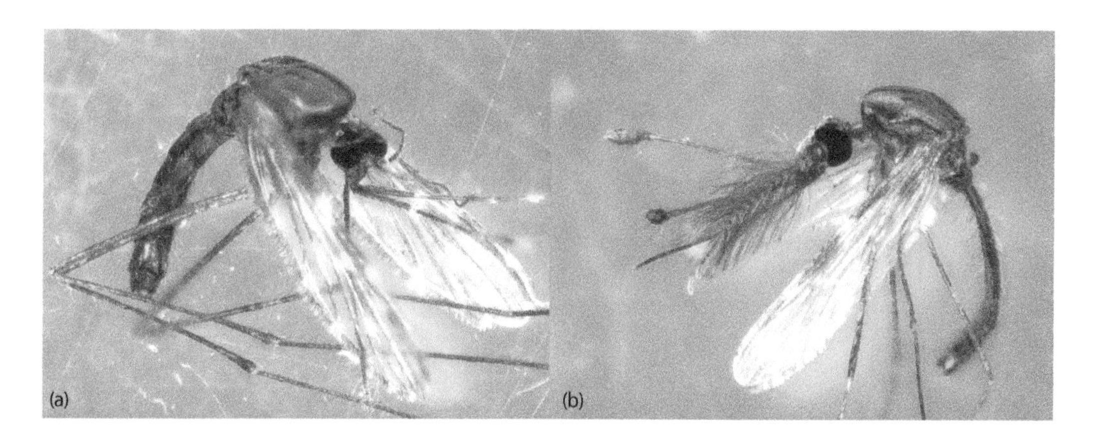

Figure 2.15 (a) Female and (b) male *Anopheles funestus*. Note the plumose hairs on the antenna and clubbed end of the palps in the male.

The fibrillae act as the male mosquitoes' 'ears' and respond to the female flight tone but not to the sound of other males. Females are also able to hear with their antennae but not as well as males. Females beat their wings approximately 400 times a second (400 Hz) but males beat their wings at 600 Hz. Despite their lower wingbeat frequency, females fly much faster than males; so in order to mate, the male has to increase his flight speed to catch up with the female once he has heard her. This may be one way that natural, or sexual, selection can act during mating.

Males are active a few minutes before females. In Furvela, Mozambique, both males and females rest inside houses. Insects can be caught, using a netting curtain placed over the open doorway, as they leave houses at sunset. Males leave a few minutes before females – it being better to be a few minutes early to mate rather than a few seconds too late (Figure 2.18). Unfed, virgin females and gravid females have similar activity patterns.

In the laboratory, virgin female *A. coluzzi* will also swarm over appropriate markers. In the wild they tend to arrive at a swarm site once the swarm has started. Both males and females use

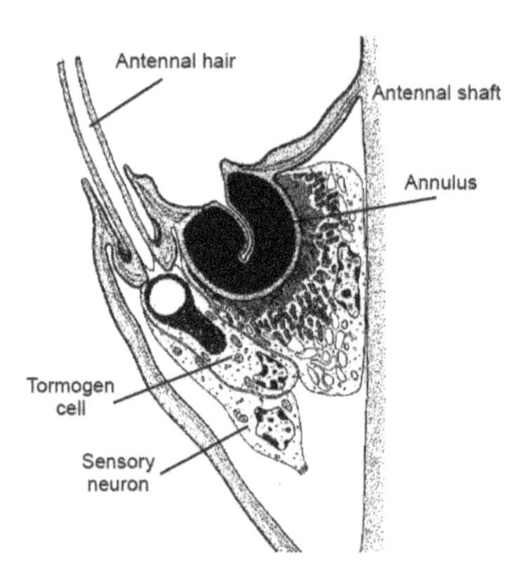

Figure 2.16 The base of a fibrillae on the male antenna showing the annulus, which swells with a change in pH. (From Nijhout, H.F. and Sheffield, H.G., *Science*, 206, 595–596, 1979.)

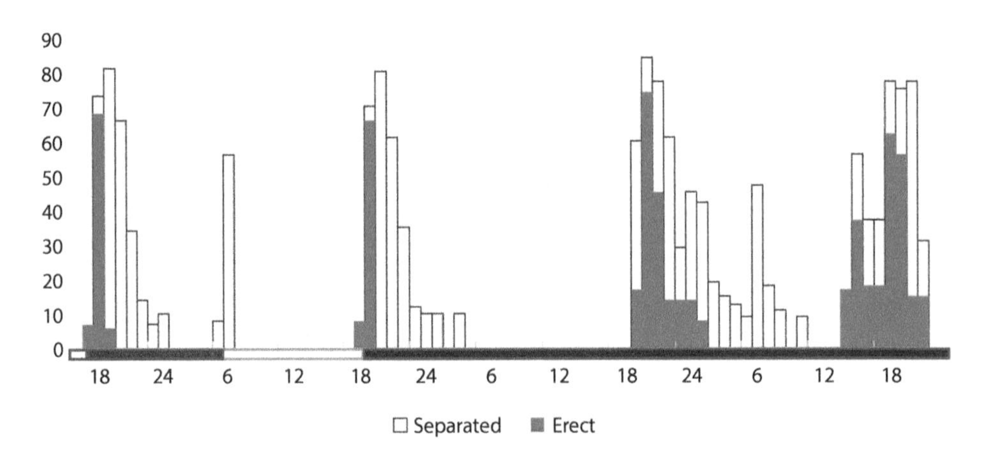

Figure 2.17 Cyclical erection of the antennae of male *A. coluzzii* continues when they are kept in constant dark. (From Charlwood, J.D. and Jones, M.D.R., *Physiol. Entomol.*, 4, 111–120, 1979.)

their hind legs for flight stabilisation, rather in the same way that the tail of a kite helps it to fly. Males have complex claws on their front legs (Figure 2.19) and simple ones on their other legs. The male flies in such a way that the claw on the front leg attaches to the outstretched hind leg of the female (Figure 2.20).

Once in copula the pair leave the swarm, the larger female pulling the male behind her. Approximately 17 seconds after copulation has been initiated the male ejaculates, depositing sperm in the female's spermatheca and a gelatinous 'mating plug', derived from the male's accessory glands, in the female's oviduct. The plug contains a number of hormones that, together with a rise in juvenile hormone (JH), help alter the female's behaviour – from mate seeking to host seeking. Females only mate once in their life, but a single male can successfully inseminate up to four females. Examination of the male's accessory glands shows whether the male has, or has not, mated. The glands are partially depleted if they have mated once but become almost fully depleted if they have mated twice.

Another way that mosquitoes may avoid mating with the wrong species could include mating at different times. Thus, in the laboratory, males of the *A. gambiae* complex respond to female flight tones

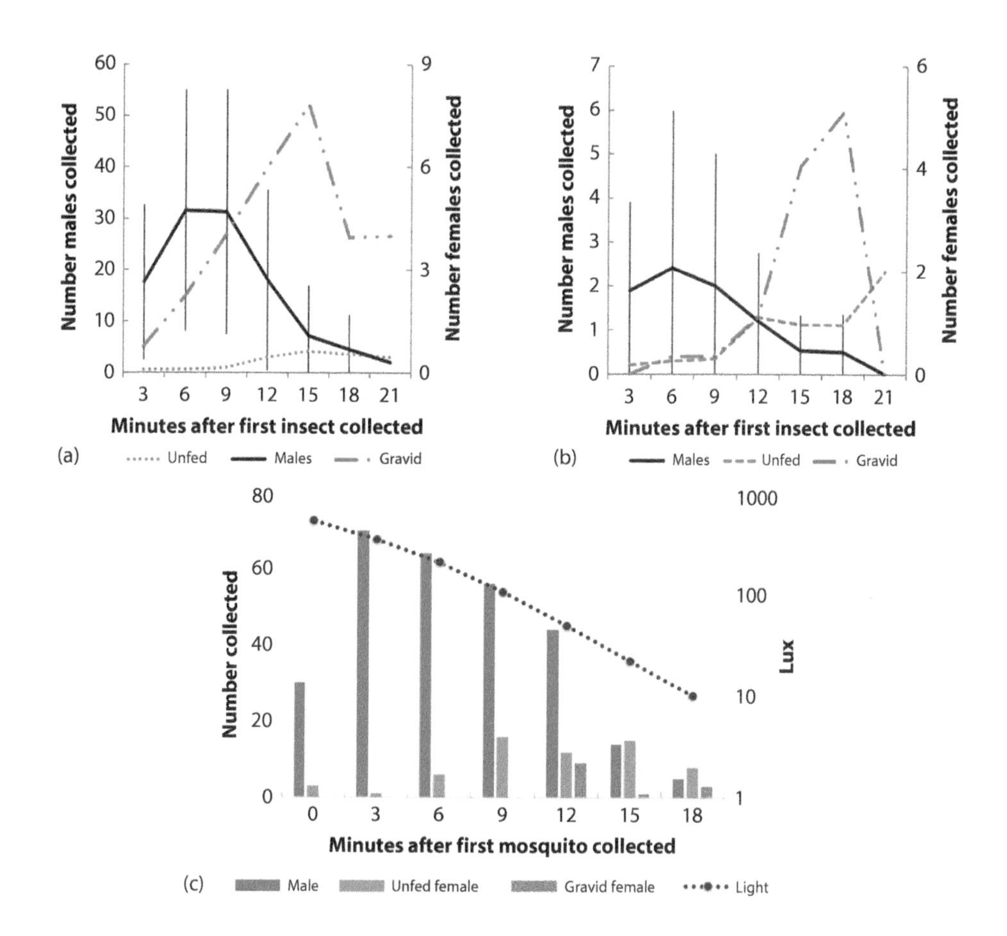

Figure 2.18 Time of collection of mosquitoes from Mozambique leaving houses relative to sunset (a) *Anopheles funestus* (b) *A. gambiae* s.l. and (c) *A. arabiensis* leaving vegetation from Eritrea. ([b] From Charlwood, J.D., *J. Vector. Ecol.*, 36, 382–394, 2011.)

Figure 2.19 The complex claw on the foreleg of a male *A. arabiensis*.

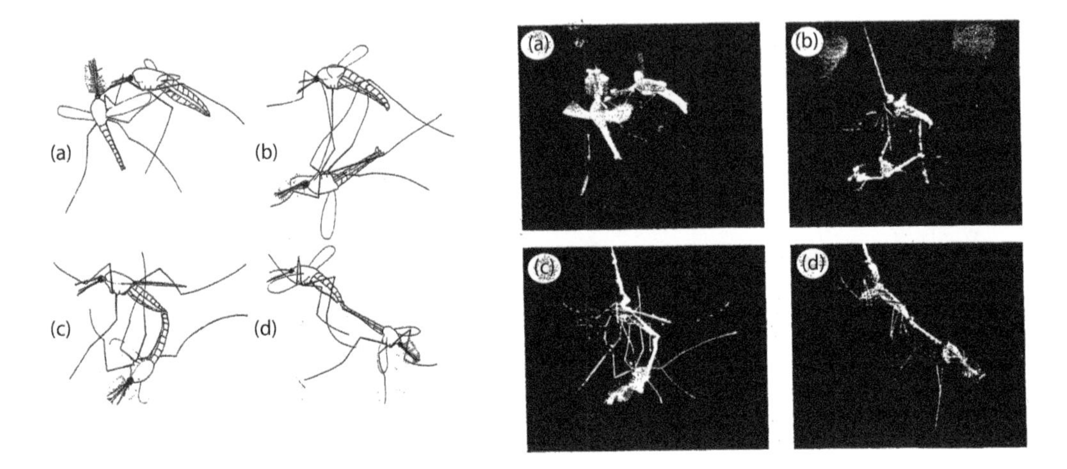

Figure 2.20 The mating sequence in *A. coluzzii*. (a) Males attach to the female using their complex claw on the front leg; (b) Using the momentum of their initial attachment, the male positions himself underneath the female; (c) Within a second or two the terminalia (claspers) of the male attach to the female; (d) The male releases his hold on the female's legs and the pair assume the end-to-end position. Note the erect antennal fibrillae of the male's antennae. (From Charlwood, J.D. and Jones, M.D.R., *Physiol. Entomol.*, 4, 111–120, 1979.)

at slightly different times (Figure 2.21a) and in both *A. coluzzii* and *A. funestus*, mating only takes place for a relatively short time – most taking place over the space of 15 minutes (Figure 2.21b and c).

A comparison of male flight activity (derived from individual insects in actographs) relative to the time of an artificial 30-minute 'sunset' and mating times after sunset observed in the field also indicates that time of mating may influence species isolation (Figure 2.22).

In Mozambique, the time that the first pair was seen in copula relative to the time of the start of swarming was a good indication of the number of pairs subsequently observed throughout the duration of the swarm (Figure 2.23).

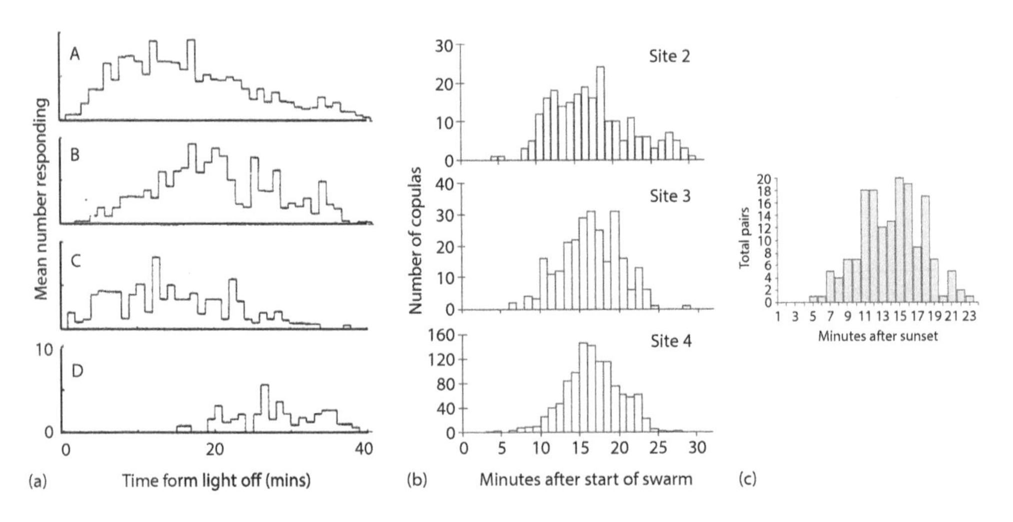

Figure 2.21 (a) Response times of different members of the *A. gambiae* complex to female flight tones following light-off in the insectary. (A) *A. coluzzii*, (B) *A. arabiensis*, (C) *A. melas* and (D) *A. merus*. (b) Number of mating pairs of *A. coluzzii* leaving three swarms in São Tomé relative to the start of swarming (c) Number of mating pairs of *A. funestus* leaving swarms relative to the start of swarming, Furvela, Mozambique. (From Charlwood, J.D. and Jones, M.D.R., *Physiol. Entomol.*, 4, 111–120, 1979; Charlwood, J.D. et al., *J. Vect. Ecol.*, 27, 178–183, 2002b; Charlwood, J.D. et al., *Malar.* J., 2, 3, 2003d.)

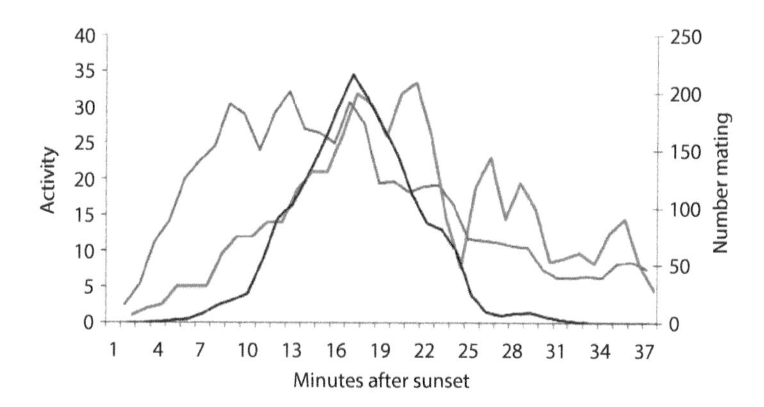

Figure 2.22 Activity patterns of male *A. arabiensis* (in blue) and male *A. coluzzii* (in red) in actographs following a sunset period of 30 minutes and mating times of *A. coluzzii* observed in São Tomé (in black).

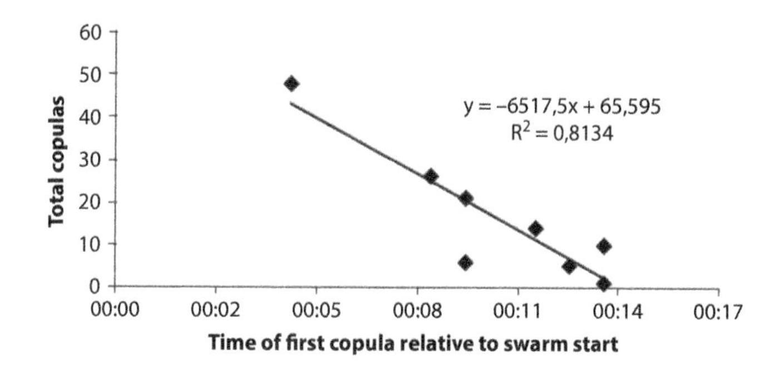

Figure 2.23 Time of the first pair of *A. funestus* seen in copula leaving a swarm relative to the total number of pairs subsequently seen during the whole swarming period. (From Charlwood, J.D. et al., *Malar. J.*, 2, 3, 2003d.)

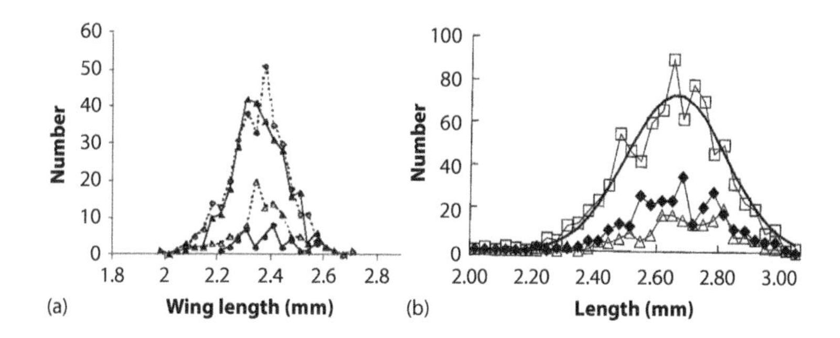

Figure 2.24 (a) Wing length of male *A. funestus* collected swarming (filled triangles), in copula (filled diamonds), and resting (open triangles) and (b) *A. coluzzii* swarming (open squares), in copula (filled diamonds) and resting (open triangles). For both species the mean sizes of each group were similar and approximated a normal distribution. (From Charlwood, J.D. et al., *Med. Vet. Entomol.*, 16, 1–3, 2002a; Charlwood, J.D. et al., *Malar. J.*, 2, 3, 2003d.)

In many species of animals, larger males obtain more mates than smaller males. In *A. funestus* and *A. coluzzii*, however, this does not seem to be the case because males caught in copula were the same size as the overall population (Figure 2.24a and b), but whether some males mate more than others remains unknown.

CHAPTER 3

The Search for the Host

BITING CYCLES

The time of day when the mosquito bites can have important consequences on disease transmission. *Aedes*, such as *Aedes albopictus* (also known as *Stegomyia albopicta*), bite during the day with a peak of biting in the late afternoon (Figure 3.1). In this way, they avoid competing with the night-biting *Anopheles*.

Among the *Anopheles*, biting activity is also distributed unevenly throughout the night. In Africa, the vectors tend to bite late at night when their preferred hosts, humans, are asleep and therefore unlikely to indulge in defensive behaviours (Figure 3.2).

Nonvectors, on the other hand, tend to bite in the earlier part of the night when their own hosts (historically game animals, more recently cattle and other domestic animals) are more accessible. Because many game animals come to waterholes in the early part of the night, a mosquito that feeds then will not necessarily have far to go to find a suitable oviposition site.

In general, despite the difference seen in spontaneous flight activity, young insects, including virgin insects, have similar biting cycles to parous ones. For example, the biting cycle of virgin and recently mated *A. coluzzii* from São Tomé and Príncipe were similar to those of parous insects, (but at the same time differed from the classic biting cycle described from the continent) (Figure 3.3).

There are some exceptions to this pattern. In Brazil, the malaria vector, *Ny. darlingi*, had biting cycles that differed between geographic locations (Figure 3.4). In particular, in Aripuãna in the Mato Grosso, the mosquito showed pronounced peaks of activity at dusk or dawn that were associated with young insects (Figure 3.5). Elsewhere in Brazil, no such peaks existed.

In species such as *Ny. darlingi*, when they switch from one activity (in this case, long-range host seeking) to another (in this case, biting) the females will spend some time 'intercurrent' resting in the vicinity of the host (Mattingly, 1965), In *Ny. darlingi* this would appear to be approximately 10 minutes because biting continued during a heavy shower but was curtailed immediately after wards (Figure 3.6). Intercurrent resting mosquitoes wave their hind legs in a circular motion. It is possible that they have sensillae on these limbs that sample the air for a different cue to that used by the antennae.

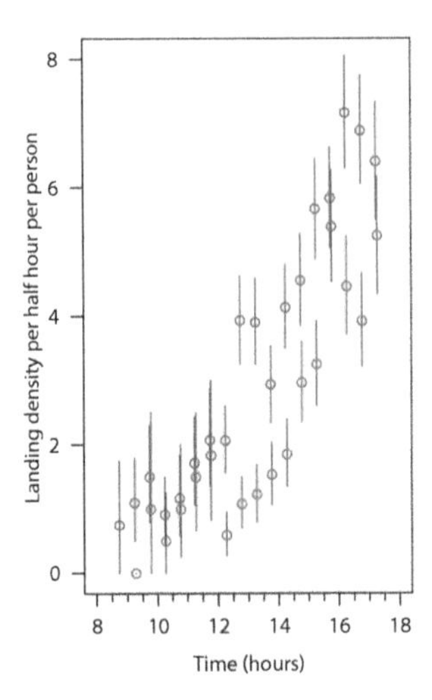

Figure 3.1 Biting activity of *Aedes albopictus* from the Ou Chra woods, Mondolkiri, Cambodia. Bars indicate 95% confidence intervals. Blue control collections, red with a pyrethroid repellent in use. (From Charlwood, J.D. et al., *Parasites Vec.*, 7, 324, 2014.)

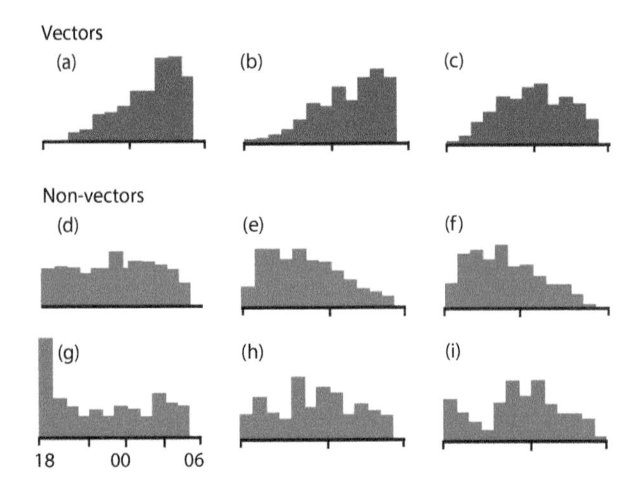

Figure 3.2 Biting cycles of African *Anopheles* vectors (a) *A. gambiae* s.l.; (b) *A. funestus*; (c) *A. nili*; non-vectors; (d) *A. coustani*; (e) *A. pharoensis*; (f) *A. wellcomei*; (g) *A. squamosus*; (h) *A. flavicosta*; (i) A. *brohieri*. (From Hamon, J. et al., *Bull. Soc. Entomol. Fr.*, 69, 110–121, 1964.)

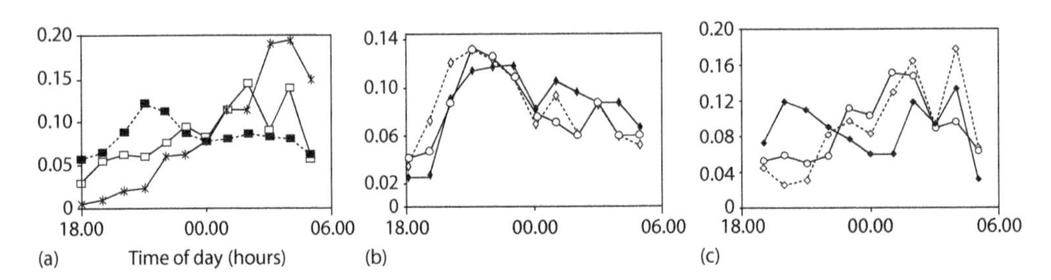

Figure 3.3 The biting cycles of *A. coluzzii*. (a) Overall cycles for the islands of São Tomé (filled square) and Príncipe (open square) and a composite figure from mainland Africa. (b) The cycles for virgin (open diamond), with mating plug (filled diamond) and mature (open circles) on the island of São Tomé and (c) Principe. (From Charlwood, J.D. et al., *Ann. Trop. Med. Parasitol.*, 97, 751–756, 2003b.)

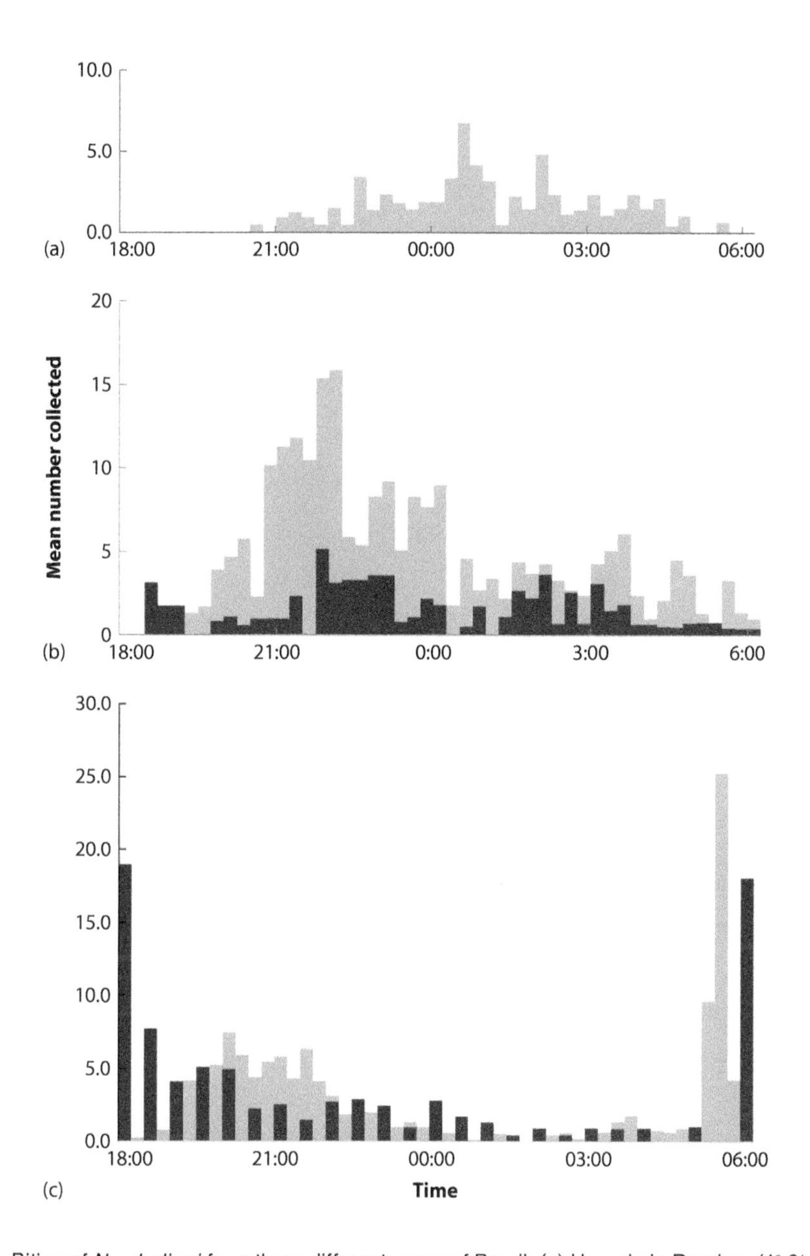

Figure 3.4 Biting of *Ny. darlingi* from three different areas of Brazil. (a) Uauaris in Roraima (4° 8′ N 64°29′ W in March 1977) (b) Km 137 on BR174, Amazonas (1°50′ S 60°30′ W – March in grey; dark grey in July 1977) and (c) Aripuanã in the Mato Grosso (10°11′ S 59°49′ W March 1977 in grey, April 1978 dark grey). The peak of activity observed in Aripuanã in the early morning was largely due to young nulliparous insects. (From Charlwood, J.D. and Hayes, J., *Acta Amazonica*, 8, 601–603, 1978; Charlwood, J.D. and Wilkes, T.J., *Bull. Entomol. Res.*, 67, 337–342, 1979.)

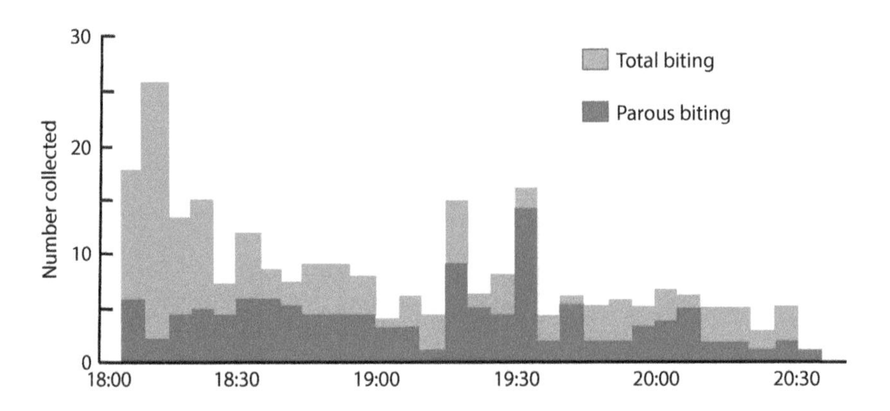

Figure 3.5 Age of *Ny. darlingi* females biting during the early part of the night from Aripuanã, Mato Grosso, Brazil (April 1978). Note the preponderance of nulliparous females during the early evening peak of biting. (From Charlwood, J.D. and Wilkes, T.J., *Bull. Entomol. Res.*, 67, 337–342, 1979.)

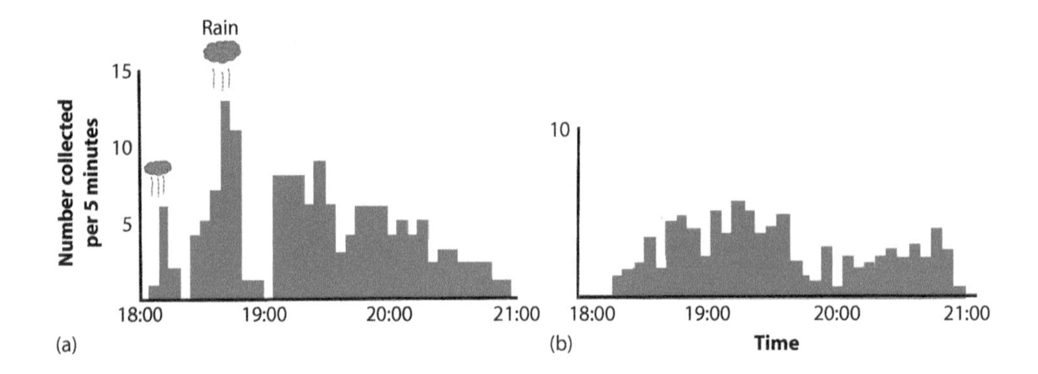

Figure 3.6 (a) Biting activity of *Ny. darlingi* at Aripuanã on one night when rainstorms occurred at 18:05–18:10 and 18:35–18:47 h; (b) Mean biting activity during the remainder of the week when no rain fell. The time of sunset has been adjusted to 18:00 h. (From Charlwood, J.D., *Bull. Entomol. Res.*, 70, 685–692, 1980.)

MOONLIGHT

Moonlight can also affect the biting cycle. In Papua New Guinea, more *A. farauti* were collected in nights leading up to a full moon than on moonless nights (Figure 3.7).

Numbers of mosquito biting were also greater during periods of full moon compared to moonless nights (Figure 3.8). Because many people spent more time outside their houses at such times, their exposure would have increased, and so, therefore, would the risk of malaria transmission.

In Mozambique, most *A. funestus* were also caught in tent-traps during periods of the night when moonlight was present (Figure 3.9).

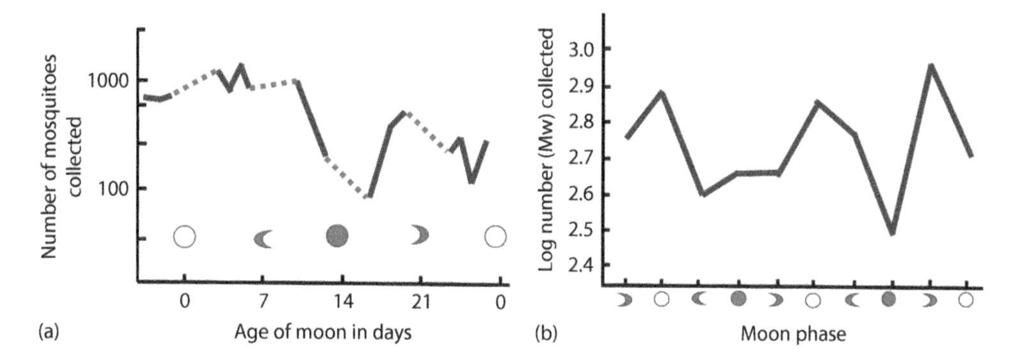

Figure 3.7 Numbers of *A. farauti* in landing collections (at different months) according to moon phase, Maraga, Madang, Papua New Guinea. (a) Over a single month and (b) over subsequent three month period. (From Charlwood, J.D. et al., *J. Med. Entomol.*, 23, 132–135, 1986d.)

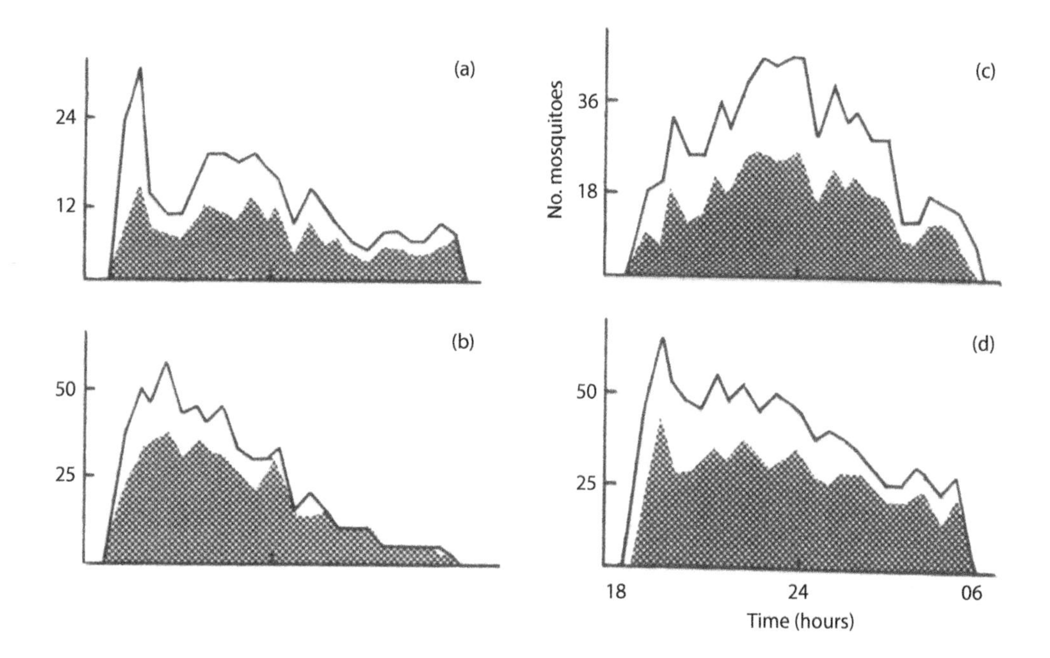

Figure 3.8 The biting cycle of *A. farauti* collected in Maraga, Madang Province, Papua New Guinea, according to moon phase. The figure is the Williams' mean of four collections (a, b, d) and six for (c). The shaded portion of the figure represents the parous fractions, and the graphs have been adjusted to a uniform height. (a) = no moon, (b) = waxing moon, (c) = full moon, (d) = waning moon. (From Charlwood, J.D. et al., *J. Med. Entomol.*, 23, 132–135, 1986d.)

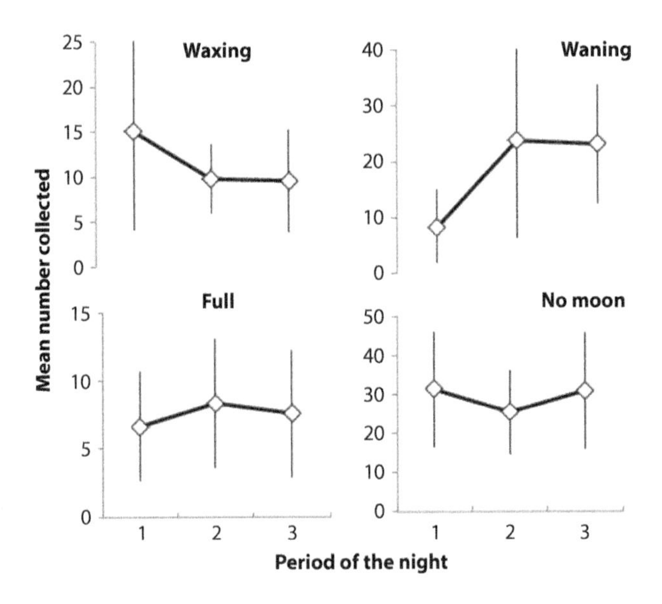

Figure 3.9 Proportion of the night's catch of *Anopheles funestus* from tent-trap collections according to phase of the moon, Furvela, Mozambique. (From Kampango, A. et al., *Med. Vet. Entomol.*, 25, 240–246, 2011.)

AGE AND BITING TIME

Although in general, the biting cycles of young and old insects are the same, among parous insects the proportion of insects that return to feed following oviposition (insects with large ovariolar sacs) may vary throughout the night (Figure 3.10).

In Ghana, numbers of part-fed and engorged *A. gambiae* s.l. (mainly *A. coluzzii*) increased during the night whilst unfed insects reached a peak just after midnight (Figure 3.11).

In this case the proportion of the collection that was parous (which was always low) decreased during the night (Figure 3.12).

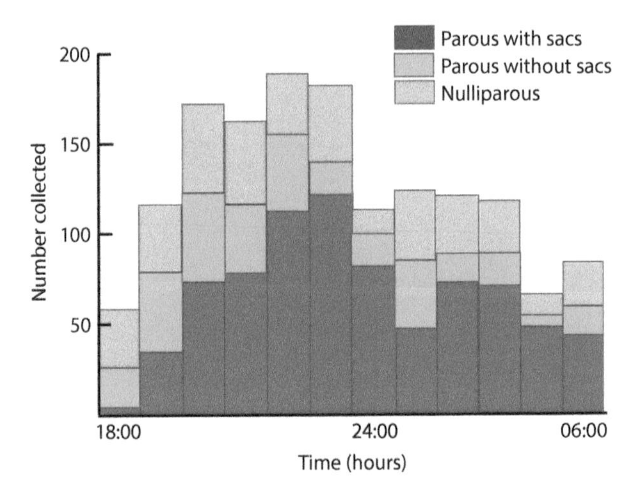

Figure 3.10 Biting activity of *A. farauti* from Madang, Papua New Guinea. Note that nulliparous insects had a similar biting cycle to parous insects but that among the latter the proportion of insects with ovariolar sacs changed with time of night. (From Charlwood, J.D. et al., *Med. Vet. Entomol.*, 2, 101–108, 1988.)

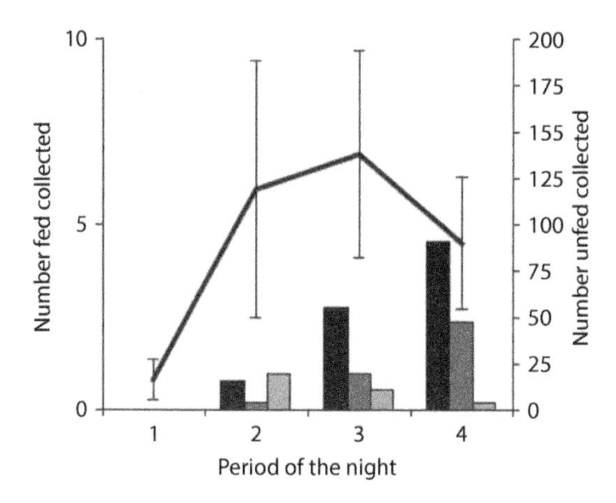

Figure 3.11 Numbers of *A. gambiae* s.l. collected in a tent-trap by period of the night, Okyreko, Ghana. (From Charlwood, J.D. et al., *Bull. Entomol. Res.,* 101, 533–539, 2011.)

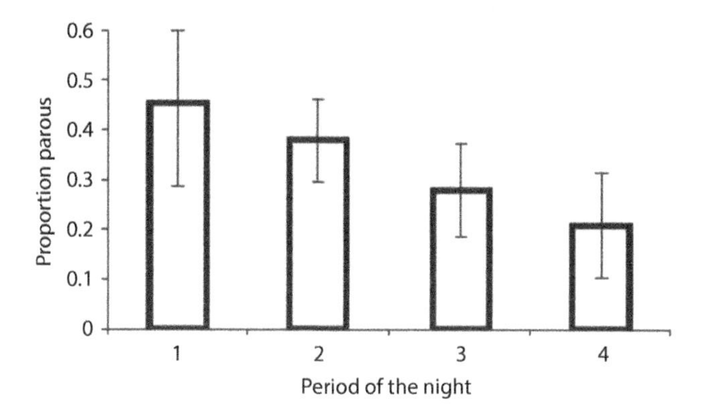

Figure 3.12 Proportion of the *A. gambiae* s.l. that were parous from tent-traps by time of the night, Okyreko, Ghana. (From Charlwood, J.D. et al., *Bull. Entomol. Res.,* 101, 533–539, 2011.)

BLOOD FEEDING

Mosquitoes primarily use olfactory cues to locate their hosts. Out of 5000 acids, alcohols, ketones and aldehydes taken from the armpits of 200 volunteers, 44 of them varied enough to give an individual chemical profile that can be read like a fingerprint. Thus, we have enough volatile molecules to satisfy even the most demanding of blood-feeding insects' location mechanisms.

The female mosquito antenna has numerous peg-like receptors (Figure 3.13), the cuticle of which is characterised by the presence of numerous small pores that permit the entry of chemicals. These sensillae are filled with a liquid that coats the entire sensillum.

The process of olfaction is shown in Figure 3.14. Volatile molecules from the host pass through the pore in the sensillum and enter the liquid space between the dendrite and the cuticle. They then form a compound with odour receptor molecules that bind to the dendrite receptor and cause a nerve impulse. After a short period, a degrading enzyme breaks down the odour molecule, freeing up the receptor, which again returns to the liquid medium of the sensilla (Figure 3.14).

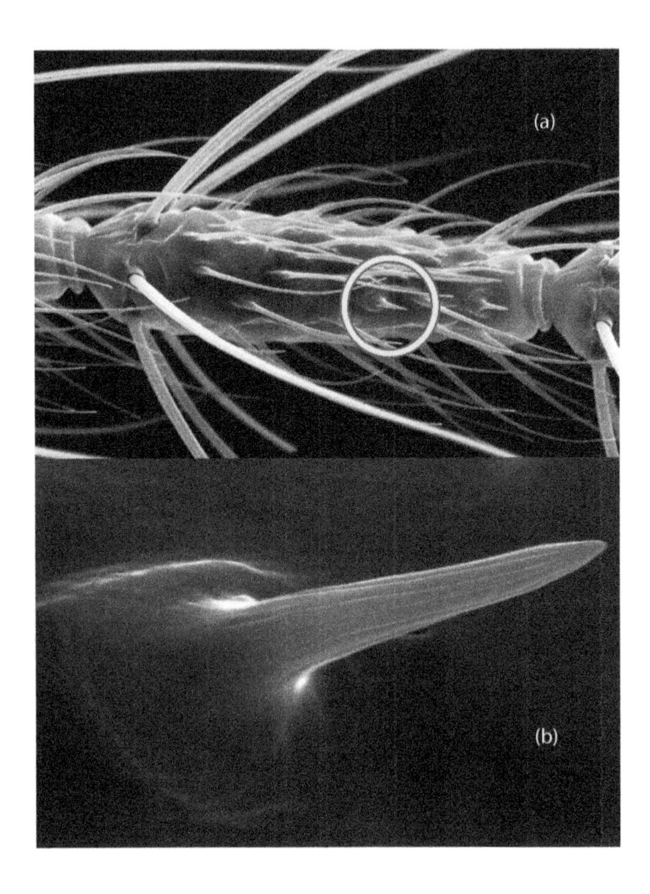

Figure 3.13 (a) Scanning electron micrograph of a segment of the antenna of a female mosquito. (b) Typical sensilla as in the yellow circle. (Courtesy of Jason Pitts.)

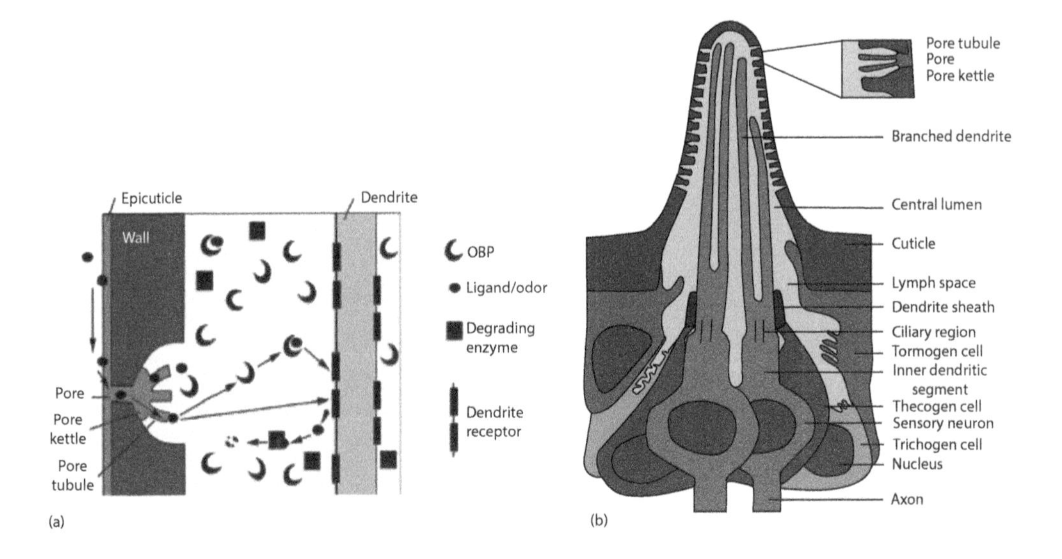

Figure 3.14 Olfactory sensillum. (a) Schematic longitudinal section through a sensillum of the single-walled type. (b) The pore region enlarged. (From Cribb, B.W. and Merritt, D.J., Chemoreception. Chapter 24/in *The Insects Structure and Function*, Chapman, R.F., Simpson, S.J. and Douglas, A.E. (Eds.), Cambridge University Press, Cambridge, UK, 2013.)

Mosquitoes disperse at random but generally upwind. Hosts give off an odour plume consisting of a variety of chemicals (lactic acid and other volatiles) and pulsed carbon dioxide. When they encounter the odour plume, mosquitoes turn towards it. If they leave the plume they make turning movements until they find it again. A single host (a human or calf) is located from approximately 30 m away (Figure 3.15).

Given the wide range of odours produced by different hosts, it is not surprising that different mosquitoes have evolved to respond to a different set of odours to locate their specific hosts. This again means that the different species will avoid competing with each other. Different hosts have different qualities of blood – corpuscle size differs between different animals as well as chemical constituents. This means that for the same amount of blood from different hosts, different numbers of eggs will be produced. Obviously, the more eggs produced the better (as far as the mosquito is concerned).

Mosquitoes that have evolved to feed primarily on humans are the ones that end up being vectors of human disease. The human blood index of vectors (the proportion of the population that feed on humans) is generally higher than 50% and, in some cases, (such as is the case for *A. funestus*) approaches 100% (Table 3.1).

When offered the choice between human and calf, vectors tend to prefer humans (Table 3.2).

In some circumstances, however, as is the case for *A. coluzzii* in São Tomé and Príncipe, both dogs and pigs are widely fed on (Table 3.3), but this is in part due to the relative inaccessibility of people due to the manner of house construction on the archipelago.

Perhaps because of the type of house, the *A. coluzzii* that do manage to feed indoors tend not to stay there to complete gonotrophic development. Thus, there is a deficit of gravid females from indoor resting collections but not from collections of mosquitoes resting underneath houses. Such collections also include a large number of male mosquitoes (Table 3.4).

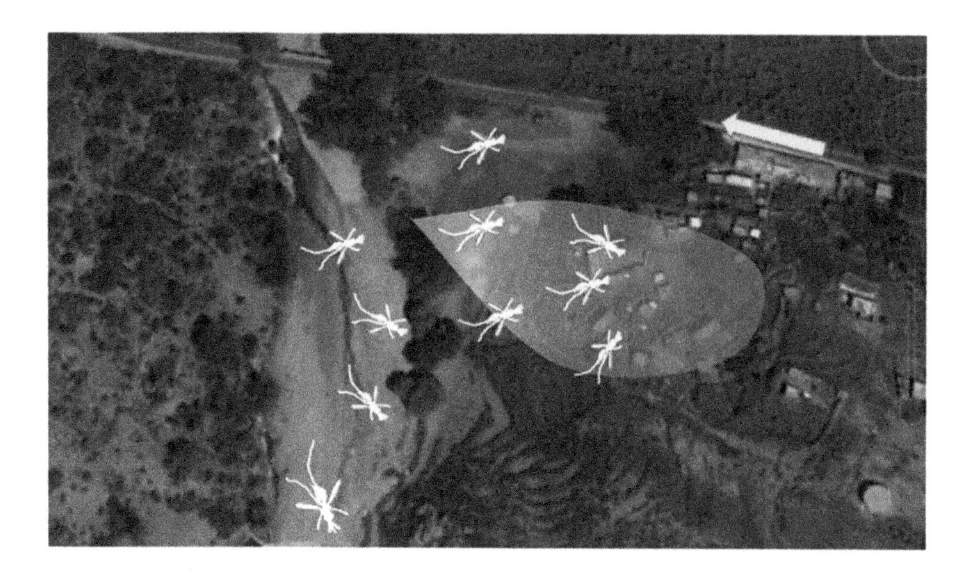

Figure 3.15 Schematic figure of the odour plume coming from a village and the tracking behaviour of mosquitoes coming from a focal breeding site (in this case, the drying river close to the hillside village of Adi Bosko in Eritrea). The yellow arrow denotes the wind direction. (After Gillies, M.T., Anopheline mosquitos: Vector behaviour and bionomics. Chapter 16 in *Malaria: Principles and Practice of Malariology*, Wernsdorfer, W.H. and McGregor, I. (Eds.), Churchill Livingstone, London, UK, 1989.)

Table 3.1 The Human Blood Index of Major, Minor and Nonvectors of Malaria from Different Parts of the Globe

Human Blood Index		
>0.5	**0.1–0.5**	**<0.1**
A. farauti[a]	*A. culicifacies*[b]	*A. algeriensis*
A. funestus[a]	*A. messeae*[b]	*A. aconitus*[b]
A. gambiae s.l.[a]	*A. quadrimaculatus*[b]	*A. annularis*
A. koliensis	*A. rufipes*	*A. mascarnsis*
A. nili[a]	*A. sergentii*[a]	*A. multicolour*
	A. sinensis[b]	*A. subpictus malayensis*
	A. sundaicus[a]	

Source: Garrett-Jones C. *Bull. World Health Organ.*, 30, 241–261, 1964.
[a] indicates major vectors
[b] other vectors

Table 3.2 Results of Trials with Man and Calf in Separate Bed-Net Traps in Upper Volta, Showing Percentage of Total Catch in Man-Baited Trap

		% of Feeds on Man
Vectors	*A. funestus*	84
	A. gambiae	77
	A. nili	65
Nonvectors	*A. flavicosta*	24
	A. pharoensis	13
	A. coustani	12
	A. squamosus	10
	A. brohieri	4
	A. rufipes	1

Source: Hamon, J. et al., *Bull. Soc. Entomol. Fr.*, 69, 110–121, 1964.

Table 3.3 Blood Meal Source of *A. coluzzii* from Riboque, São Tomé

			Resting Collections			
	Light Trap Collections		Indoors		Outdoors	
Host	**Number**	**%**	**Number**	**%**	**Number**	**%**
			Single Feeds			
Human	379	87.3	161	93.4	87	20.6
Dog	11	2.5	1	0.5	185	43.8
Pig	0	0.0	10	5.2	84	19.9
Chicken	1	0.2	0	0.0	0	0.0
Goat	0	0.0	0	0.0	0	0.0
Rat	0	0.0	0	0.0	0	0.0
Total	391	90.0	172	89.1	356	84.3
			Mixed feeds			
	22	5.1	21	10.9	45	10.7
			Negative/other			
	21	4.8	0	0.0	21	5.0
Total	434	100.0	193	100.0	422	100.0
HBI	0.92		0.94		0.27	

Source: Sousa, C.A. et al., *J. Med. Entomol.*, 38, 122–125, 2001.

Table 3.4 ***Anopheles coluzzii*** **Captured Resting Indoors in 52 Rooms and in 33 Outdoor Resting Sites**

	Females					Males
Sites	**Unfed**	**Engorged**	**Half-Gravid**	**Gravid**	**Total**	**Total**
Indoors (PSC)	9	113	4	0	126	6
Outdoors (OR)	269	172	91	169	701	575

Source: Sousa, C.A. et al., *J. Med. Entomol.*, 38, 122–125, 2001.
Abbreviation: PSC = Pyrethrum spray catch; OR = outdoor resting.

Figure 3.16 Representation of the odour plume coming from a house. Odours travel farther than carbon dioxide (which when pulsed acts as an attractant). Once inside, convection currents from the body will also provide cues for the mosquito. On coming to the wall of the house, nonvectors turn aside but vectors go upwards.

The factors that make a mosquito likely to be a vector of disease are not just restricted to host choice. For night-biting insects, such as *Anopheles*, the ability to enter houses is also important. Many species will approach the odour plumes that emanate from the eaves and other openings of houses (Figure 3.16). Should there be openings close to ground level, then a variety of species will enter but if, as is usually the case, the openings are those between the roof and walls (either at the eaves or at the gable ends of houses) then it is only the vectors that enter. In other words, when hitting the wall of the house, nonvectors turn sideways whilst vectors turn upwards.

In addition to the mosquito undertaking active host seeking, outdoor resting mosquitoes may take the opportunity to feed should the mountain (the host) come to Mohammed (the mosquito). If the mosquito does not obtain a meal (or at least probe a host) within a short time of activation, it will desist. If it manages to probe, it will persist. Thus, *Ochlerotatus cantans*, a crepuscular biter, resting in woods during the day in England, attacked a host that arrived into the woods. They were marked with a dot of paint before being brushed off. Mosquitoes that were felt to probe were marked with one colour and those that were brushed off before probing with another. Those that probed returned and attempted to feed a number of times before eventually returning to rest. Nonprobers returned to rest straight after being brushed off the host.

Invitation Effects

During feeding, as a result of the metabolic processes involved, a mosquito may produce volatile compounds. Host-seeking females may perhaps use these compounds to find the host.

On the Wirral in England, groups of 25 hungry *O. cantans* were placed on the legs of a bait and a piece of paper was placed between the skin and the cages so that on one leg they had access to feed but not on the other. A 1.2 × 1.2 m piece of card was held between the host's legs. Two collectors caught and counted mosquitoes on each leg. After a dozen replicates, significantly more mosquitoes were collected on the leg that had feeding insects on it compared to the one that didn't (a mean of 16 per leg compared to 6.8 on the control leg). This implies that there was some sort of 'invitation' effect that helped unfed insects locate a suitable site on the host for feeding.

A similar situation was observed among *Aedes albopictus* in Cambodia. The site of biting on two hosts was noted. Relative to surface area, most mosquitoes showed a preference for landing on the head (Figure 3.17). Landings (within three minutes of each other) on a collector were about 2.5 times

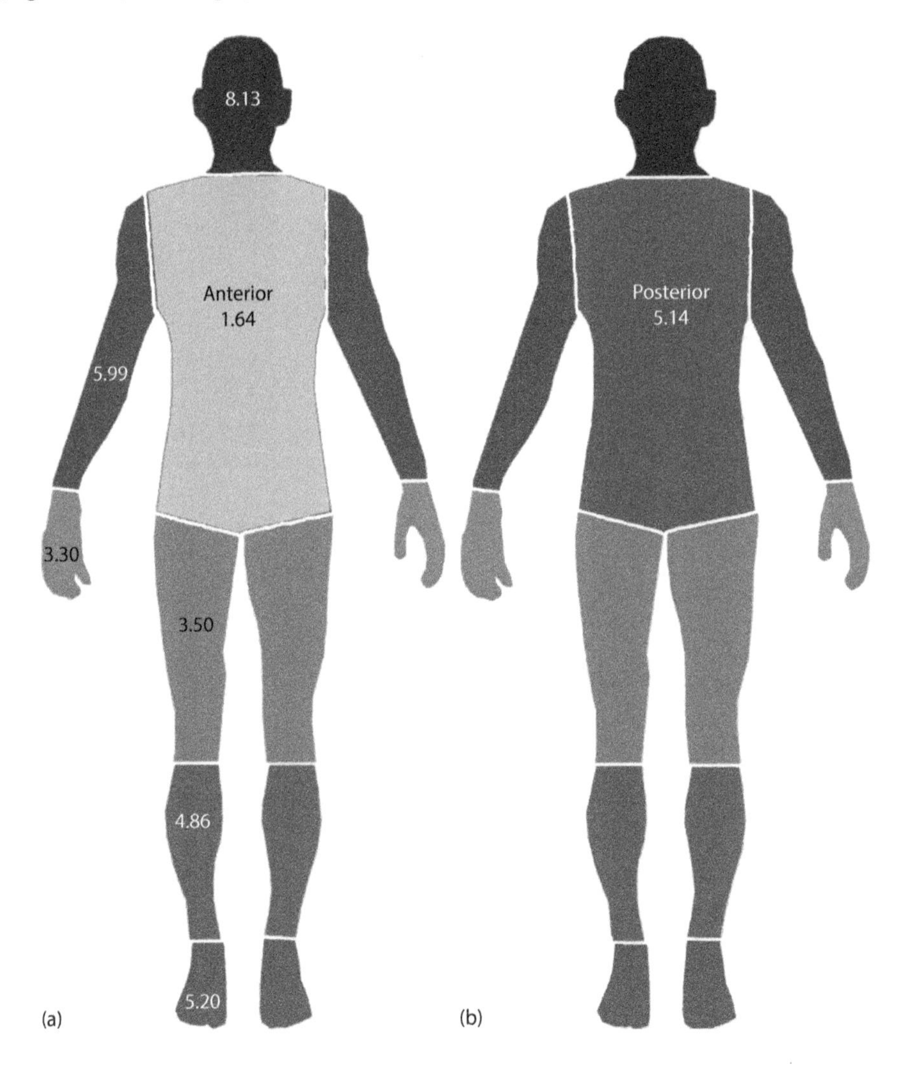

Figure 3.17 Average landing density (dm^{-2}) of *Aedes albopictus* on body regions, Okyerko, Cambodia. The head was the most attacked part of the body. (From Charlwood, J.D. et al., *Parasites Vec.*, 7, 324, 2014.)

more likely to be on the same body part than on a random body part, weighted for landing site preference. This again suggests that there may be an 'invitation' effect with successfully feeding insects signalling to other host-seeking females. The possible advantage of such a system has been questioned, but it is possible that the effect is only present in species where siblings may be feeding at the same time. It may also be useful for species in which only a limited part of the host is available to be bitten. Sheep for, example, are bitten close to their noses because the rest of the body is covered with wool.

Some mosquitoes have very definite feeding sites. In his book *Mosquitoes*, Gillett (1971) reported that *Sabethes bellasori*, a canopy-frequenting, monkey-feeding, mosquito, famous for the large, brilliantly coloured paddle-like scales on its legs, bit on the nose of the host. Indeed, the four that I caught up a tower in a forest close to Bélem in Brazil, all landed on my nose.

Once the host has been located, it is time to feed. The mouthparts of the female mosquito consist of several tubes protected by a labium (Figure 3.18).

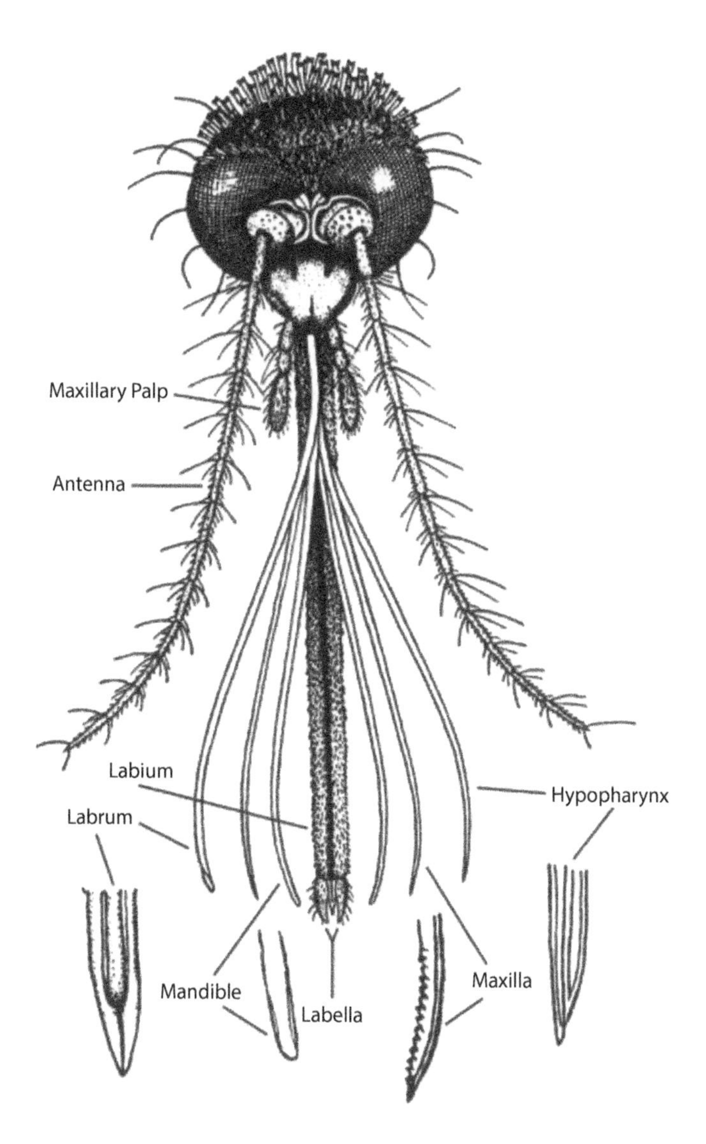

Figure 3.18 Mouthparts of a female mosquito. (From Matheson, R. *Handbook of the Mosquitoes of North America*, 2nd ed., Comstock, Ithaca, NY, 1944.)

The uppermost structure, the labrum, is slender, pointed and grooved along its ventral surface. In between the labrum and a 'lower gutter' (labium) are five needle-like structures; namely a lower pair of toothed maxillae, an upper pair of mandibles, which usually lack teeth (although in *Anopheles* they are very finely toothed), and finally a single hollow stylet called the hypopharynx. The tubes are held together by surface tension of the saliva and together are called the fascile. The different parts have different functions – pumping saliva into the wound to prevent coagulation and to anaesthetise the biting site, as well as a tube to suck up the blood/saliva mixture (Figure 3.19).

Muscles on either side of the head alternately contract to push the fascile into the skin. Once in, it can search for a suitable blood vessel so that ingestion can start. The process of feeding is basically 'spit and suck' as seen in the action potentials of nerves in the proboscis (Figure 3.20). Mosquitoes can ingest between two and four times their body weight.

When stretch receptors in the abdomen fire (giving feedback to the brain), then feeding stops. If the nerves between abdomen and head are cut, then the insect will continue feeding until it explodes!

The internal anatomy of a female mosquito is shown in Figure 3.21. Most of the thorax is taken up with the flight muscle, although it is here that the salivary glands reside. The abdomen contains the digestive tract (stomach, crop and hindgut) as well as the reproductive system and Malphigian tubes (responsible for excretion). In mosquitoes, the crop is a separate sac that is located off the side of the oesophagus. Sugar meals end up in the crop whilst the blood passes directly into the midgut of the mosquito.

The foregut of mosquitoes contains a series of teeth-like structures, the cibarial armature, that protrude into the lumen of the gut. In *Anopheles* mosquitoes, the armature is well developed but it is less so in culicines. One effect of the armature is that it mechanically kills microfilariae (the infective stage of filariasis). In *Anopheles*, the transmission of filariasis increases as the number of

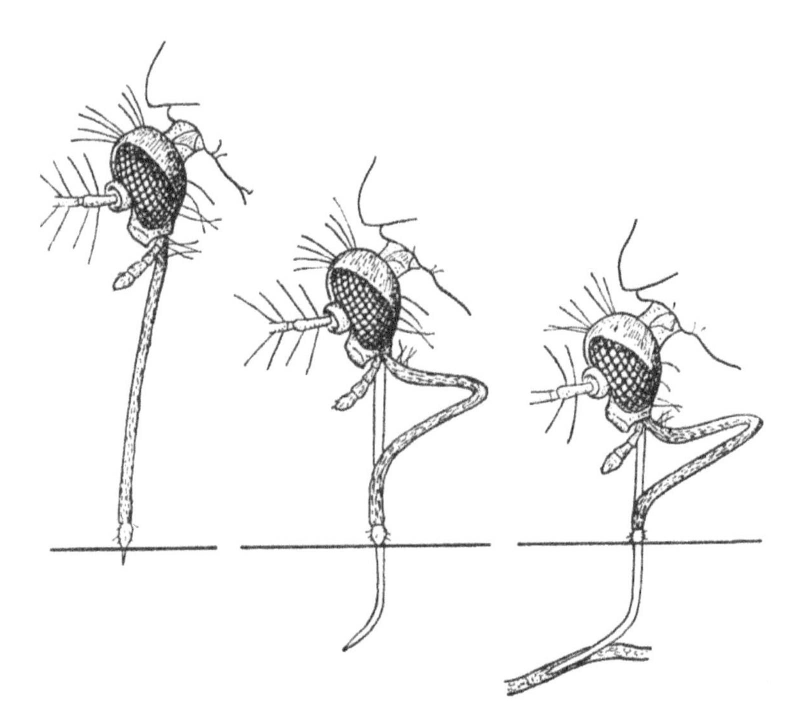

Figure 3.19 Feeding of a *Culex* mosquito.

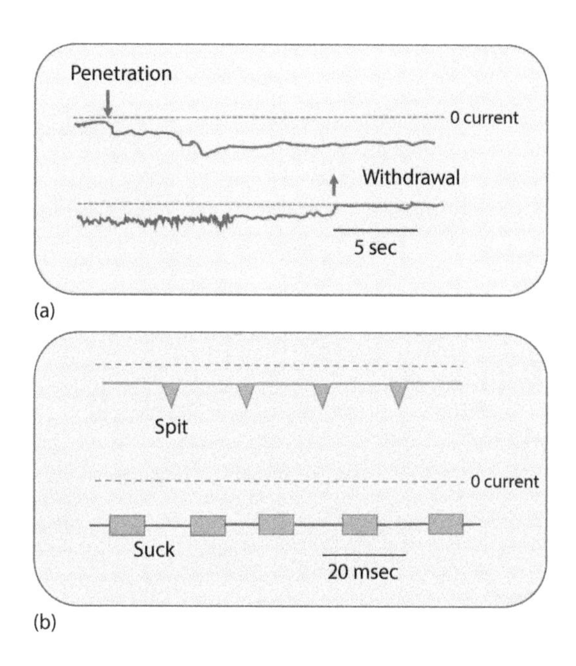

Figure 3.20 'Spit' and 'Suck' changes recorded during the feeding by individual females of *Aedes aegypti* on a mouse. Mosquito and mouse formed part of an electric circuit, which was closed when the mosquito's mouthparts penetrated the skin. Downward displacement of the trace indicates increases in current. Upward displacements indicate decreases in current. (From Kashin, P., *J. Insect Physiol.*, 12, 281–284, 1966.)

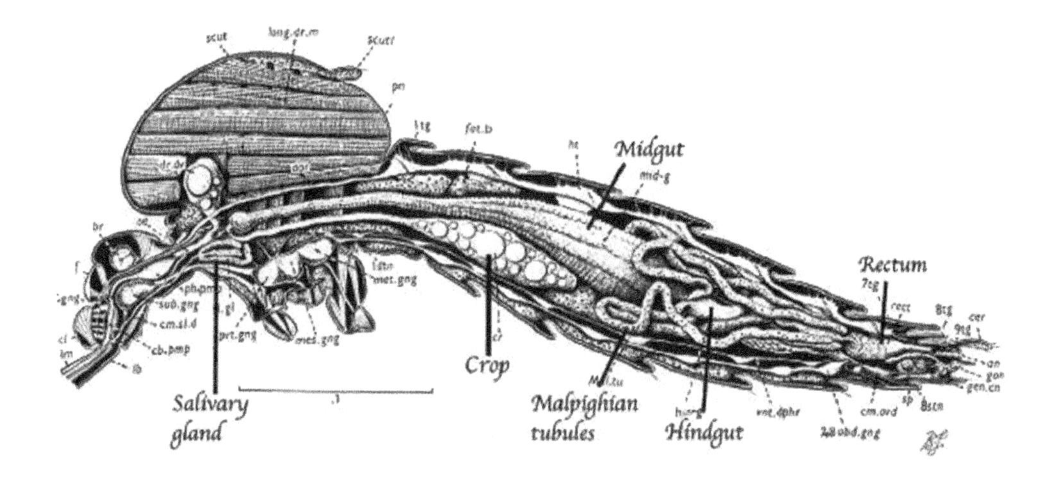

Figure 3.21 The internal anatomy of a blood-feeding insect. Above a generalised schema and below the mosquito. (From Jobling, Courtesy of the Wellcome Museum.)

microfilaria increases (a process known as facilitation) whilst in *Aedes* vectors survival of microfilariae decreases with increasing density (a process known as limitation).

As soon as it has a blood meal in the stomach and the stretch receptors have fired, the mosquito secretes a peritrophic matrix (PM). (The larva also produces a similar matrix but, because they are slightly different, they are given the name PM-1 for adults and PM-2 for larvae.) This is a chitinous

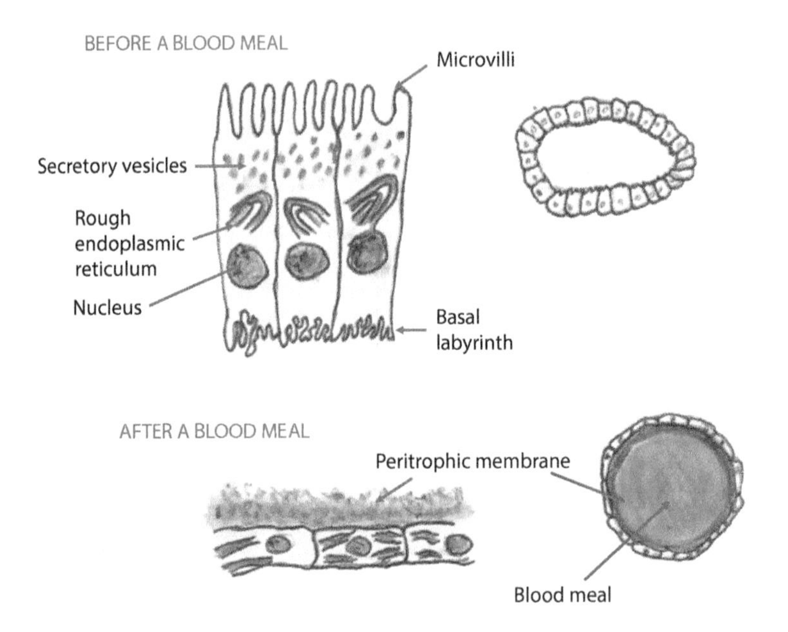

Figure 3.22 Epithelial cells lining the gut wall before and shortly after a blood meal. (Reprinted from *Biology of Disease Vectors*, 2nd ed., Devenport, M. and Jacobs-Lorena, M., The peritrophic matrix of haemotophagous insects, Chapter 22, Copyright 2005, with permission from Elsevier.)

matrix that allows small molecules to pass through but may help in the prevention of the penetration of bacteria.

In some species, the PM may also prevent the ookinete of the malaria parasite from penetrating the gut wall. Basically, vesicles in the epithelium of the gut pass into the gut itself where they bind together to make the PM (Figure 3.22).

Blood is generally a very fluid, and hot, medium and is a very dilute form of food. Thus, one of the first things to occur when a mosquito feeds is that excess liquid is excreted. This excretion may also help cool the insect.

The stages of blood meal digestion include the following:

- Removal of excess water – may also help temperature equilibrium
- Breakdown of vertebrate blood cells (haemolysis)
- Hydrolytic degradation of macromolecules in the blood meal (digestion)
- Absorption of small molecules into the midgut epithelial cells and subsequently into the haemocoel

Haemolysis breaks down cells to release proteins and other nutrients, making them accessible to the digestive enzymes. It may be achieved mechanically by the cibarial armature, or biochemically by haemolytic factors including small peptides and free fatty acids.

The *Anopheles* mosquito not only sucks up blood but also injects parasites during the bite when feeding. It also injects a cocktail of bioactive molecules including enzymes in the saliva. Some of these salivary compounds are essential to the *Plasmodium* life cycle. They have substantial anti-haemostatic, anti-inflammatory and immunomodulatory activities that assist the mosquito in the blood-feeding process by inhibiting several defence mechanisms of the human host. Many of them elicit strong immune responses, shown by the swelling and itching that accompany a mosquito bite.

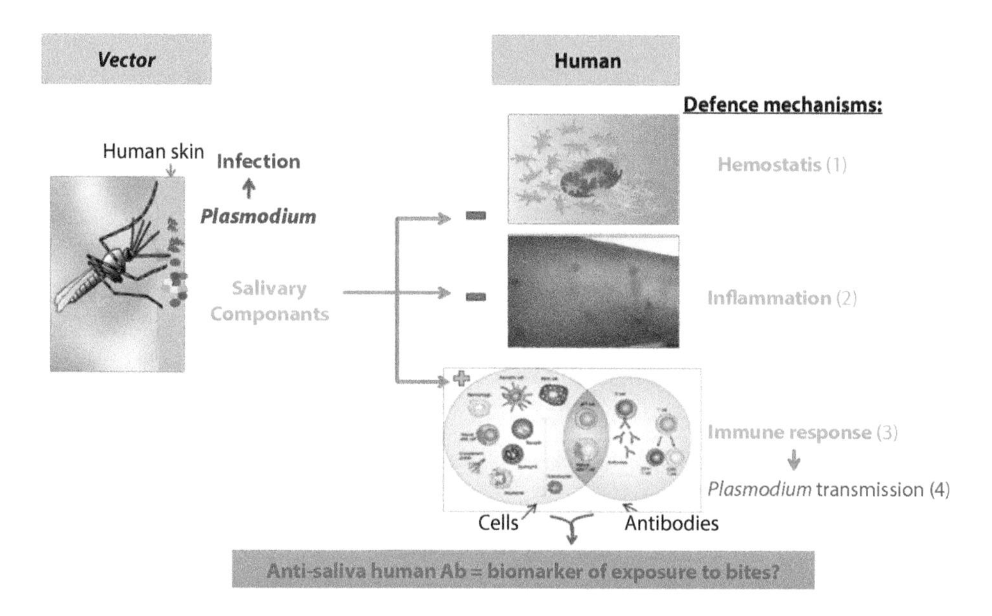

Figure 3.23 Effects of *Anopheles* saliva on haemostatic, inflammatory and immune reactions of the human to the bites of the vector. (From Drame, P.M. et al., New salivary biomarkers of human exposure to malaria vector bites. Chapter 23, in *Anopheles Mosquitoes—New Insights into Malaria Vectors*, Manguin, S. (Ed.), InTech, Rijeka, Croatia, 2013.)

Specific acquired cellular and/or humoral responses are developed by human individuals when they are bitten by mosquitoes. These immune responses may play several roles in pathogen transmission and disease outcomes (Figure 3.23). It remains possible that in areas of seasonal transmission the increased amount of biting at the start of the mosquito season elicits a response that is detected by circulating malaria parasites in asymptomatic people. The parasites may then start producing gametocytes, rapidly enhancing transmission.

The prevalence of gSG6 protein, found in the saliva of *Anopheles*, can be used as a marker of exposure to biting females. It appears to be specific to *Anopheles* and because of its synthetic nature reproducibility of the immunological assay is to be expected. Thus, levels of the antibody (determined by ELISA) differed in groups before and after control with insecticide was attempted (Drame et al., 2003).

Such an assay may be useful in the future as an indicator of recent exposure in areas where mosquito density falls to very low levels (Figure 3.24). It is still in relatively early stages of development and may not work outside of Africa. In a recent trial in Cambodia there was no difference seen between control and intervention villages despite a significant difference in *A. dirus* densities (Figure 3.25).

Once the mosquito has fed, and if the blood meal is sufficient, then host seeking is inhibited and egg development initiated (Figure 3.26). Anophelines are generally gonotrophically concordant and, with the possible exception of an initial pre-gravid feed, each blood meal results in the development of a full batch of eggs.

The inherent risks in host seeking and feeding differ according to the time of day that the insect feeds. Some of the advantages and disadvantages of feeding at different times of the day are outlined in Table 3.5. Host defensive mechanisms are particularly a factor for day-biting insects.

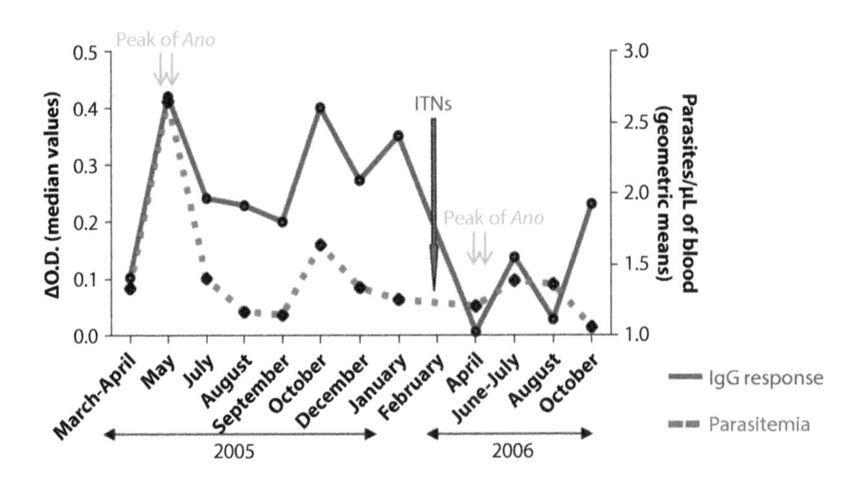

Figure 3.24 gSG6 levels and levels of parasitemia before and after insecticide-treated net distribution. (From Drame, P.M. et al., New salivary biomarkers of human exposure to malaria vector bites. Chapter 23, in *Anopheles Mosquitoes—New Insights into Malaria* Vectors, Manguin, S. (Ed.), InTech, Rijeka, Croatia, 2013.)

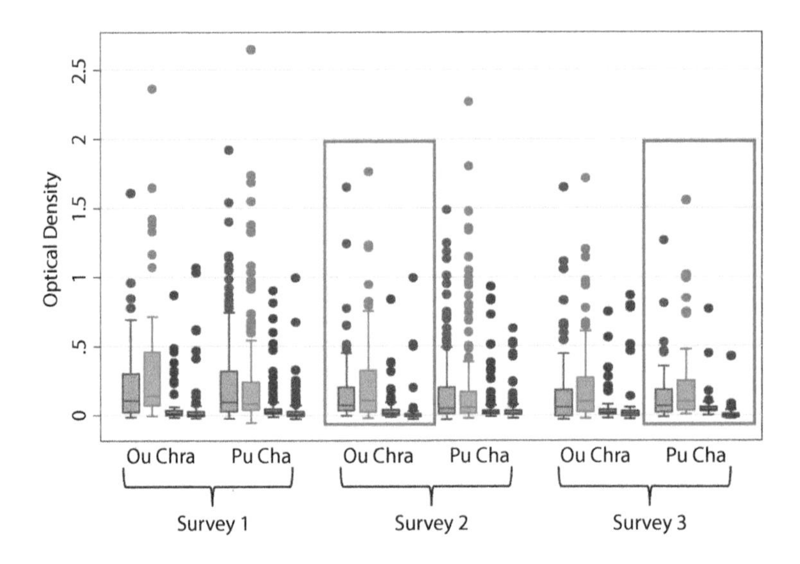

Figure 3.25 gSG6 levels in control and intervention villages (using spatial repellents) from Mondolkiri, Cambodia. (From Charlwood, J.D. et al., *Malar. World J.*, 8, 11, 2017a.)

As the numbers of mosquitoes attacking a host increase, then so might defensive behaviour on the part of the host. This may be an individual response (such as more active swatting) or a community response. In some situations (as was the case in Papua New Guinea before wide-scale distribution of nets), people may be more likely to use mosquito nets as density increases. This may reduce feeding success among the mosquitoes. In Tanzania, where the vectors were late-biting *A. gambiae* and *A. funestus*, there was, however, no evidence of decreased feeding success as numbers of mosquitoes increased (Figure 3.27). Should this have been the case then there would have been a

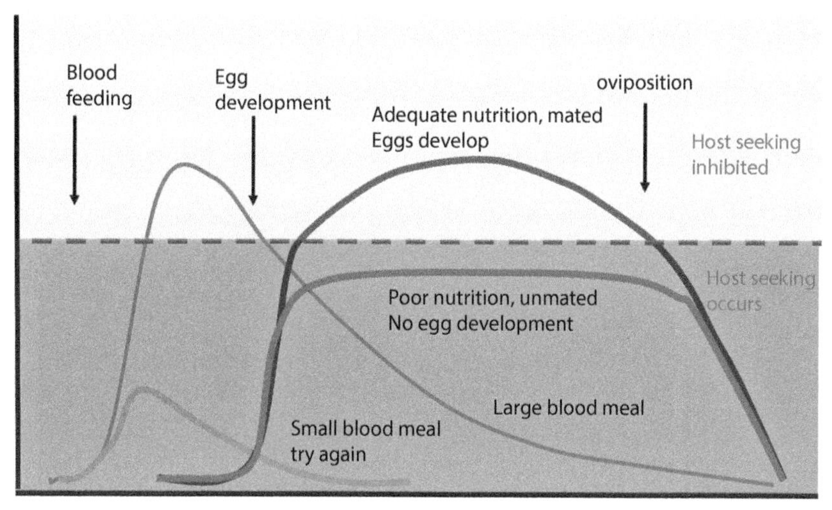

Figure 3.26 Factors resulting in the inhibition of host-seeking behaviour in blood-fed mosquitoes. (Reprinted from *Biology of Disease Vectors*, 2nd ed., Klowden, M.J. and Zwiebel, L.J., Vector olfaction and behaviour, Chapter 20, Copyright 2005, with permission from Elsevier.)

Table 3.5 Advantages and Disadvantages of Feeding During the Day or Night

	Day	Night	Crepuscular
Disadvantages	1. Greater risk of desiccation	1. Poor visual clues (especially colour)	1. Short window of opportunity to feed
	2. Greater wind turbulence	2. Low wine speed and hence poor directional clues in host-odour plumes	2. Predators still active
	3. Greater risk from predators	3. Greater background levels of atmospheric carbon dioxide	3. Sit and wait strategies not feasible
	4. Host mobile (disadvantage for odour-responding insects?)	4. Host less mobile, so sit-and-wait strategies less feasible	
	5. Greater risk from defensive behaviour of active host	5. Cold temperatures at high altitudes	
	6. Hot and low humidity in low tropical areas		
Advantages	1. Good visual clues	1. Less risk of desiccation	1. Visual cues available
	2. Higher wind speeds providing good directional clues in odour plumes	2. Host more likely to be at rest so reduced risk from host defensive behaviour	2. Hosts may congregate
	3. Reduced background levels of atmospheric carbon dioxide	3. Less risk from predators	3. Wind speed and humidity optimal
	4. Host mobile, making a sit-and-wait strategy feasible	4. Less atmospheric turbulence and hence more continuous odour plumes	4. Temperature optimal

Source: Gibson, G. and Torr, S.J., *Med. Vet. Entomol.*, 13, 2–23, 1999.

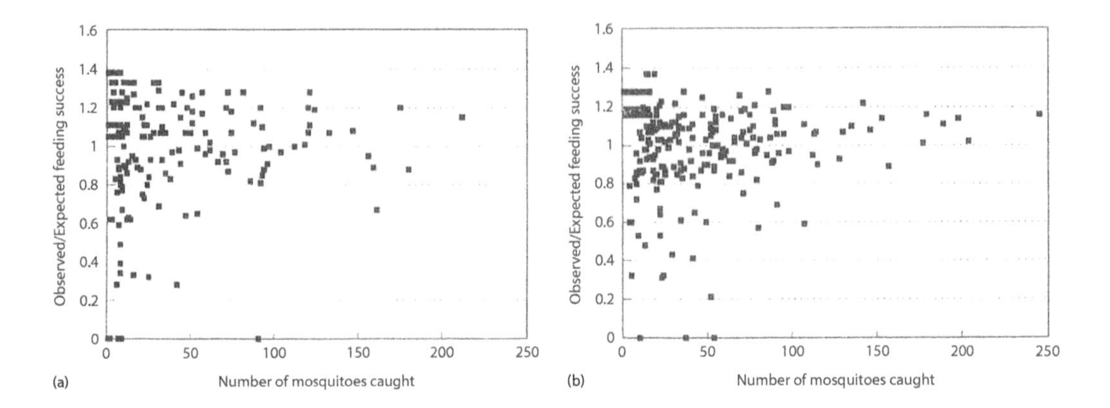

Figure 3.27 Relative feeding success by mosquito density from Namawala, Tanzania. (a) *A. gambiae* s.l. (b) *A. funestus*. (From Charlwood, J.D. et al., *Acta Tropica*, 60, 3–13, 1995a; Charlwood, J.D. et al., *Ann. Trop. Med. Parasitol.*, 89, 327–329, 1995b.)

declining trend in the feeding success as numbers increased. Relative feeding success was, however, relatively constant. This is probably because people, who at the time of high densities were occupied in preparing their fields for harvest, and who did not have mosquito nets for protection, were just too tired to resist.

Immediately following a blood meal, the abdomen of the mosquito is distended with blood and flight is limited. In Papua New Guinea, the majority of fed females dispersed less than 50 m. In this case they had fed on a unique host (a buffalo) that had been introduced into the village (Figure 3.28).

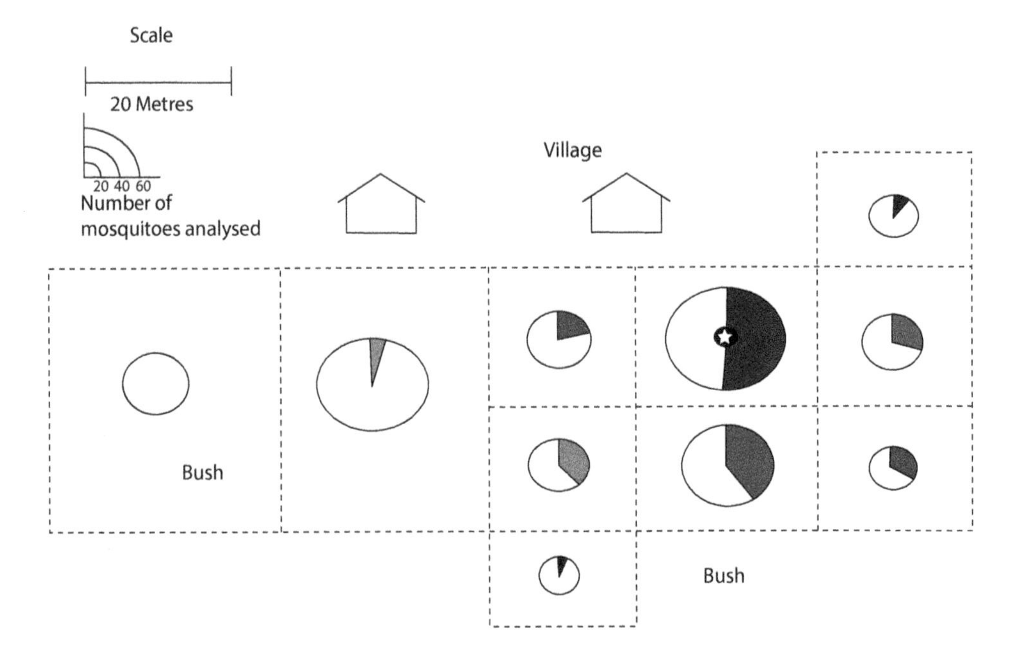

Figure 3.28 Map of recapture areas in Maraga village. The circles represent the number of engorged females of *Anopheles farauti* analysed, the shaded portion being the proportion of buffalo-fed mosquitoes in each area. The star denotes the position of the buffalo. (From Charlwood, J.D. et al., *Bull. Entomol. Res.*, 75, 463–475, 1985b.)

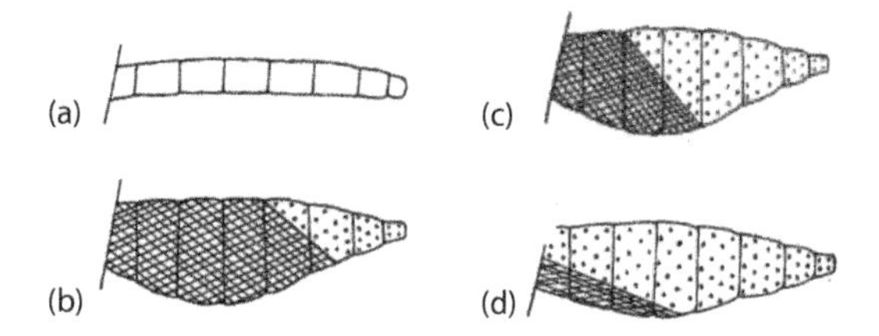

Figure 3.29 Appearance of a female mosquito's abdomen: (A) unfed; (B) recently engorged; (C) semi-gravid; (D) gravid. Ready-to-lay semi-gravid insects are commonly seen during the cooler months of the year when gonotrophic development takes three, instead of two, days.

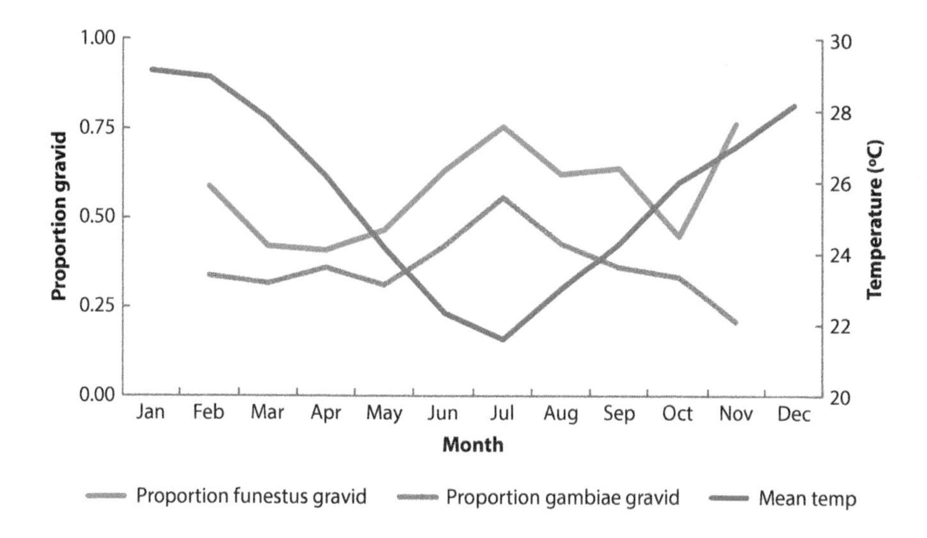

Figure 3.30 Proportion of *A. funestus* and *A. gambiae* s.l. semi-gravid and gravid in resting collection and temperature, Furvela, Mozambique. (From Charlwood, J.D., *Peer J.*, 3099, 2017a.)

Unfed mosquitoes have a thin, cigar-shaped abdomen (Figure 3.29A). On taking a blood meal, the abdomen of the mosquito becomes distended with blood (Figure 3.29B). In the warmer months of the year it takes two days for the mosquito to develop eggs and so insects in resting collections are either unfed (and likely to be newly emerged), freshly fed or gravid (Figure 3.29D). At temperatures below 25°C, development may take three days in which case insects will also be semi-gravid with half of the abdomen taken up with the blood meal (now a darker red/black in colour) and half with developing eggs (Figure 3.29C).

The effect of temperature on the different proprtions of *A. funestus* and *A. gambiae* s.l. in resting collections from Mozambique is shown in Figure 3.30. The higher proportions of semi-gravid and gravid insects at the lower temperature is an indication that the time taken to develop eggs has increased from two to three days.

CHAPTER 4

Dispersal

Knowledge of the flight range of mosquitoes is of paramount importance in control programmes, as it is essential to know the width of 'barrier zones' needed to prevent the infiltration of adults into an area where control measures are being assessed. Dispersal of mosquitoes is likely to be influenced by several extrinsic and intrinsic factors, including among others: meteorological conditions, local topography, physiological status of mosquitoes, body size, population density, and availability of oviposition and resting sites.

Saltmarsh *Aedes* such as *Ochlerotatus vigilax*, which emerge by the coast, have been collected more than 90 kms inland and as much as 160 kms from the shore. *Anopheles pharoensis* was collected during the Second World War in the desert, 45 kms from the nearest breeding sites and more recently in Israel, 280 kms from the nearest known source. These insects included infected females, indicating that the movement was not restricted to young females (as it often is in migrating insects).

Passive transport is also important. *Anopheles* carried aboard aircraft to non-malarious countries such as France, Switzerland, Belgium and Britain, are responsible for the phenomenon known as airport malaria, where people living close to airports, who have never left the country, get malaria from these illegal immigrants. Of greater impact was the transport of *Anopheles gambiae* s.l. from Dakar in West Africa to Recife in Brazil in the 1940s. This resulted in major epidemics of malaria in subsequent years and required a massive effort for elimination to be achieved.

Dispersal is commonly thought to result from flights made at random with air movement being important. Migratory flights may, however, be an integral part of the biology of the species. Migration is often associated with younger insects. Migration may involve movements of a whole population of animals, and is often seasonal, with the individuals or their offspring frequently returning to their parent's place of origin.

Although physiologically capable of several kilometres of flight, the distance that most mosquitoes will disperse during their life is much less. Indeed, most dispersion is leptokurtic – in other words, a small number of insects move farther than might be expected from a normal distribution but the great majority move much less. The distance that most insects move can be considered their ambit. For *A. aegypti* this may only be a few hundred metres throughout their life, and for most malaria vectors the ambit is determined more by the environment than by any innate tendency of the mosquito to disperse. In areas where suitable oviposition sites and feeding sites are in close proximity, dispersal will be considerably less than in areas where they are widely separated.

Dispersal can be measured by examining the decline in numbers of insects away from unique breeding sites. This has been done for *A. funestus* in Southern Mozambique. Resting collections were made from 16 randomly selected houses each day for eight days. Mosquitoes were collected by two collectors working for 20 minutes per house using manual aspirators and torches. The decline in numbers of *A. funestus* with increasing distance from the breeding site was fitted to linear and exponential curves, with correlation coefficients of 0.276 and 0.23, respectively. The exponential model predicted that fewer than five female *A. funestus* would be collected during a 20-minute collection period at distances of 300 m or more from the larval habitat. Resting densities of young pre-gravid females were highest close to the larval habitat.

Dispersal of *A. arabiensis* from a restricted breeding site is also implicit in data from Namawala, Tanzania, where female *A. arabiensis* were caught in light traps in diminishing numbers away from the edge of the Kilombero Valley where the larvae were found (Figure 4.1).

Sabatinelli et al. (1986) described a similar situation from Ouagadougou (Burkina Faso) where numbers of *A. gambiae*, and malaria prevalence, decreased with distance from a water reservoir. In the Gambia the numbers of mosquitoes collected also decreased with distance from the Gambia River (Clarke et al., 2002). In this case, though, there was an inverse relationship between prevalence and distance – implying that older females were dispersing farther than younger ones.

In Brazil, the malaria vector *Nyssorhynchus darlingi* (formerly *Anopheles darlingi*) was the only mosquito collected away from the edge of the finger lake that acted as the habitat for the larvae of a variety of zoophilic *Anopheles* (Figure 4.2). The forest around the collection site had recently been clear-felled and burnt, which may have influenced dispersal of the non-vectors.

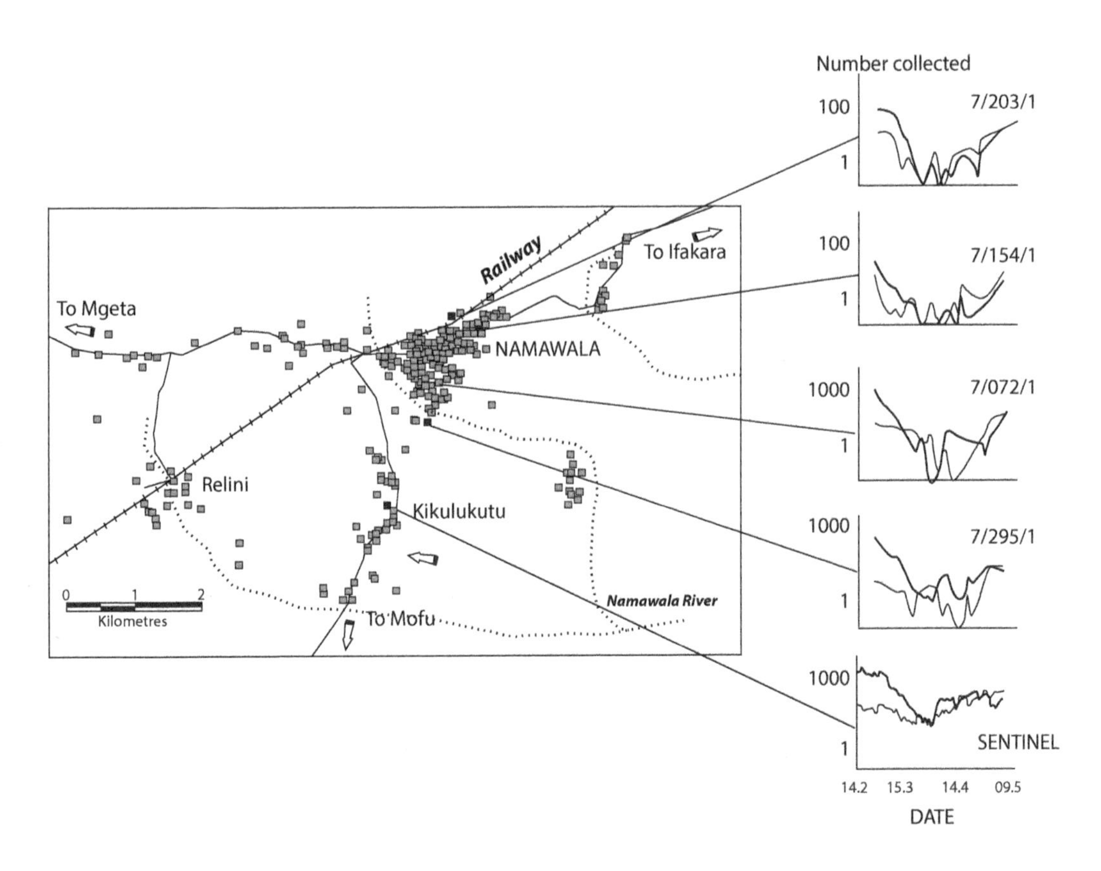

Figure 4.1 Numbers of *A. arabiensis* (thick line) and *A. funestus* (thin line) collected at different distances from the edge of the Kilombero Valley where breeding was taking place. Note the log scale. Thus, in the locations farthest from the Valley, maximum numbers of *A. arabiensis* were <100 a night whilst closer to the Valley edge they were > 3000 a night. Dispersal of *A. arabiensis* was nonrandom but related primarily to the distribution and numbers of houses. (From Charlwood, J.D. et al., *Bull. Entomol. Res.* 85, 37–44,1995c.)

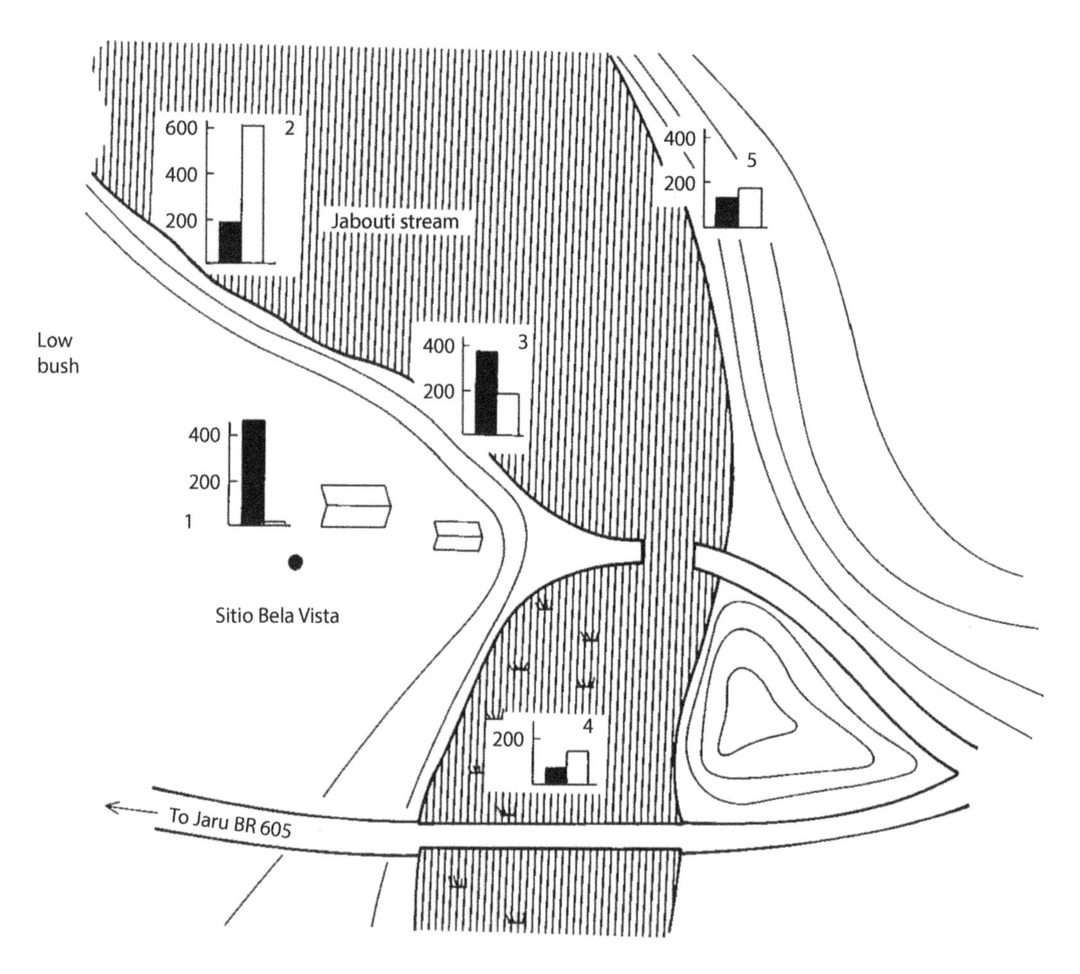

Figure 4.2 Distribution of *Ny. darlingi* (solid histogram) and other *Anopheles* (clear histogram), collected by human landing catch, from the breeding site, Rondonia, Brazil. (From Charlwood, J.D. and Alecrim, W.D., *Ann. Trop. Med. Parasitol.*, 83, 601–603, 1989.)

ESTIMATING DISPERSAL BY CAPTURE-RECAPTURE TECHNIQUES

The most widely used method of determining mosquito dispersal is by capture-recapture (also known as Mark-Release-Recapture, MRR) experiments. These involve either marking and releasing a known number of laboratory-reared insects or collecting mosquitoes from the field, counting and marking them and then releasing them (Figure 4.3). The choice depends on the objectives of any particular experiment and the availability of material.

These kinds of experiments can provide a tremendous amount of data but depend on obtaining a sufficient number of recaptured insects for their conclusions to be robust. In these experiments dispersal cannot be separated from mortality because either way the insects are lost from the population. Laboratory reared insects have the advantage that the insects released are all of a known age but colonised mosquitoes may have atypical survival rates and dispersal. The age of field collected insects, on the other hand, prior to release is not known.

Marking can be carried out in a number of ways. Perhaps the easiest way is to dust the insects with fluorescent powder that glows under ultraviolet (UV) light (Figure 4.4). Large numbers of

Figure 4.3 Male *Anopheles funestus* marked with yellow fluorescent powder on their release, Furvela, Mozambique.

Figure 4.4 Two recaptured anophelines marked with red fluorescent powder are visible in this picture illuminated with a UV light. (From Benedict, M.Q. et al., *Vector-Borne Zoonotic Dis.*, 18, 39–48, 2018.)

insects can be marked together through the use of this technique. A variety of colours allows a number of different treatments or insects to be used before the insects are released. For example, insects can be blood-fed before release and their dispersal, survival and recapture probability can be compared to mosquitoes released unfed. In this way the cost of blood-feeding to the insect, if any, can be determined. The most useful colours are yellow, blue, magenta, red and orange.

There are a number of assumptions in these kinds of experiments:

1. The mark should not affect the survival of individuals.
2. It should not cause abnormal behaviour that changes the chances of it being caught.
3. The mark should be retained (at least for the duration of the recapture period).
4. Marked individuals must mix homogeneously in the populations before being recaptured.
5. The chance of collecting marked and unmarked individuals should be the same.
6. Sampling must take place at discrete time intervals, and sampling time must be relatively short in relation to the total length of the study.

In many cases insects are released at a central point and recapture sites are placed in concentric circles around it. It should be remembered that as the circumference of the circle increases then so should the sampling effort. To estimate the relative numbers of marked mosquitoes present in each annulus, the density at each distance from the release point is multiplied by the area of the annulus at that distance.

Gillies (1961) marked 132,000 laboratory-reared *A. gambiae*, either topically with paint or with radioisotopes, and recaptured 1019. It became apparent that dispersal had to be measured over a greater distance, but this was not feasible with existing procedures without drastically reducing the proportion of houses sampled. To avoid this difficulty marked mosquitoes were released at a peripheral release point, situated 1–1.5 miles from the centre of the sampling area (Figure 4.5).

In this way although some mosquitoes escaped from the recapture area almost immediately, those flying towards the centre could be caught up to 3.2 kms (2 miles) from the peripheral release point. A straight line was fitted to the recapture data when the log percentages of the total females recaptured were plotted against distance on an arithmetic scale. The mean flight ranges recorded for males and females released at the centre were 0.83 and 1.0 km (0.52 and 0.64 miles), respectively, while for females released at the periphery this range was 1.6 kms (0.98 miles). A few adults were caught at the maximum range of 3.2 kms from the release point (Figure 4.6a). Gillies (1961) concluded that the dispersal of *Anopheles gambiae* was nonrandom but related primarily to the distribution and numbers of houses. The survival rate of the released mosquitoes from these experiments fit an exponential distribution (Figure 4.6b).

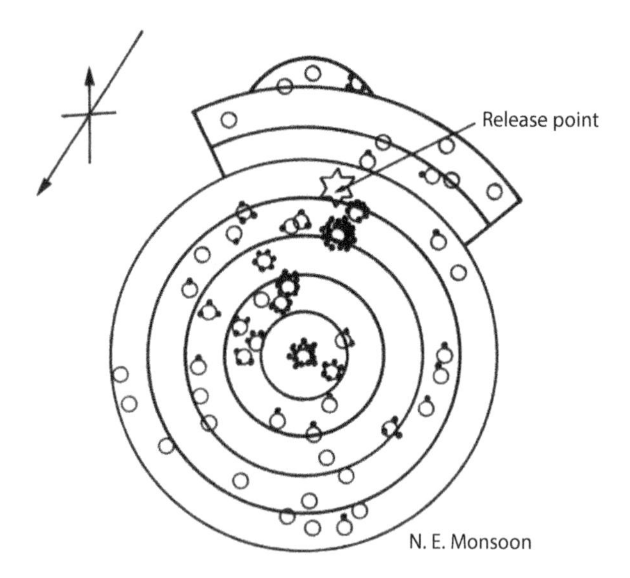

Figure 4.5 Pattern of release and recapture data from Muheza in Tanzania. The dots represent recaptured mosquitoes and the open circles represent houses. (From Gillies, M.T., *Bull. Entomol. Res.*, 52, 99–127, 1961.)

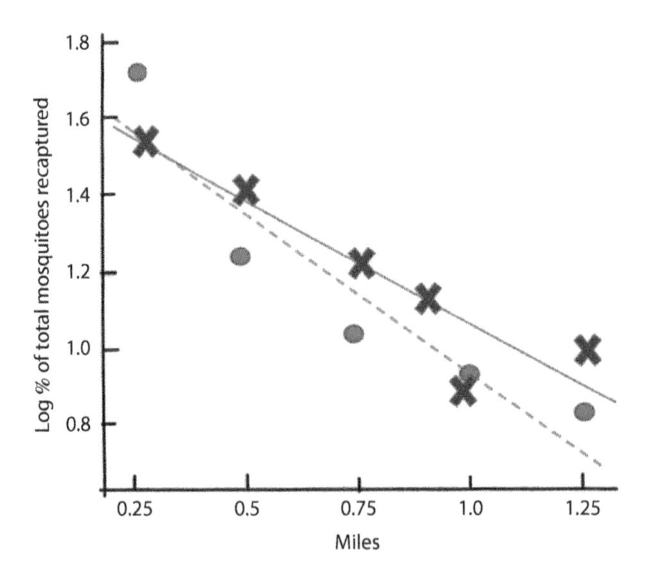

Figure 4.6 (a) Dispersal and (b) female survival data of *A. gambiae* from the recapture data of Gillies (1961). In (a) the solid line represents males and the broken line females; *p* is the probability of survival through one day.

In Namawala, southeast Tanzania, the dispersal and survival of *Anopheles funestus* and *Anopheles gambiae* s.l. was also studied using mark-release-recapture methods. Resting mosquitoes were collected from houses between 08:00 and 10:00 h using oral aspirators. Mosquitoes were marked by a small quantity of DayGlo™ fluorescent powder that was blown into a plastic bag containing a 15 cm/side cage in which the mosquitoes were held. Following release, resting mosquitoes were collected, over a period of two weeks, from 11 houses. A CDC light-trap was operated daily from a sentinel house. A total of 4262 *Anopheles funestus* and 645 *Anopheles gambiae* s.l. were released over 2 days and 184 and 48, respectively, were recaptured over 10 days. Dispersal of the marked mosquitoes among the three catching and release sites is presented in Figure 4.7. For the *A. funestus* there was a significant net movement from Area 1 to Area 3, but this was not the case for the *A. gambiae* s.l., which moved in each direction equally. The proximity of Area 3 to rice fields, and hence oviposition sites, may have been a factor in determining the apparent preference for this area, especially by *A. funestus*.

Dispersal of *A. punctulatus*, a member of the *A. punctulatus* complex, and one of the principal vectors of malaria and filariasis in Papua New Guinea, was studied in the Sepik and Madang areas of the country. For the experiment in the Sepik, mosquitoes collected resting were released from the site A in Figure 4.8.

For the first release, anophelines collected in landing and resting catches were used. For the second release, only those obtained resting were used. Mosquitoes from the landing catch were blood-fed before being marked. Mosquitoes that did not feed, and the unfed and gravid ones from the resting catches, were discarded. Marking, with fluorescent dye, was performed in the manner described by Charlwood et al. (1986a). Separate colours were used for each release, which took place in an unused house in area A (Figure 4.8) at 07:00 hours on two consecutive mornings. The house was close to the road, in a small hamlet in a valley approximately 200 m wide. On either side of the valley the ground rises steeply, so that all collecting sites more than 500 m distant were also considerably higher than the point of release.

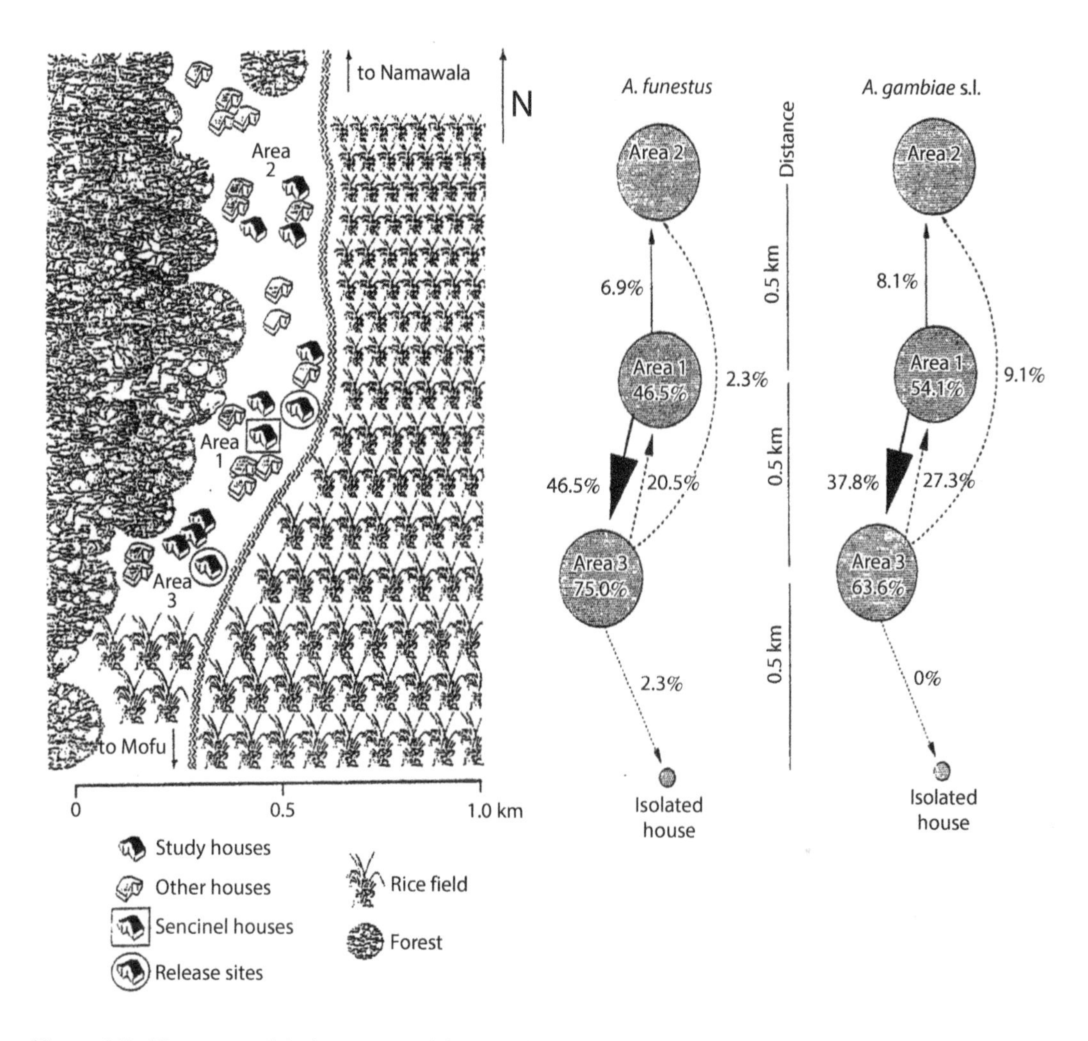

Figure 4.7 Movement of *A. funestus* and *A. gambiae* s.l. in Kikulukutu Namawala, Tanzania. Solid arrows show mosquitoes released in Area 1; broken arrows show mosquitoes released in Area 3. (From Takken, W. et al., *Bull. Entomol. Res.*, 88, 561–566, 1998.)

The house resting population of *A. punctulatus* was subsequently sampled in five areas, designated A–E in the figure, up to 1.8 km away from the release site. Collectors each searched the same three or four houses between 06:00 and 08:00 hours on the seven mornings after the second release. Torches were used to locate the mosquitoes and aspirators to catch them. The number of collectors working in each area varied. Thus, in the furthermost area one man collected whereas in the area closest to the release, but with the same number of houses, two men collected.

Three hundred and thirty-two *A. punctulatus* were released on the first day and 565 on the second. Recapture rates from the two releases were similar. Thirty-one (9.3%) and 51 (9.0%) were subsequently recaptured, the majority (78.0%) in the first four days after release. The single mosquito collected the day after release had undeveloped ovaries and had presumably been released unfed or returned to refeed. Apart from this one mosquito, the minimum time before recapture was two days. It was assumed that there was no pre-gravid stage and that only one blood mean was required per oviposition cycle. Thus, this was equivalent to the minimum duration of the oviposition cycle using the techniques described by Charlwood et al. (1986a); the overall oviposition cycle length was estimated to be 2.90 days.

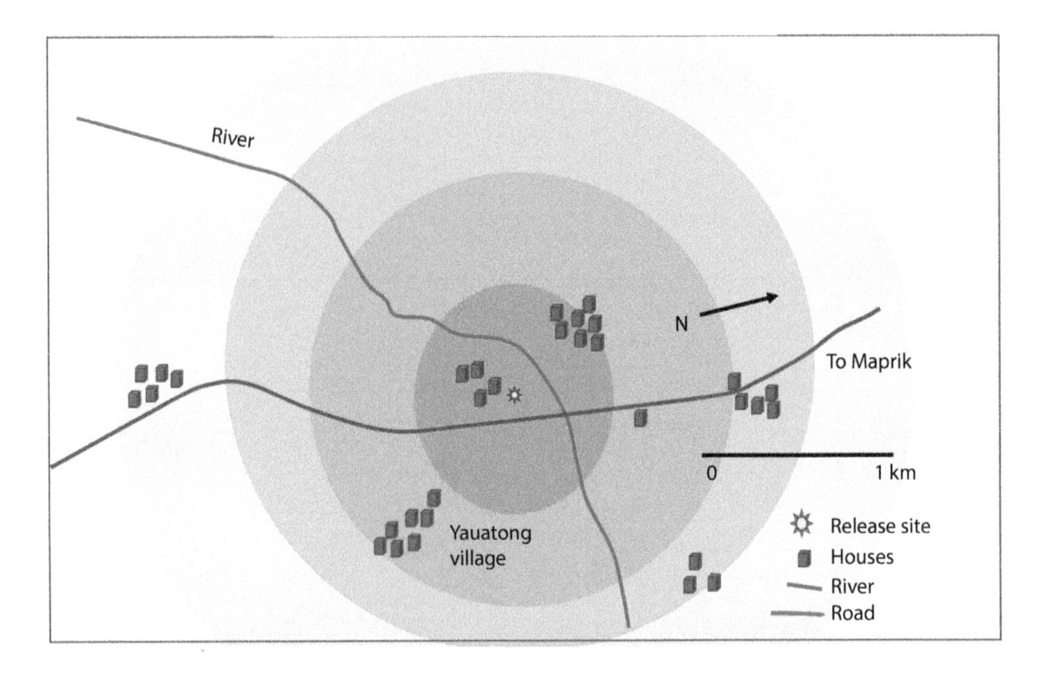

Figure 4.8 Release pattern of *A. punctulatus* from the Sepik, Papua New Guinea. (From Charlwood, J.D. and Bryan, J.H., *Ann. Trop. Med. Parasitol.*, 81, 429–436, 1987.)

Marked mosquitoes were collected in all of the recapture areas. There was an exponential decline in the number collected per collector as the distance between release and recapture area increased ($r = -0.90$, $l = -3.6$, $p = 0.037$). The time between release and peak recapture varied according to recapture area, being longest in the areas farther from the release site. The overall daily survival rate of the recaptured mosquitoes determined by regression (Service, 1976; Charlwood et al., 1986a) was 0.77. Survival rates were similar in infected and uninfected females ($l = 0.10$).

DISPERSAL OF MALES

Female mosquitoes have to locate five resources: a place to mate, a source of blood, a place to rest, a place to oviposit and (perhaps) a source of sugar. Male mosquitoes, on the other hand, only need to discover a place to mate, a source of sugar and a place to rest. Because newly emerged females appear to disperse over a limited area, it would seem reasonable that males, too, should limit their dispersal to places where such females are likely to be found. This appeared to be the case in Furvela, Mozambique, but this may not be the case elsewhere. Rather, males appear to follow Ross's model and disperse at random away from potential places to meet young females. In the collections from Boane, in Southern Mozambique, males tended to disperse farther than females, but it was not determined if they returned to the proximity of the larval habitat to swarm and mate. In Gillies' (1961) experiments it was a male mosquito that was recaptured the farthest from the release site.

Male mosquitoes, that were collected as they exited houses at dusk in Furvela, were marked with fluorescent powder and recapture rates from different houses determined. Numbers recaptured declined with distance such that the mean distance dispersed was 175 m (Figure 4.9). Five males were also recaptured swarming.

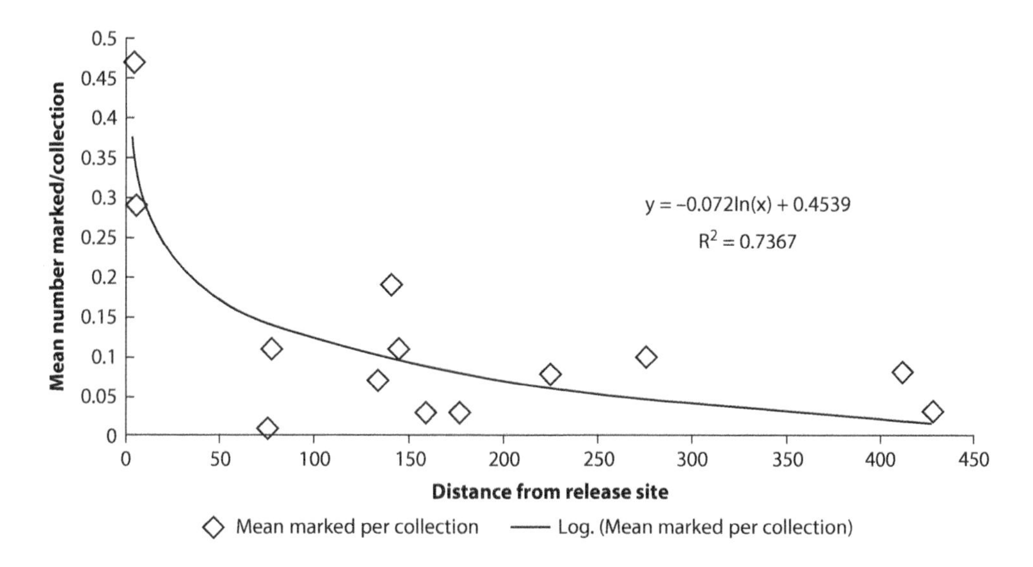

Figure 4.9 Recapture rates of *Anopheles funestus* males collected exiting houses, Furvela, Mozambique. (From Charlwood, J.D., *J. Vector. Ecol.*, 36, 382–394, 2011.)

The source of sugar for males is likely to differ between different environments. The environment occupied by *A. coluzzii* and *A. arabiensis*, in Burkina Faso at least, is drier than that occupied by *A. gambiae.* The plants from these environments will have different scents. There is, therefore, likely to be selection in males for plant (sugar) location (especially for plants that produce volatiles at night) just as there is among females for host location.

INDOOR/OUTDOOR MIXING

Should there be distinct indoor- and outdoor-biting populations of mosquitoes, then this will affect estimates of transmission and also will influence ways to control them. One way of determining whether there are such discrete populations is to mark insects collected biting indoors and outdoors with different colours and to see whether they are preferentially recaptured indoors or outdoors. This was done in Papua New Guinea with three members of the *A. punctulatus* complex. The results indicated that there was mixing of the mosquitoes and so such discrete populations did not occur there (Table 4.1).

Among some of the other results obtained using this technique (in addition to estimates of survival and the usual oviposition cycle duration discussed elsewhere), the effect of transporting *A. farauti*, a mosquito like *A. funestus* that breeds in permanent bodies of water, from Agan to Maraga (and vice versa), (Figure 4.10) was investigated.

Results (Table 4.2) indicated that transported mosquitoes dispersed more than non-transported mosquitoes – some of them indeed returning to their original village of capture. This led to the idea that the mosquitoes may have some kind of memory of the site from where they may have either emerged or oviposited.

Table 4.1 Numbers of Mosquitoes Recaptured Indoors or Outdoors According to the Location of Their Original Capture Site, Butelgut and Umuin Villages, Papua New Guinea

			Site of Recapture		
		Site of Capture	**Indoor**	**Outdoor**	
A. koliensis	Butelgut	Indoor	2	5	$X^2 = 0.10$ n.s.
		Outdoor	4	4	
A. punctulatus	Butelgut	Indoor	8	10	$X^2 = 0.11$ n.s.
		Outdoor	5	6	
A. koliensis	Umuin	Indoor	3	3	$X^2 = 0.22$ n.s.
		Outdoor	9	9	
A. farauti	Umuin	Indoor	14	20	$X^2 = 0.25$ n.s.
		Outdoor	27	52	

Source: Charlwood, J.D. et al., *Papua New G Med. J.*, 29, 19–27, 1986.

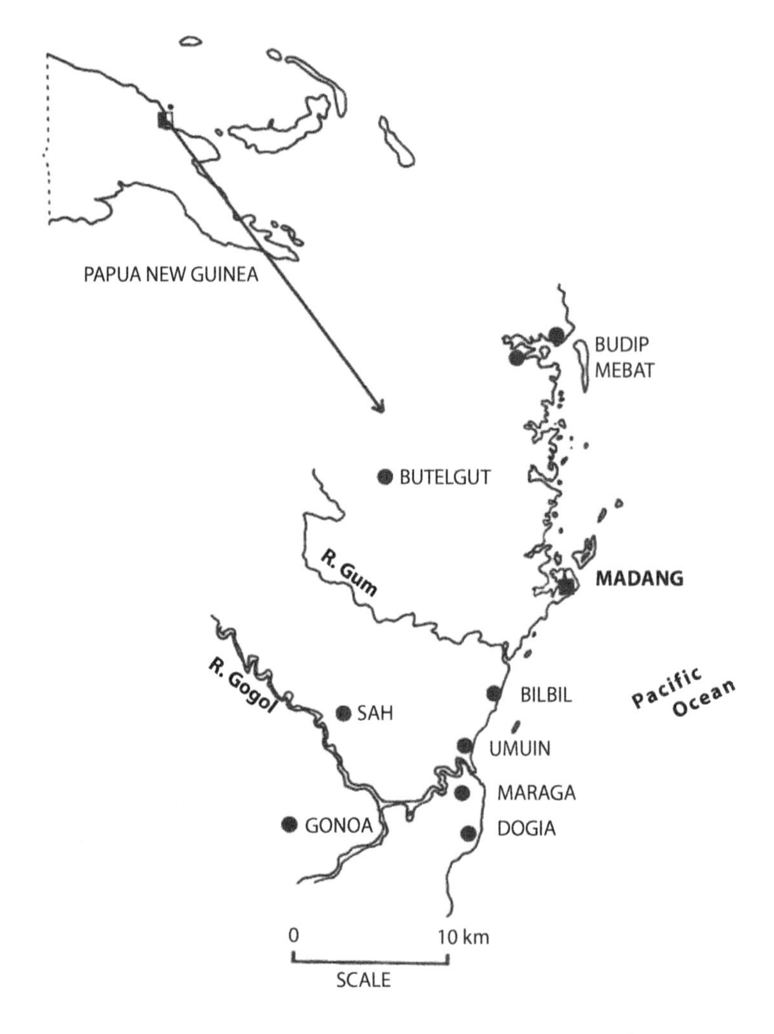

Figure 4.10 Map of the villages in which studies were performed in Papua New Guinea. Note that Maraga and Agan are separated by the estuary of the Gogol River. The village of Butelgut is located 25 km to the north of Yagaum. (From Charlwood, J.D. et al., *Med. Vet. Entomol.*, 2, 101–108, 1988.)

Table 4.2 Recapture Rates of *A. farauti* Released in Their Village of Capture or Taken and Released to a Nearby Village (Agan and Maraga Villages) (for Location See Figure 3.13)

	Released in Agan	Released in Maraga
Non-displaced		
Place of origin	Agan	Maraga
No. released	1306	2073
Recaptured where released	140 (10.7%)	170 (8.2%)
Recaptured in other village	0	0
Displaced		
Place of origin	Maraga	Agan
No. released	1916	1058
Recaptured where released	142 (7.4%)	44 (4.2%)
Recaptured in other village	1 (0.05%)	6 (0.6%)

Source: Charlwood, J.D. et al., *Med. Vet. Entomol.*, 2, 101–108, 1988.

CHAPTER 5

Population Dynamics

As noted by Darwin, animals can all potentially produce more offspring than actually survive to reproduce in the next generation. The rate at which a species can reproduce when there are no constraints on not doing so is known as R_0, the intrinsic rate of increase. As described by Klowden (2007), 'consider the reproductive potential of the female mosquito, *Aedes aegypti*. She can produce about 125 eggs from each blood meal she ingests, and those eggs can develop to adults in about 10 days after she feeds. If one female mosquito began to reproduce in the spring with a single blood meal and all her offspring survived and likewise reproduced, by the end of the summer there would be more than 7×10^{21} mosquitoes. Their biomass at the end of these three months would be more than 1×10^{19} g, about 33,000 times the weight of the total human population'.

Long-term trends in population biology are the domain of evolution and genetics (the 'unseen hand' in ecology). Whilst the way that energy is partitioned in the way that it is (e.g., in the mosquito into eggs or energy for sustaining the adult female) is the domain of optimisation theory (trade-offs).

Why populations are what they are at any given moment, in particular population dynamics in space and time, is the domain of ecology. There are a number of factors that will influence the population dynamics of animals in general. As their name implies, density-independent factors, especially meteorological factors, will act on the population irrespective of density, whilst density-dependent factors will have an increasing effect as numbers increase. In the absence of these, populations will grow up to the point that the environment can sustain them, the, so-called, carrying capacity. How much density-dependent factors, such as predators, operate depends on the stability of the environment. Thus, mosquitoes that breed in tree holes (a stable larval habitat) may be affected by density-dependent factors but others, such as *A. gambiae* s.l., that depend on temporary habitats are more affected by density-independent factors. Typical patterns of population fluctuations, with and without controlling factors, are shown in Figure 5.1.

A growing population is a young population. The proportion of insects that have taken an infectious blood meal and survived through the extrinsic cycle is low in such populations and so, therefore, is the sporozoite rate (Figure 5.2). As the population declines, this proportion increases and the risk per bite also increases.

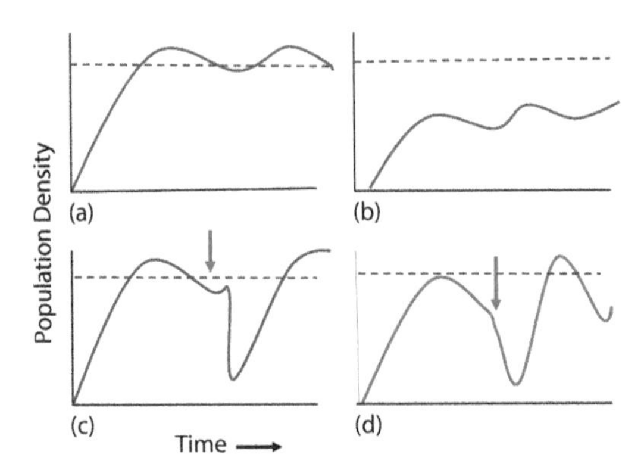

Figure 5.1 Population cycles with and without controlling factors: (a) No external mortality factor – the population oscillates around the carrying capacity of the environment, (b) Growth with density dependence below the carrying capacity, (c) Growth with a single density-dependent factor, (d) Population with density-dependent and density-independent mortality factors. (Reprinted from *Biology of Disease Vectors*, 2nd ed., Hemingway, J., Biological control of mosquitoes, Chapter 43, Copyright 2005, with permission from Elsevier.)

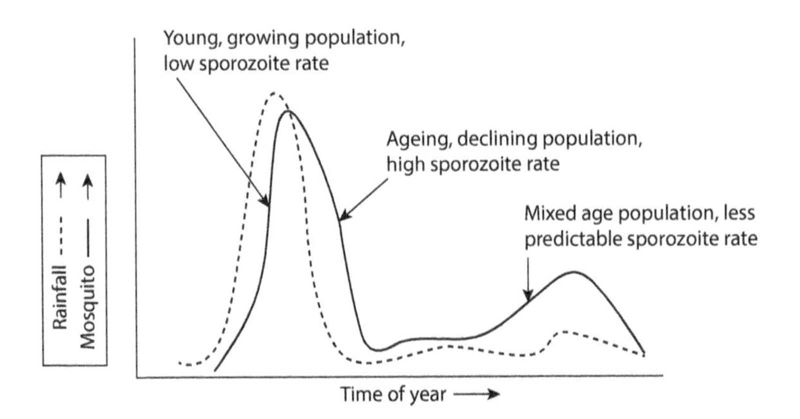

Figure 5.2 Risks associated with different patterns of vector numbers. (From Billingsley, P.F. et al., Rapid assessment of malaria risk using entomological techniques: Taking an epidemiological snapshot. Chapter 6, in *Environmental Change and Malaria Risk: Global and Local Implications*, Ur Frontis, Wageningen, the Netherlands, 2003.)

RAINFALL AND TEMPERATURE

The two most important factors that influence dynamics of the main African malaria vectors are rainfall and temperature. In general, *A. gambiae* s.l. numbers increase after the rain, sometimes in an explosive manner. Thus, in Namawala in May following a short period of rain, numbers of *A. arabiensis* in light traps increased from less than 100 a night to more than 1600 a night over a period of three days (Figure 5.3). The parous rate at the time dropped from 80% to less than 5%, indicating that the increase in numbers was due to newly emerged insects. In a short while the population returned to previously lower levels. This pattern of 'boom and bust' is typical of temporary pool breeders and the generalised pattern associated with transmission risks. The implication from

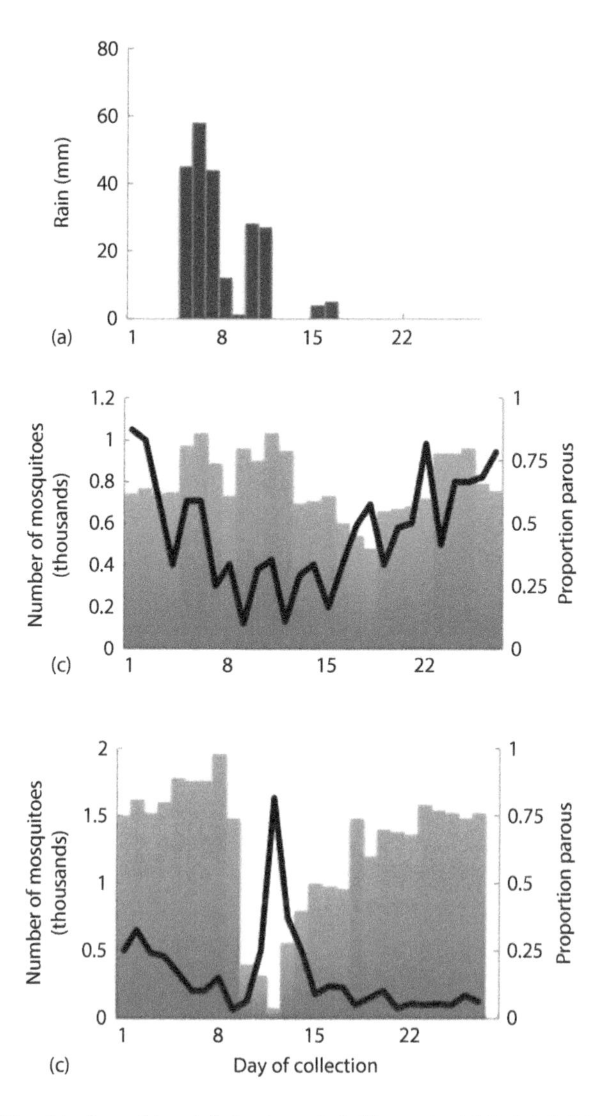

Figure 5.3 'Boom and bust' in *A. arabiensis* following rain in Tanzania. (a) Rainfall, (b) numbers of *A. funestus* and (c) numbers of *A. arabiensis* collected in Kikulukutu at the edge of the Kilombero Valley. The red histograms represent the proportion of the population that was parous. (From Takken, W. et al., *Bull. Entomol. Res.*, 88, 561–566, 1998.)

data like this is that larval survival was much higher than is usually described. At the same time, numbers of *A. funestus* decreased from 1000 a night to 200, and parous rates increased, indicating that in this species larval populations were depressed by the rain.

The *A. funestus* larvae occurred in clean permanent pools that were muddied by the rain. This had a depressing effect on numbers emerging, whereas the rain created numerous temporary pools used by the *A. arabiensis*.

A similar picture of *A. gambiae* s.l. numbers fluctuating with rainfall but the *A. funestus* being independent of rain was also seen in other villages in the Kilombero Valley of Tanzania monitored

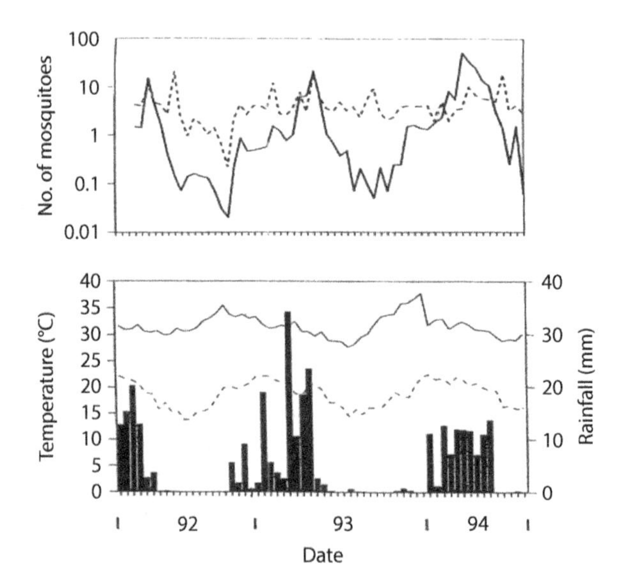

Figure 5.4 Two-week period variation in entomological and meteorological data. Top graph: mean mosquito densities. The values plotted are the expected numbers caught in light traps adjusted to the mean for the sampled neighbourhoods ···· *Anopheles funestus* — *A. gambiae* s.l. Bottom graph: histogram = daily rainfall, — = daily maximum temperature ···· = daily minimum temperature. (From Charlwood, J.D. et al., *Am. J. Trop. Med. Hyg.*, 59, 243–251, 1998.)

over a number of years (Figure 5.4). These collections were undertaken before the wide-scale utilisation of Long Lasting Insecticide-treated Nets (LLINs) in the area.

Numbers of *A. gambiae* s.l. collected as they exited houses at sunset from the village of Furvela in southern Mozambique were also more closely related to mean air temperatures than were *A. funestus* (Figure 5.5). This probably reflects the kind of water that acts as the larval habitat of the

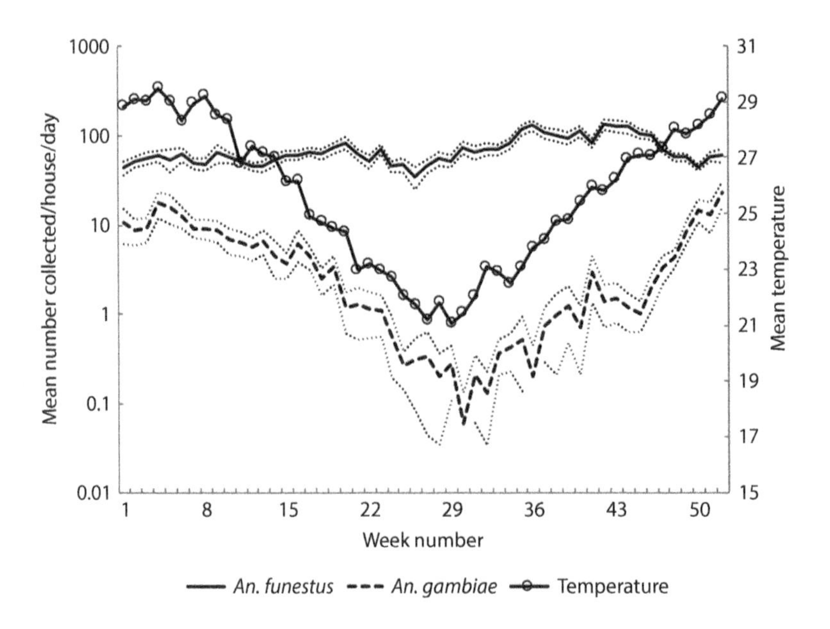

Figure 5.5 Numbers of male *Anopheles funestus* and *A. gambiae* s.l. in exit collections from Furvela, Mozambique, and mean air temperature. (From Charlwood, J.D., *J. Vector. Ecol.*, 36, 382–394, 2011.)

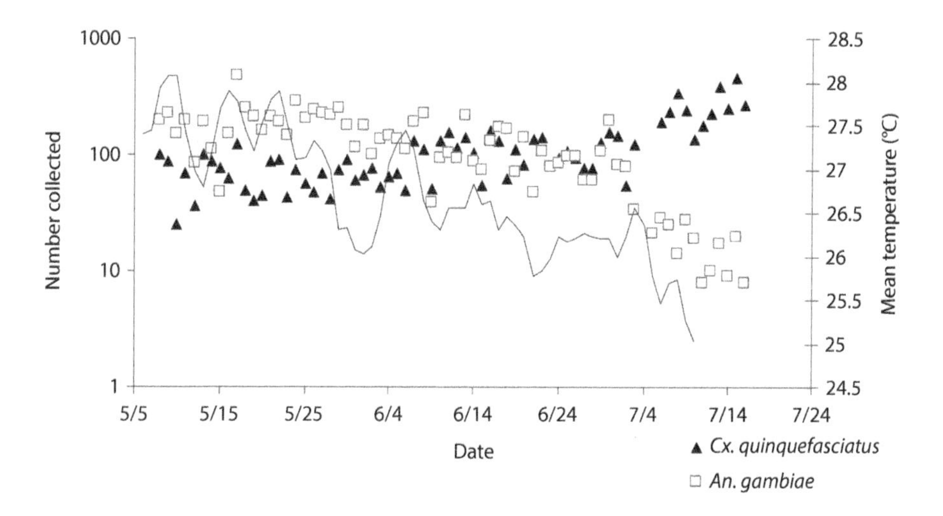

Figure 5.6 Numbers of *A. coluzzii* and *Cx. quinquefasciatus* collected in a sentinel light trap and mean daily temperature, São Tomé. (From Charlwood, J.D., et al., *Malar. J.*, 2, 7, 2003c.)

two species. *Anopheles gambiae* is well known for its preference for sunlit pools whilst *A. funestus* occupies more permanent, shaded habitats in which temperature fluctuations of the water are likely to be more stable than they are for the former species.

A similar effect of decreasing temperature on numbers of *A. coluzzii* and *Cx. quinquefasciatus* in light-traps from houses was seen in São Tomé. Once again, the species with the more permanent larval habitat increased as temperatures dropped but numbers of *A. coluzzii* decreased (Figure 5.6).

In addition to affecting the larvae, rainstorms can have an effect on the adult population. Figure 5.7 shows the proportion of *A. funestus* with mating plugs (indicative of recent mating) and, in parous mosquitoes, the proportion with sacs (indicative of recent oviposition) following an evening of heavy rain compared to the four nights before and after the rain. Numbers were temporarily depressed but rapidly resumed to the levels seen before the storm.

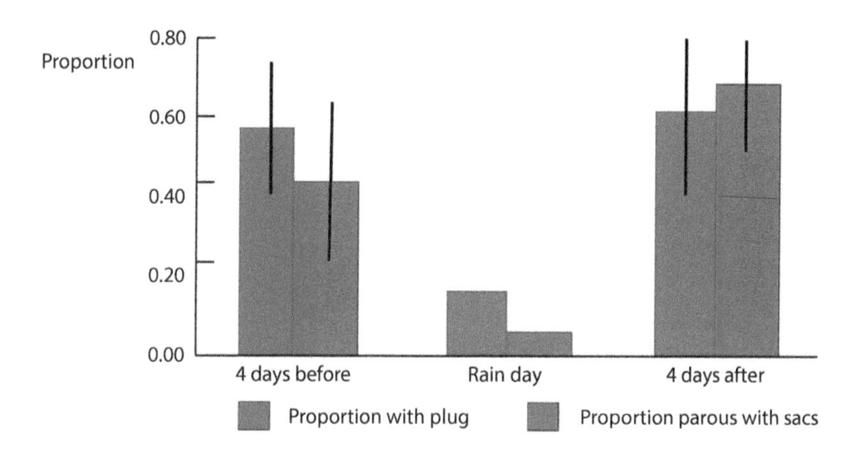

Figure 5.7 Proportion of newly emerged *Anopheles funestus* with mating plugs and parous females with sacs on the four days before and after a rainstorm. (From Charlwood, J.D. and Braganca M., *J. Vector Ecol.*, 37, 1–5, 2012a.)

DRY SEASON SURVIVAL

African malaria vectors often have to contend with survival through a long dry season, often with a complete absence of potential oviposition sites. How they manage to do this depends to a certain extent, once again, on the environment. One way for mosquitoes to cope with stresses of temporarily unsuitable conditions is to aestivate (during hot and dry periods) or hibernate (during cold periods). Aestivating mosquitoes may become gonotrophically discordant and take multiple blood meals per egg batch. The first indication that females would do this was provided by Omer and Cloudsley-Thompson (1970). They found gravid, nulliparous, but blood-fed, *A. arabiensis* resting in *turkel*s from Sudan. More recently, an *A. gambiae* female marked and released before the dry season in Mali was recaptured seven months later, indicating that this species can also aestivate in the manner described earlier (Adamou et al., 2011; Lehmann et al., 2010). The peninsular of Linga Linga in Mozambique also has a long dry season. During this time *Anopheles funestus* can still be caught in light traps; however, the proportion of insects that are gravid increases significantly to the point where they are the dominant type collected.

Thus, there is evidence that all three major African vectors can survive by aestivation. In the Sudan, where water was available despite the high temperatures, breeding continued normally and the insects were gonotrophically concordant. In Mali, although present, the *A. arabiensis* apparently did not aestivate (Adamou et al., 2011, Lehmann et al., 2010) and in Mozambique, where suitable habitat was available, the population continued without any sign of gonotrophic discordance.

Elsewhere, the species may be restricted to isolated 'refugia' populations that expand their range during the wet season. This was the case in the Kilombero Valley of Tanzania where *A. gambiae* was restricted to villages close to the escarpment of the Udzungwa Mountains with annual rainfall in excess of 1600 mm. *A. funestus* occurred in sites where seasonal streams had stopped flowing and so became seasonal lakes and *A. arabiensis* was found in the valley itself where it fed on the abundant game and bred in water left behind in hippopotamus footprints (Figure 5.8).

During the wet season the valley floods, the game moves to higher ground (although much of the game has now been killed by poachers) and the *A. arabiensis* becomes an endophilic, anthropophagic mosquito.

When the rains start, the runoff from the nearby fields muddies the water used by *A. funestus* and the population declines. It only increases again towards the end of the wet season when the rice fields are reasonably mature and large expanses of potential larval habitat are available. Once the rain starts, humidity rises and suitable larval habitat becomes available *A. gambiae* will rapidly expand its range. It is, however, not certain that aestivating insects of all three species do not occur in other areas of the Kilombero Valley. In Eritrea, *A. arabiensis* is the dominant vector. Given that the pattern of malaria transmission, and the climate, is similar to that observed in the Sudan, it is likely that the species also aestivates in Eritrea.

Southeast Asian anophelines are physiologically adapted to humid conditions, having wider spiracles than African ones. They may invest in large numbers of eggs per batch to the detriment of longevity and thereby vectorial capacity. In Cambodia, despite long periods without rain, none of the species examined in a recent study showed evidence of gonotrophic discordance (Charlwood et al., 2016b).

Anopheles arabiensis appears to have adapted to urban habitats in some parts of its range as shown by the distribution of both *A. arabiensis* and *A. gambiae* in the environs of Benin City in the late 1970s (Figure 5.9).

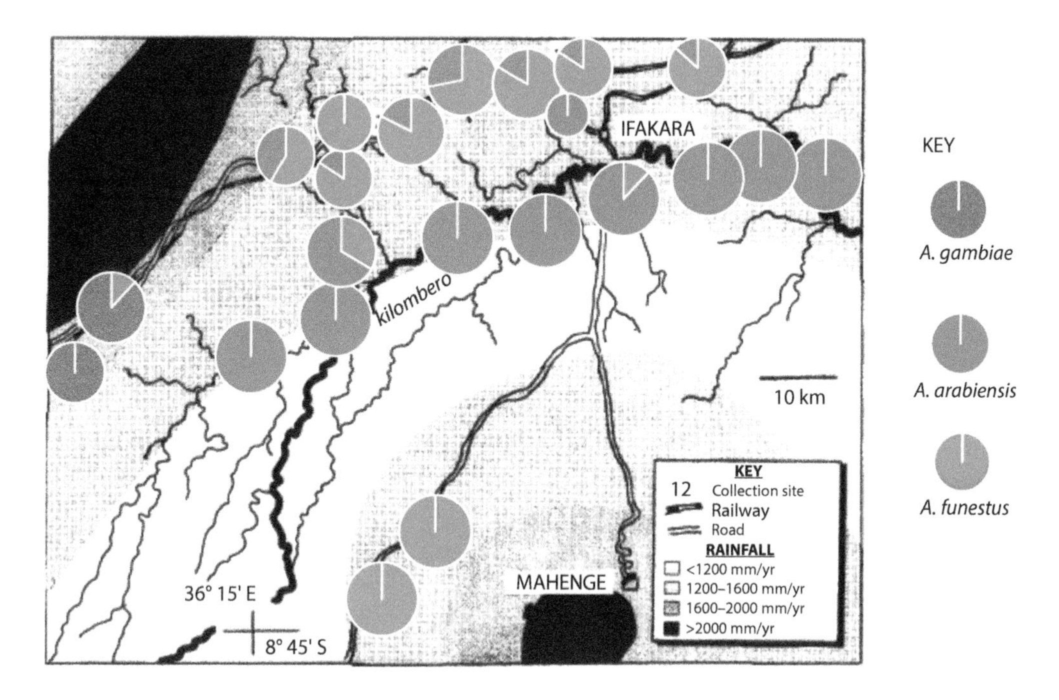

Figure 5.8 Distribution of malaria vectors during the dry season in the Kilombero Valley, Tanzania. (From Charlwood, J.D. et al., *Am. J. Trop. Med. Hyg.*, 62, 726–732, 1999.)

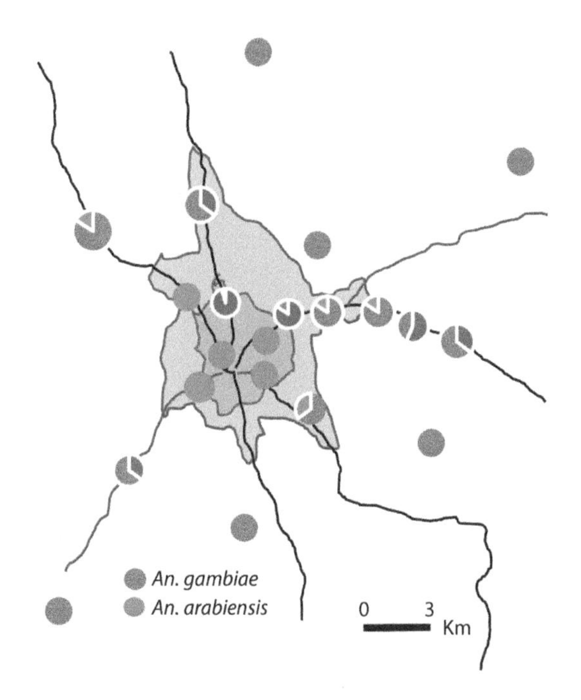

Figure 5.9 The distribution of *A. gambiae* (in black) and *A. arabiensis* in and around Benin City. The stippled area shows the limit of urbanisation. (From Coluzzi, M. et al., *Parassitologia*, 16, 107–109,1974.)

CHAPTER 6

Mapping

Mosquitoes vary as much in space as in time. The first computer-generated maps of *A. gambiae* s.l. and *A. funestus* density were produced using light-trap data from Namawala (Figure 6.1). The techniques were much the same as those still used today, but each iteration took several hours to complete instead of the minutes that it would take today. Densities were also restricted to information from houses (Figure 6.2a and b).

There are now many further examples of density maps of mosquitoes available and journals devoted to this aspect of vector biology. In general, highest densities occur close to the edge of villages, as was observed in the village of Massavasse in Mozambique. Massavasse is a 1 × 2 km village in the middle of an irrigation project where rice is the main crop. There are a number of different water bodies in and around the village that provide suitable habitat for a variety of mosquito species (Figure 6.3).

It is often the case that mosquitoes will fly close to boundaries or edges between open areas and denser vegetation. One explanation for the epidemiological pattern observed in Brazil, where the boundary is between rainforest and open areas where the forest has been cut down, is that the *Ny. darlingi* are dispersing along this boundary for long distances. Indeed, one marked specimen was caught 7.2 kms from the site of release. In landing collections numbers of *Ny. triannulatus* were also several folds higher close to the edge of a forest compared to the interior of the forest (1193 of 1243 collected at the forest edge) or the middle of an adjacent airstrip (757 of 855 being collected at the forest edge) (Charlwood and Wilkes, 1981).

The mosquitoes in Massavasse were also most common towards the edge of the village both in light-trap collections and tent-traps (Figure 6.4). Tent-traps have the advantage that they only differ in the attractiveness of the host in the tent. Tent-traps can also be placed at predetermined locations that do not depend on there being houses available. In Massavasse, tent-traps were located at the centroid of the 16 rectangular areas used to subdivide the village.

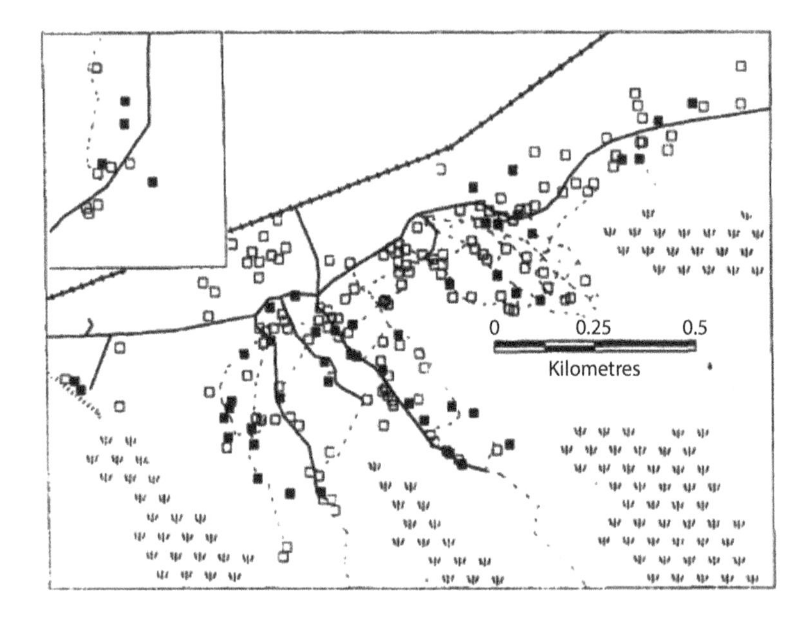

Figure 6.1 Map of Namawala, Tanzania. The filled circles represent sampled houses and the open circles other houses in the village. The inset shows the hamlet Kikulukutu (where capture-recapture studies were performed) close to the Kilombero Valley.

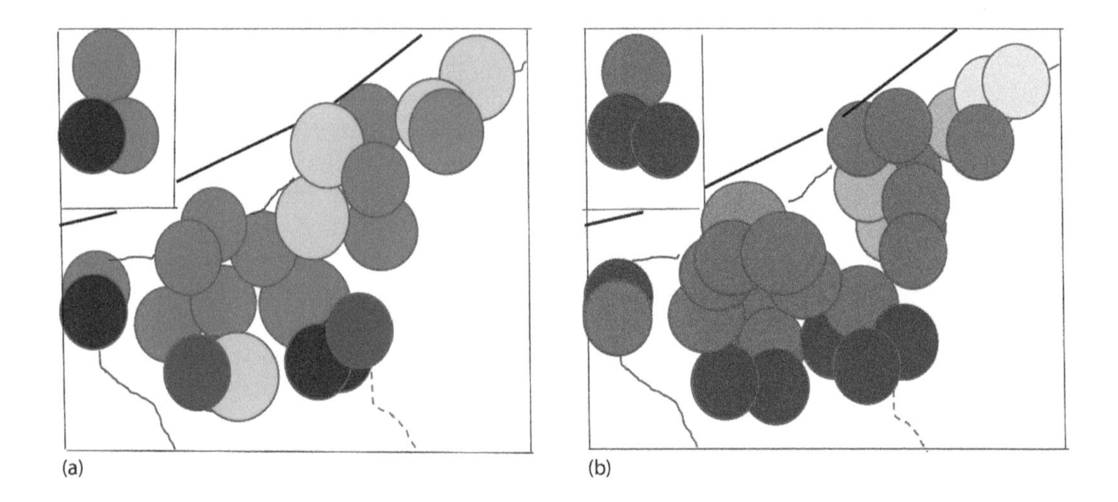

Figure 6.2 Density maps of (a) *A. gambiae* s.l. and (b) *A. funestus* from Namawala, Tanzania. (From Smith, T.A. et al., *Acta Trop.*, 59, 1–18, 1995.)

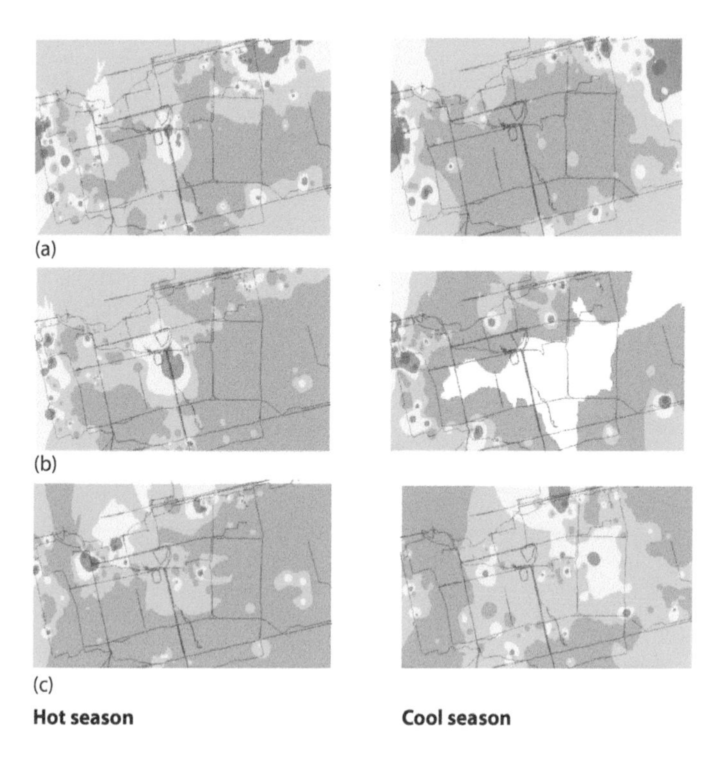

Figure 6.3 Density maps of mosquitoes from light-trap collections, Massavasse, Mozambique. The colours show an increasing density from green to red. (a) *A. funestus*, (b) *A. arabiensis*, (c) *Cx. quinquefasciatus*. (From Charlwood, J.D. et al., *Geospatial Health*, 7, 309–320, 2013b.)

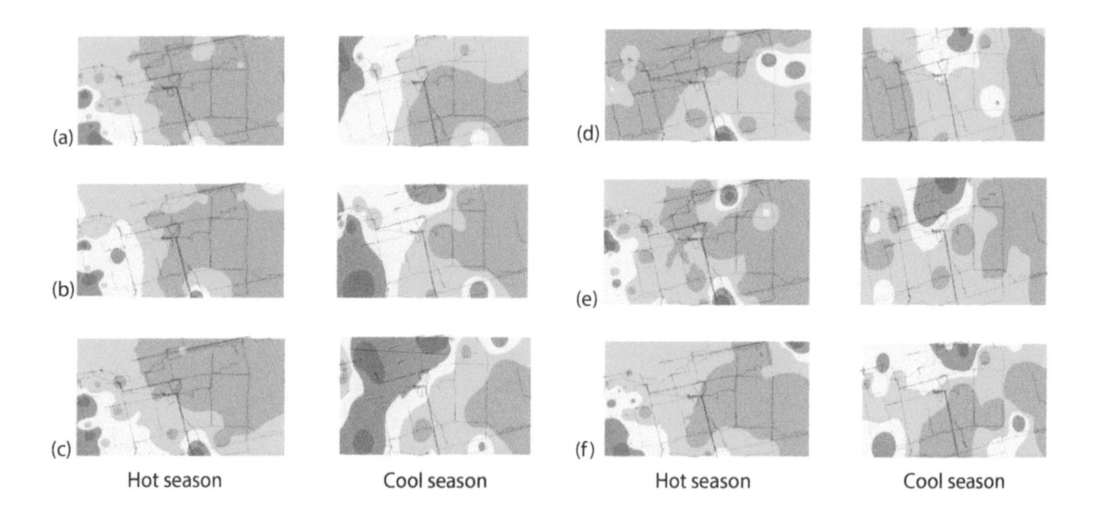

Figure 6.4 Map of mosquito densities from Massavasse, Mozambique, obtained using tent-trap collections. The colours show an increasing density from green to red. Note that densities were, in general, greatest towards the edge of the village. (a) *A. funestus*, (b) *A. arabiensis*, (c) *A. tenebrosus*, (d) *Cx. quinquefasciatus*, (e) *Cx. tritaeniorhynchus*, (f) *Ms. africana*. (From Charlwood, J.D. et al., *Geospatial Health*, 7, 309–320, 2013b.)

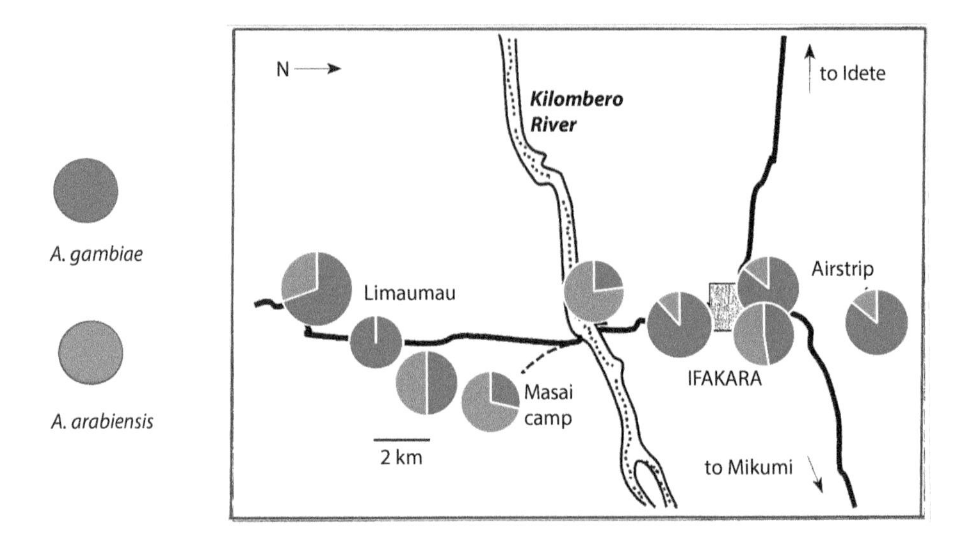

Figure 6.5 Distribution of larvae of *A. gambiae* (blue) and *A. arabiensis* (orange), determined by PCR, close to the town of Ifakara, Tanzania. (From Charlwood, J.D. and Edoh, D., *Tanzania J. Med. Entomol.*, 33, 202–204, 1996.)

With the advent of PCR technology it is now possible to identify all larval stages of members of the *A. gambiae* complex. This enables the relative density of the different larval instars in different habitats to be determined. This was done in areas close to the town of Ifakara, Tanzania (Figure 6.5). Here, it was found that the *A. arabiensis* were predominant close to the cattle camps of the Masai and Mangati.

CHAPTER 7

Vectorial Capacity

The simple question 'How many infectious mosquitoes would be expected to come from a single infectious mosquito after just one generation of the parasite?' requires a relatively complicated answer.

It was soon realised by Ronald Ross, the man who discovered that it was transmitted by mosquitoes, that malaria was a disease that could be described mathematically. He developed a model, later refined by George MacDonald, that captures many of the components of transmission. Ross's original models were refined by a number of others, notably the mathematician Lotka. They were originally designed to predict the impact of larval control (or source reduction, SR) on malaria transmission. They were developed before the discovery and use of DDT for adult control and did not include a number of biological parameters (either because they were considered unimportant or because the parameters could not at that time be measured). It was George MacDonald who realised that, because of the long delay required for the parasite to complete sporogony in the mosquito, longevity of the mosquito is a potential weak link in malaria transmission.

The Ross-MacDonald model of vectorial capacity (C), the expected number of infectious bites that will eventually arise from all the mosquitoes that bite a single person on a single day, still forms the basis of most of the more complicated models of transmission.

The classic equation of MacDonald is: $C = \frac{m\,a^2 p^n}{-\ln p}$

where m = the number of mosquitoes biting an individual human, a = the number of blood meals per day per vector on individual humans (the human blood index), p = the daily survival rate and n = the extrinsic incubation period of the parasite (i.e., the time to complete sporogony, which is the time between the ingestion of gametocytes in one blood meal to the time when sporozoites are injected in a subsequent blood meal). The human blood index is raised to the power of two because two blood meals on humans are required for transmission (one meal for the insect to acquire the parasite and one to transmit it). The duration of the extrinsic cycle varies between species of malaria and is dependent on temperature. The most sensitive parameter in the model is the daily survival rate of the mosquito (Figure 7.1, courtesy of Prof. Tom Smith).

This still serves as the basic model from which numerous variants have been developed over the past 50 years. Most of the recently developed models include such things as superinfection (the possibility that a person can be infected by a second parasite before the first infection has been cleared) and immunity on transmission. The adverse effect of increased mosquito mortality on the overall fecundity of the population, giving rise to fewer eggs being laid and fewer larvae developing in succeeding generations, is also included in a number of recent models (Brady et al., 2016; Reiner et al., 2013; Smith et al., 2012). However, it remains true that reducing adult survival remains the most effective single factor to reduce transmission.

Ross began his search for evidence of mosquito involvement in malaria transmission in 1894 in India. At that time, he, primed by Patrick Manson who had earlier discovered the involvement of mosquitoes in filariasis transmission, was under the impression that mosquitoes did not survive egg laying and that the disease was transmitted in the water used for oviposition. (Indeed, transmission by drinking

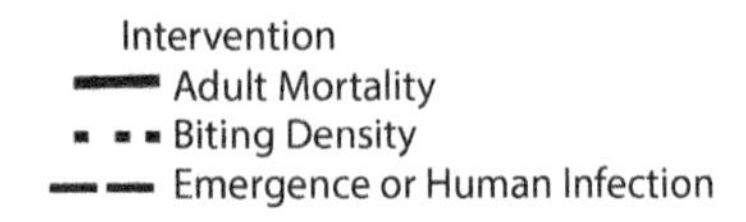

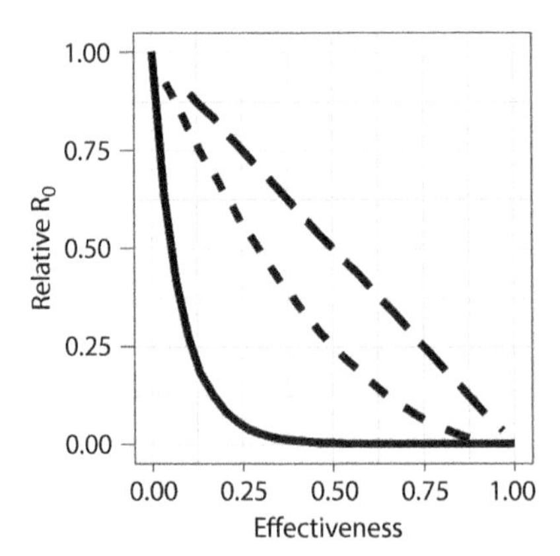

Figure 7.1 Sensitivity of R_0 to changes in mosquito density, biting rate and mosquito survival calculated for the Ross-MacDonald model (based on Koella 1991). MacDonald's formula for R_0 is

$$= \frac{ma^2bcp^n}{-t\log(p)}$$

where the different parameters are m = the number of female mosquitoes per human host, a = the number of bites per mosquito per day, b = the probability of transmitting infection from an infectious mosquito to a human (per bite), c = the probability of transmission of infection from an infectious human to a mosquito (per bite), τ = the rate of recovery of humans from infectiousness, p = the daily survival of adult mosquitoes and ω = the duration of the extrinsic cycle (the time required for the development of sporozoites from infection of the mosquito). Changes in parameter values are represented as the efficacy of shown factors relating to the original setting (e.g., an efficacy of 50% corresponds to a multiplication of m, or $1/\omega$ by 0.5; m or $1/\tau$ for the basic reproductive number linearly); therefore, this efficacy corresponds to a 50% decrease in reproductive number. Biting rate, a, enters the reproductive number quadratically, so that an efficacy of 50% in reducing this leads to reduction in R_0 to $0.5^2 = 0.25$ times its original value. Efficacy in adulticiding of 50% corresponds to a 50% reduction in survival per unit time. This enters the formula for R_0 as a power function, so decreases in this lead to the largest changes in R_0. (From Smith, T.A. et al., *Cold Spring Harb. Perspect. Med.*, doi:10.1101/cshperspect.a025460, 2017.)

contaminated water was a popular belief even until late last century.) It was only much later that the role of the bite took precedence and that survival of the mosquito was a critical factor in transmission.

Given that malaria transmission is extremely sensitive to the survival/mortality rate of adult mosquitoes, considerable effort has gone into measuring mortality rates in the field because it is only insects that have survived the extrinsic incubation period following an infective feed that will be vectors. Survival rates have been estimated by dissection of the females' ovaries, by capture-recapture experiment, by estimation of increasing infection in mosquitoes maintained after collection and by the decline in numbers when recruitment to the population has ceased. Daily survival can be calculated by knowing the number of oviposition cycles that a mosquito completes and the duration of each cycle. The oviposition cycle (also known as the gonotrophic cycle) refers to the time taken between one blood meal and the next, or one egg laying and the next. Together survival rate and oviposition interval have a major impact on the proportion of mosquitoes that might be vectors, a change from a two-day cycle to a three-day one producing a fourfold increase in the potential numbers of vectors whilst a 10% increase in survival per cycle will also have a fourfold increase (Table 7.1).

Table 7.1 Effect of Changing Survival Rates and Oviposition Cycle Lengths on Proportion of a Mosquito Population That Will Survive to Become Vectors of Malaria[a]

	Survival Rate per Blood Meal		
Mean Oviposition Interval (days)	**0.4**	**0.5**	**0.6**
2	0.01	0.03	0.12
2.5	0.02	0.07	0.22
3	0.04	0.13	0.32
3.5	0.07	0.19	0.43
4	0.11	0.25	0.54

Source: Charlwood, J.D. et al. *Bull. Entomol. Res.* 76: 211–227, 1986a.
[a] Assuming a 12-Day Extrinsic Development Time of the Parasite.

OVIPOSITION (GONOTROPHIC) CYCLE DURATION

There are two main ways to measure the duration of both the oviposition cycle and survival in the mosquito: by dissection of the females ovaries and by capture-recapture experiment. The best, perhaps, is by a capture-recapture experiment in which case insects at a particular phase of the cycle are released and then recaptured at the same phase in subsequent cycles. This has been done in many places including Papua New Guinea with members of the *A. punctulatus* complex; in Brazil with *Ny. darlingi* and in Tanzania with *A. gambiae* and *A. funestus*. The decline in numbers of marked mosquitoes recaptured over time can form the basis of survival rate estimates, either by simple regression or by a more complicated series of models that take into account the cyclical nature of recaptures.

Different species have different oviposition cycle lengths. Thus, in Papua New Guinea, the members of the *A. punctulatus* complex had cycles that differed both between species and by village (Table 7.2).

More recently, mosquitoes from Cambodia were shown, by dissection, to have cycles that differed between species but did not vary between locations of collection site. Smaller mosquitoes (such as *A. aconitus*) tended to have shorter cycles compared to larger species (such as *A. barbirostris*) (Figure 7.2).

The cycle may also be affected by ambient illumination. In Papua New Guinea *A. farauti* were blood fed, marked and released at different phases of the moon. They returned more rapidly on nights when there was illumination from the moon before the sun had set (i.e., at the time that the mosquitoes were in search of an oviposition site) (Table 7.3).

Table 7.2 Estimated Oviposition Cycle Duration of the Three Papua New Guinean Vectors Based on MRR Experiments

	A. punctulatus	*A. koliensis*	*A. farauti*
Maraga	–	–	3.0 (0.67)[a]
Umuin	–	2.9 (1.00)	2.1 (0.35)
Mebat	2.7 (0.66)	2.4 (0.50)	–
Butelgut	3.7 (1.00)	3.2 (0.60)	–

Source: Charlwood, J.D. et al., *Bull. Entomol. Res.*, 76, 211–227, 1986a.
[a] Standard deviations in brackets.

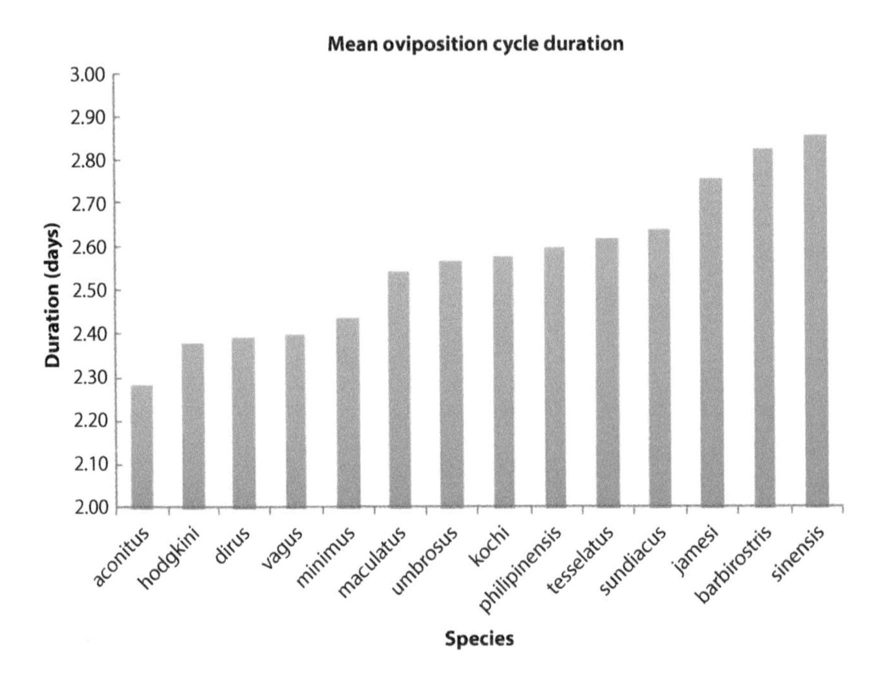

Figure 7.2 Oviposition cycle lengths, estimated by dissection, of *Anopheles* from Cambodia. (From Charlwood, J.D. et al., *Med. Vet. Entomol.*, 2016a; Charlwood, J.D. et al., *Malar. J.*, 15, 356, 2016b.)

Table 7.3 Estimated Oviposition Cycle Lengths of *A. farauti* and *A. koliensis* Marked and Released in Agan Village, Papua New Guinea, at Different Phases of the Moon

Experimental	A		B		C		D
Release date	17/11		24/11		30/11		7/12
Moon phase	4		1		2		3
A. farauti (days)	2.35		2.26		2.04		2.40
P		N.S.		0.02		0.01	
A. koliensis (days)	2.50		2.22		2.00		2.25
P		N.S.		N.S.		N.S.	

Source: Birley, M.H. and Charlwood, J.D., *Ann. Trop. Med. Parasitol.*, 8, 415–422, 1989.
Abbreviation: A = No moon, B = New moon, C = Full moon, D = Waning moon.

MOSQUITO SURVIVAL DETERMINED BY DISSECTION

The alternative to capture-recapture experiment is to examine the ovaries of females that come to feed and to determine if they have returned shortly after oviposition or with a delay. This requires an understanding of the generalised life history of the adult female mosquito.

A generalised schema of a female mosquito's life history follows the pattern described below (Figure 7.3). As we have seen, after emergence females may take a pre-gravid meal before or after mating. Under normal circumstances a pre-gravid meal is only taken by newly emerged insects. Subsequent blood-meals give rise to egg development.

The various categories of mosquito in the above schema can be determined by dissection of the ovaries. Dissection also provides information on the time spent between oviposition and re-feeding (i.e., the duration of the oviposition cycle).

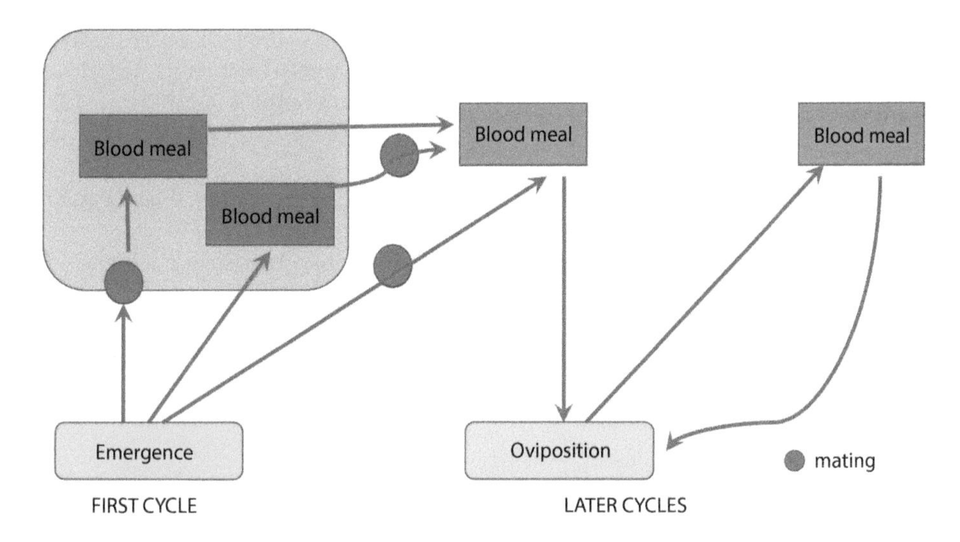

Figure 7.3 A generalised schema of a female mosquito's life history after emergence.

The ovaries of culicines differ slightly from those of anophelines. Culicines have three spermathecae whilst anophelines have only one (Figure 7.4). Anophelines have ampullae (where the egg stays for a moment to be fertilised before being laid). The ovaries of both sets of mosquitoes are heavily invested with tracheae because both require a great deal of oxygen (=energy) during the process of egg maturation.

The paired ovaries are basically hollow tubes to which are attached numerous ovarioles. The ovarioles of a newly emerged insect are very small, and the terminal follicle (egg) is without yolk (that's no yolk). Species such as *A. gambiae* s.l. and *A. funestus* may take a pre-gravid blood

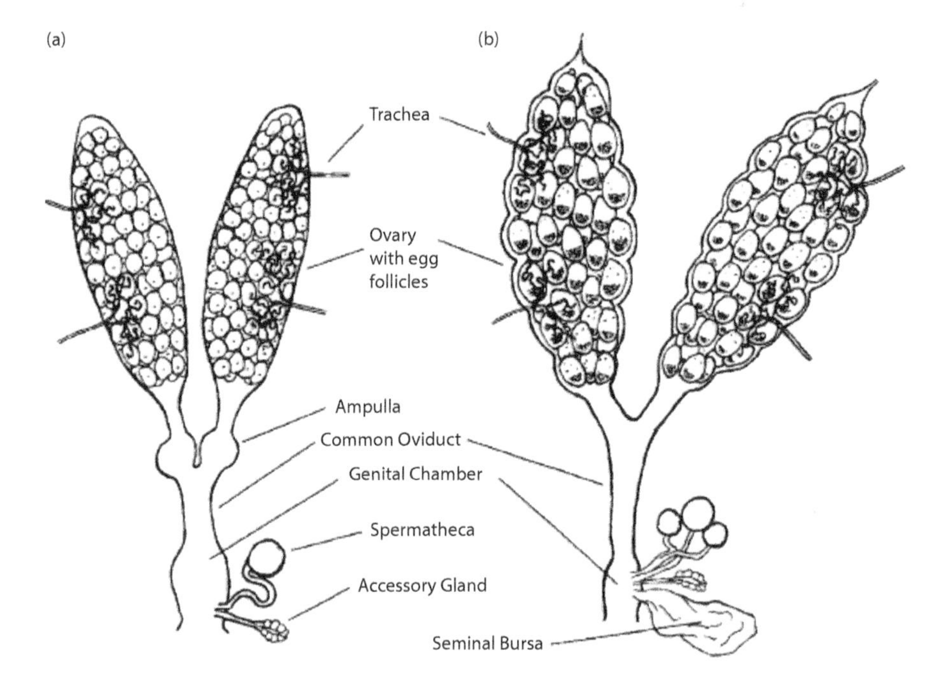

Figure 7.4 Ovaries of (a) *Anopheles* and (b) *Culex*. (Reprinted from *Medical and Veterinary Entomology*, Foster, W.A. and Walker, E.D., Mosquitoes (Culicidae) Chapter 12, Copyright 2002, with permission from Elsevier.)

meal on their second or third night after emergence. Although this is used primarily for adult maintenance some of the meal is directed towards initial egg development and so the follicle develops from what would previously be classified as a Stage I ovariole to a Stage II, as in Figure 7.5. Behind this primary follicle lies a second ovariole that will develop into the next egg once the first follicle has been laid, and behind this lies the germarium.

It is from the backward budding of the germarium that future follicles (and hence future eggs) will be produced. The stages of egg development are shown in Figure 7.6.

Following the maturation of the first batch of eggs, irreversible changes occur in the ovaries of female mosquitoes. The tightly packed and coiled tracheolar system, characteristic of nulliparous

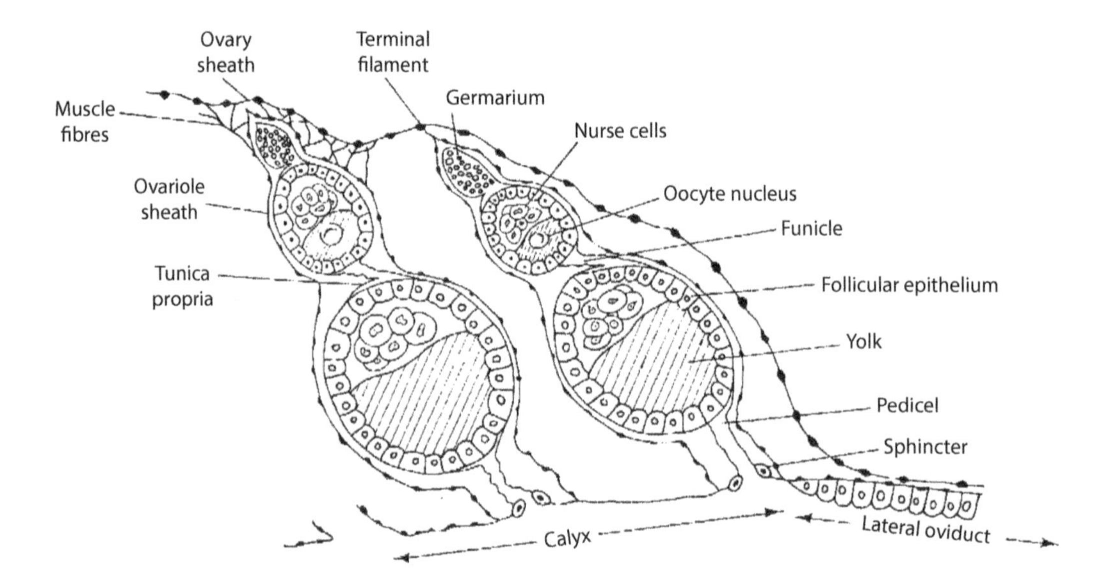

Figure 7.5 Ovariole structure. (From Detinova, T.S., Age-grouping methods in Diptera of medical importance with special reference to some vectors of malaria, Monograph series WHO no 47, 216 pp. 1962.)

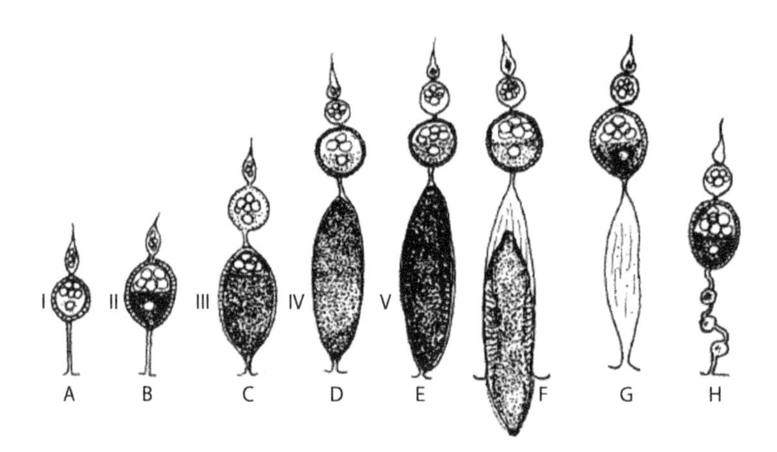

Figure 7.6 Stages (From I to V) in the maturation and laying of an anopheline egg. (From the WHO, *Manual on Practical Entomology in Malaria*, Offset publication 13 World Health Organization, Geneva, Switzerland, Vols. 1+2, 1975.)

insects, becomes stretched and uncoiled as the eggs develop and never return to their previous state (Detinova, 1962). The tracheolar system can be seen in ovaries that are dissected in distilled water and allowed to dry. Once dry, the ovary can be examined under a compound microscope. The 'dry' technique is simple and has been widely used.

It is easier to see these changes in the trachaeal system of culicines compared to anophelines. Compare Figure 7.7a and b with c and d.

The developing follicle was classified into five stages by Christophers early in the 1900s and these continue to be used today. Basically, the ovariole of a newly emerged insect without yolk is classified as Stage I. The Stage II ovariole is known as the 'resting stage' because without a blood meal there will be no further development. But once a blood meal has been taken, development through to Stage V when the egg is completely formed, and ready to be laid, takes approximately two days depending on temperature. Stage III ovarioles are the ones that have the best cells for polytene chromosome preparation. Once the egg has been laid, the stretched pedicel remains as a sac but this decreases in size over the subsequent 24 hours (Figure 7.8).

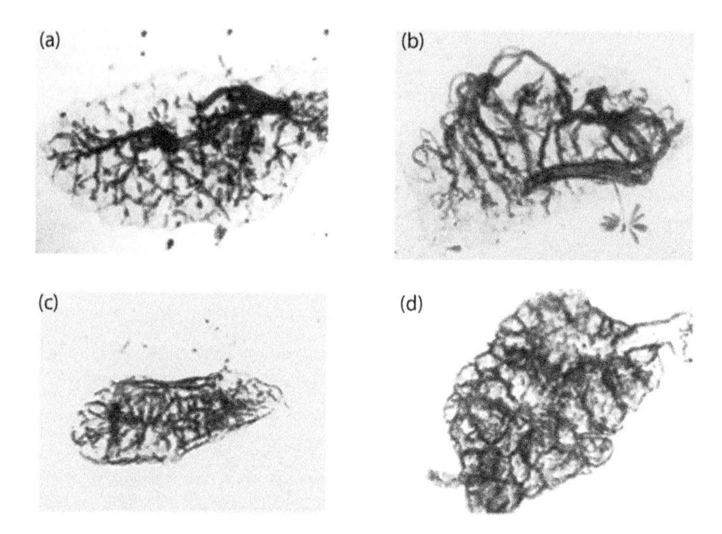

Figure 7.7 Dried ovaries of (a) a nulliparous *Culex quinquefasciatus*, (b) a parous *Cx. quinquefasciatus*, (c) a nulliparous *A. arabiensis* and (d) a parous *A. arabiensis.* Note the coiled tracheoles in the nulliparous specimens. (From Charlwood, J.D. et al., *Peer J.*, 5155, 2018.)

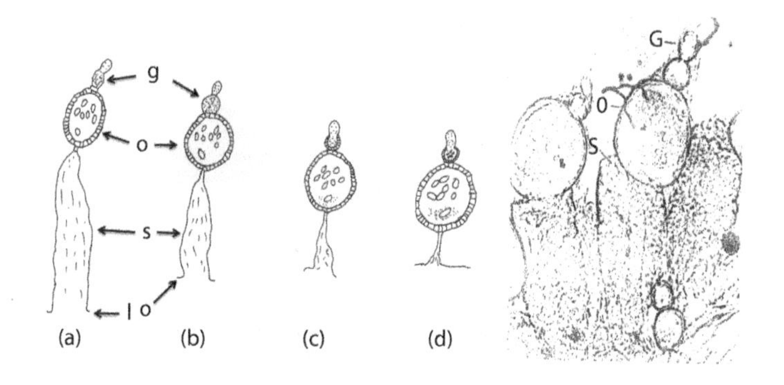

Figure 7.8 Large ovariolar sacs seen in an *Armigeres milnensis* and shrinkage of sacs seen in mosquitoes in the 24 hours following oviposition. (From Charlwood, J.D and Galgal, K., *Aust. J. Entomol.*, 24, 313–320, 1985; Charlwood, J.D. et al., *Peer J.*, 5155, 2018.)

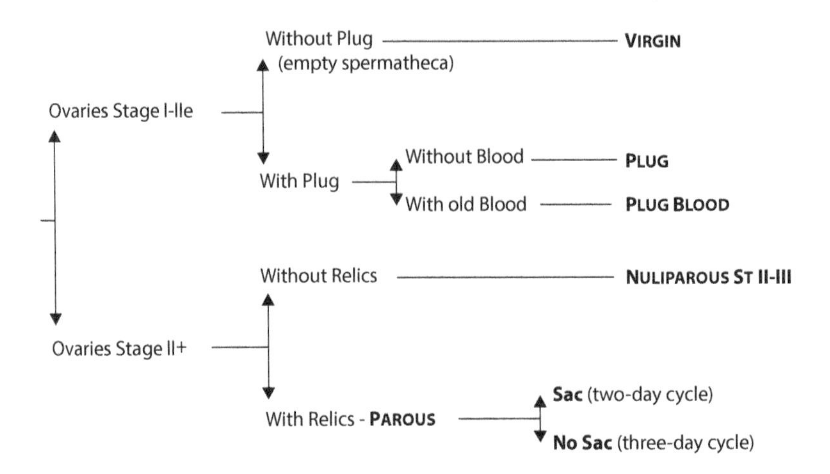

Figure 7.9 A generalised schema of the information obtained by dissection of the female anophelines ovaries.

Dissection of the ovaries, therefore, enables the researcher to determine the time between oviposition and the time of capture. Given that egg maturation is a temperature-dependent process, which in the tropics generally lasts two days, this enables the duration of the oviposition cycle to be estimated (as in the schema outlined in Figure 7.9). The dissection also enables an estimate of survival rate to be determined because it separates Nulliparous insects from Parous ones. This dissection is relatively simple to do but, like advanced age grading described below, it does require transmitted light (light coming from under the specimen).

The dissection can also reveal the exact number of ovipositions that a mosquito has completed, but this is more difficult. It requires a considerable degree of technical skill and a certain 'knack'. Previously, it was thought that serial ovulations give rise to serial dilatation in each ovariole. However, it is now recognised that ovarioles that give rise to an egg result in a single basal dilatation without a reformed stalk for each ovulation. Reliable indicators of gonotrophic age can be derived from ovarioles that do not develop eggs, in particular 'dwarf' ovarioles that never develop eggs. The dissection basically requires a complete examination of the ovary. When found, the number of dilatations on the stalk is an indication of gonotrophic age (Figure 7.10).

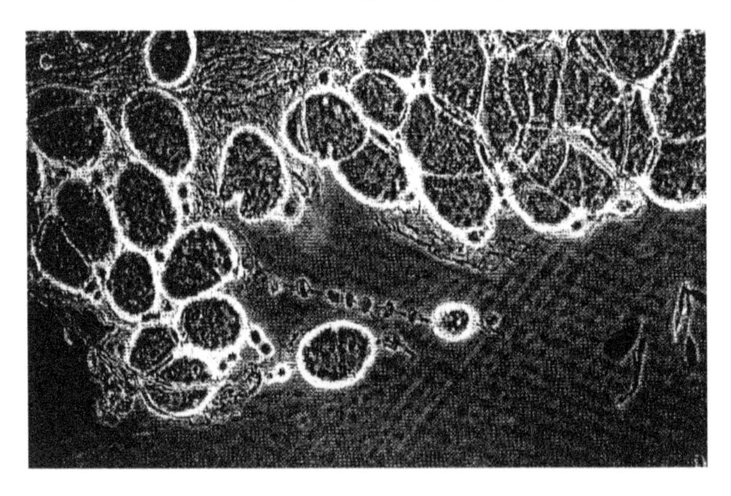

Figure 7.10 A seven-parous *A. farauti* from Papua New Guinea as identified by the dilatations on a dwarf ovariole (note the difference in size of the ovariole from a standard ovariole).

A further refinement is that of the oil injection technique, developed by the Vietnamese entomologist Traung Hoc. In this case, a Pasteur pipette is drawn out to form a short length of very thin tubing. A piece length of rubber tubing and a matchstick are added to the end. The rubber tube is filled with paraffin oil and the pipette is introduced into the common oviduct and the rubber tubing squeezed to inject the oil. The ovary is then cut lengthwise along the ventral surface using dissecting needles. This spreads the ovary out onto the slide. A cover slip is placed over the spread ovary, which is then examined with a compound microscope and the different types of follicles counted.

In parous females, three different types of functional ovarioles were observed (Figure 7.11): (1) ovarioles with dilatations and a basal body (diagnostic ovarioles); (2) ovarioles possessing only a granular basal body; and (3) ovarioles with a granular basal body and a variable number of dilatations.

In addition to estimating the absolute gonotrophic age of the mosquito, the number of ovarioles in the different categories can be counted (Figure 7.12). This enables an estimate of the number of eggs laid in each oviposition cycle to be determined.

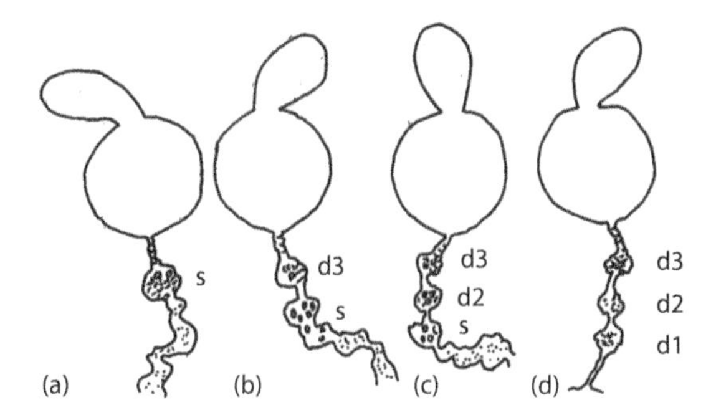

Figure 7.11 Possible appearance of individual follicles in a three-parous mosquito (a) With a sac, (b) A sac and a single dilatation (c) A sac and two dilatations and (d) three dilatations. (From Hoc, T.Q. and Charlwood, J.D., *Med. Vet. Entomol.*, 4, 227–233, 1990.)

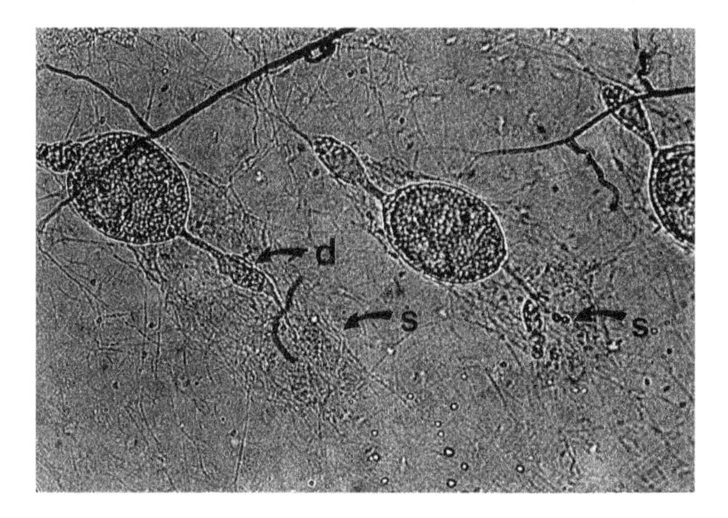

Figure 7.12 Photomicrograph of a two parous *Ochlerotatus cantans*, one ovariole having a sac and a dilatation and another having just a single sac.

SURVIVAL RATE ESTIMATION

It is important to know if mosquitoes are more likely to die as they get older or if the factors that are likely to kill them do so independent of mosquito age (as MacDonald assumed). If they are more likely to die as they age, then simple models of malaria will overestimate the vectorial capacity of the insect and they need to take this into account when assessing the possible effects of any intervention. Should survival be independent of age, and the duration of the gonotrophic cycle be a constant, then dividing the population into two age classes (Nulliparous and Parous) is sufficient and an exponential function can be used to estimate survival.

In this case the daily survival rate (p) is estimated by

$$p = \sqrt[u]{M}$$

where u is the mean duration of the oviposition cycle of the population and M is the parous rate.

There are a considerable number of assumptions associated with this estimation. The most important are that the population is stable and is not changing over time and that all age classes are sampled in proportion to their relative density in the population. Mosquito populations are, however, rarely stable and for parous rates to be useful by themselves, samples should be made over a complete population cycle (normally a year).

One way of eliminating the need for a stable population is to determine survival rates by time series analysis according to the method developed by Birley and colleagues (Birley & Rajagopalan, 1981; Charlwood et al., 1987; Holmes & Birley, 1987). For this estimate, insects need to be sampled for a minimum of 30 consecutive nights. As with parous rates, the method assumes that survival is independent of age and that nulliparous and parous insects are sampled in proportion to their relative density in the population. It also assumes that sampling provides a consistent estimate of relative density. Under these circumstances the number of parous insects sampled on one night (M_t) is equal to the total number sampled one oviposition cycle earlier (T_t) multiplied by the survival rate S.

$$M_t = S.T_{t-\mu}$$

The equation predicts that a cross-correlation of M_t and T_t will have a peak at a time lag μ days. This model is independent of fluctuations in recruitment rate and S is estimated by treating the equation as a linear regression through the origin. When the oviposition cycle is distributed among two or more days an equivalent multiple regression may be used and the mean oviposition cycle estimated. Failing this, the cycle can be determined by the proportion of parous insects that have or do not have ovariolar sacs.

Another complication occurs when the mosquito goes through a pre-gravid phase and takes two blood meals to complete the first oviposition cycle. Two methods that have the equivalent assumptions of the parous rate have been developed. The simplest is that of Garrett-Jones and Grab (1964). They produced a series of graphs of daily survival rate (proportion of the mosquito population surviving through one day) by parous rate according to five different feeding rhythms (Figure 7.13), viz:

Feeding on day

1. 2//5,7,9,11
2. 2//6, 9, 12,15
3. 2,4//6,8,10,12
4. 2,4//7, 9, 11,13
5. 2,4//8,11,14,17

where // is the time of first oviposition.

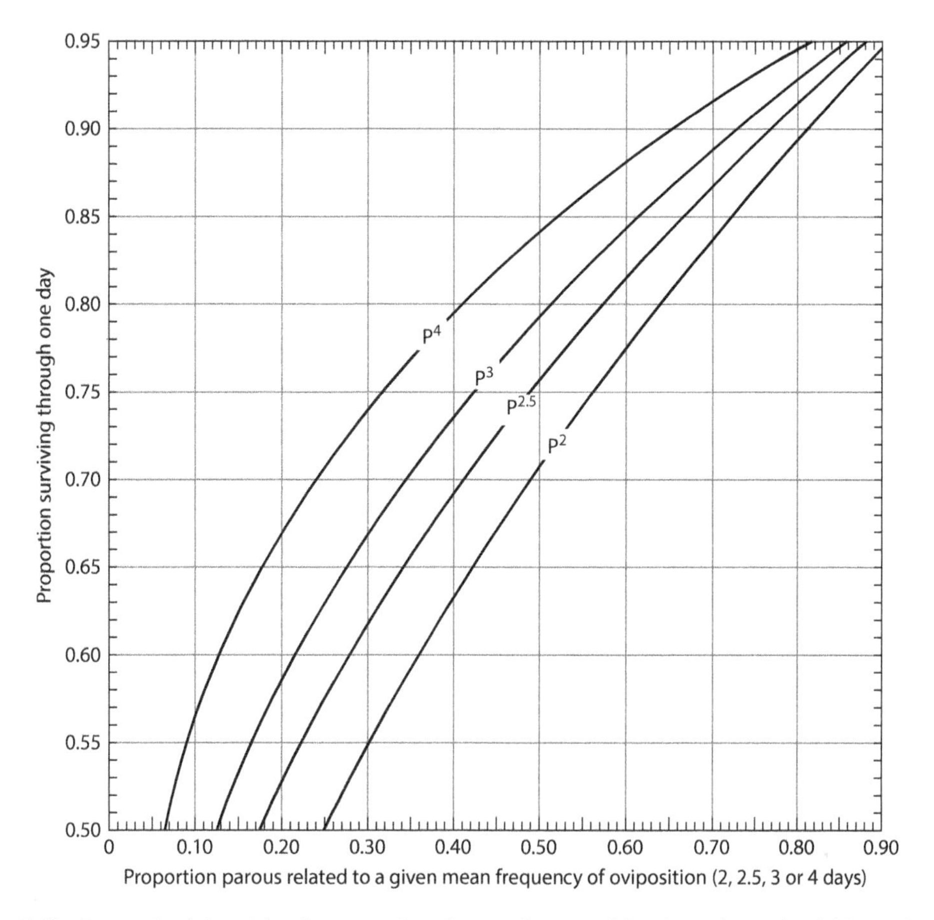

Figure 7.13 Curves for determining the proportion of mosquitoes surviving through one day (*p*) from observed proportions parous related to various irregular rhythms of feeding and oviposition. (From Garrett-Jones, C., *Bull. World Health Organ.*, 30, 241–261, 1964.)

The last three of these rhythms are equivalent to insects that take a pre-gravid feed. Thus, assuming that the development of the eggs is constant, if the time taken to return to feed (i.e., the sac stage) is known then the survival can be easily calculated. The assumptions that apply to all parous rate data (age-independent survival and equivalent sampling of the different age groups) also apply, of course, to this met hod.

In general, numbers of mosquitoes recaptured in capture-recapture experiments show an exponential decline (even accounting for the cyclical nature of recaptures) indicating that survival is independent of age. Results from advanced age-grading dissection are more ambiguous.

The number of mosquitoes in each age class N*x* (e.g., 1-, 2-, 3-parous, etc.) are plotted as in (N*x*) against age to produce a survivorship curve. If an approximately straight line is obtained, then the simple exponential model is probably appropriate – that is, mortality is independent of age, and a straight line can be fitted by eye or by linear regression of In (N*t*) against age. The slope provides a good estimate of mortality rate (Figure 7.14).

If the points of the survivorship curve do not fall on a straight line but follow the Gompertz function on a curve that is concave below, it indicates that mortality probably increases with age. Should survival be independent of age then the age specific mortality rate will be a flat (horizontal) line; if

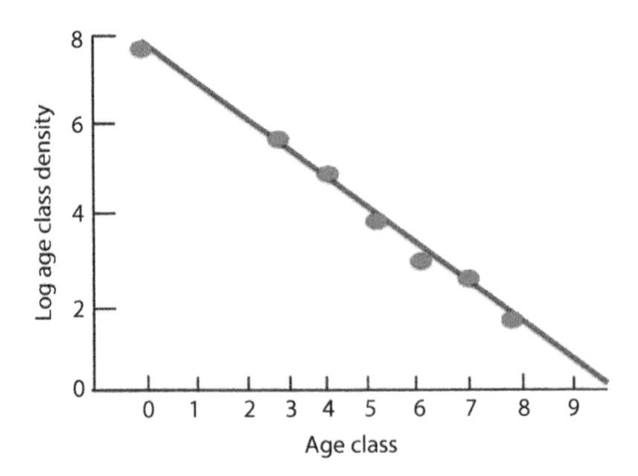

Figure 7.14 Age distribution of *A. farauti* from Papua New Guinea determined by dissection. (From Charlwood, J.D. et al., *J. Animal Ecol.*, 54, 1003–1016, 1985a.)

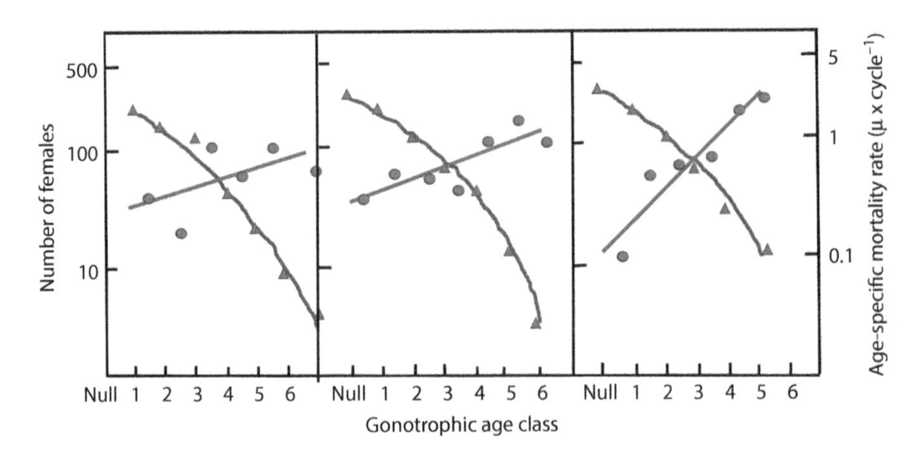

Figure 7.15 Age distribution of *A. farauti* from three villages (Bilbil, Umuin and Maraga) from Papua New Guinea. Note the higher age-specific mortality rate (the straight line) from Maraga. If survival were independent of age, this line would be horizontal.

survival increases with age then it will have an upward slope, the gradient of which is equivalent to the force of mortality with age.

In one series of dissections from Papua New Guinea, survival of *A. farauti* followed the exponential pattern (Figure 7.14), but in a further series of experiments survival followed the Gompertz function in all three villages (Bilbil, Umuin and Maraga), where collections were simultaneously undertaken (Figure 7.15). Survival was similar in the first two villages (Bilbil and Umuin), which shared a similar ecology but were lower in Maraga where breeding sites were located considerably farther from the village. Thus, in this PNG example, the age-specific mortality rate in Maraga was higher than in the other two villages.

Dissection data from Muheza, Tanzania, obtained in their classic study of *A. gambiae* by Gillies and Wilkes (1965) also indicated that survival was age dependent (Figure 7.16).

A further estimate of survival from Tanzania was available when recruitment to a large population of *A. arabiensis* ceased following an absence of rain of several weeks. The population was

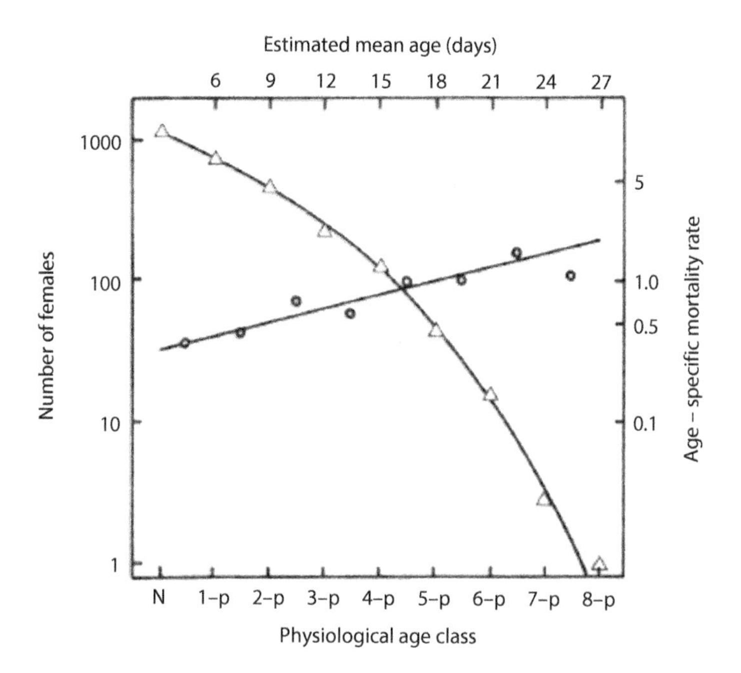

Figure 7.16 Survival of *A. gambiae* from Muheza, Tanzania, follows a Gompertz distribution with an increasing force of mortality as the insects get older. (From Clements, A.N. and Paterson, G.D., *J. Appl. Ecol.*, 18, 373–399, 1981.)

followed on a daily basis using a light-trap from a sentinel house. Over a month-long period, numbers declined from more than 3000 mosquitoes a night to fewer than 10 a night in a regular fashion. At the same time the sporozoite rate increased from less than 0.5%–2% (Figure 7.17).

Table 7.4 gives estimates of daily survival rates determined by a number of different methods (dissection, the decline mentioned above and mark-release experiments).

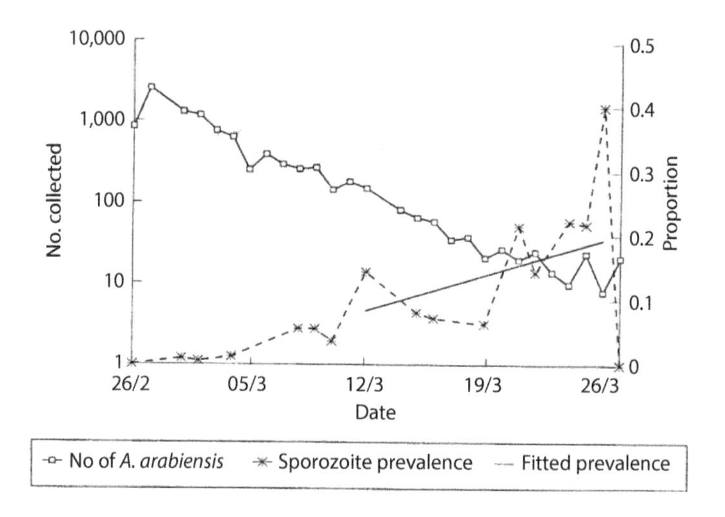

Figure 7.17 The decline in *A. arabiensis* and increase in sporozoite rates observed following an absence of rain in Namawala, Tanzania. The daily survival rate estimate was derived from the regression of numbers caught with days of collection. (From Charlwood, J.D., *Parasitol. Today*, 11, 184–185.)

Table 7.4 Estimates of Survival Rates of *Anopheles gambiae* s.l. and *A. funestus* from Namawala and Michenga Villages, Tanzania, Calculated Using Various Methods

	Namawala		Michenga
	Anopheles gambiae s.l.	*Anopheles funestus*	*Anopheles gambiae* s.l.
Estimates of survival per day (p)			
P_d	0.827 (0.004)	–	–
P_c	0.813	0.645-0.730	–
$P_M = M^{(I/u)}$	0.839 (0.025)	0.813 (0.013)	0.77 (0.018)
Estimates of survival per gonotrophic cycle (P_u)			
M	0.623 (0.004)	0.611 (0.005)	0.49 (0.01)
$P_r = (R/D_r)^{0.5}$	0.427 (0.030)	0.665 (0.064)	–
$P_n = (R/D_n)^{0.5}$	0.604 (0.060)	0.849 (0.114)	–
Estimates of survival per extrinsic incubation period (P_E)			
$P_{E,d} = P_d^E$	0.110	–	–
$P_{E,c} = P_c^E$	0.091	0.006-0.026	–
$P_{E,M} = M^{E/u}$	0.130	0.091	0.046
$P_{E,r} = S/D_r$	0.152 (0.021)	0.163 (0.030)	–
$P_{E,n} = S/D_n$	0.304 (0.061)	0.367 (0.069)	0.324 (0.041)
$P_{E,R} = SM^2/R$	0.327 (0.041)	0.132 (0.018)	–

Source: Charlwood, J.D. et al., *Bull. Entomol. Res.*, 87, 445–453, 1997.
Abbreviations: R = immediate oocyst rate; S = sporozoite prevalence; D_r = delayed oocyst rate (resting catch); D_n = delayed oocyst rate (net-with-holes catch); M = parous rate.

The take-home message from the above table is that estimates of daily survival are relatively similar but that when extrapolated to survival per extrinsic incubation period of the parasite these small differences have an enhanced effect.

The gonotrophic cycle length can be estimated by determining the proportion of parous females returning to feed with and without sacs. If one assumes that egg development takes two days and that females oviposit on the night that they are gravid, then insects with sacs have a two-day cycle and those without a three-day cycle. The mean cycle length is then:

$$\mu = \frac{\left[(\text{n Sac} \times 2) + (\text{n No} - \text{sac} \times 3)\right]}{(\text{n Sac} + \text{n No} - \text{sac})}$$

where μ is the mean feeding frequency of parous insects in days.

Temperature affects the post-oviposition behaviour of mosquitoes. In Tanzania, Gillies and Wilkes (1963) found that at temperatures above 26.5°C *Anopheles funestus* developed eggs in two days but returned to feed with a delay of a day, and that at temperatures below 26.5°C egg development took three days but the insects returned to feed on the night that they oviposited, meaning that the cycle took three days irrespective of temperature. In Ghana, the proportion of *Anopheles coluzzii* with ovariolar sacs also increased as temperatures decreased (Figure 7.18).

Some caution should always be observed when considering sac stages, however, because they are a dynamic phenomenon and will shrink rapidly over time. For example, in Muleba, the sacs of *Anopheles arabiensis* that died soon after they had been collected were larger than those insects that remained alive through the night and were only killed just before dissection (Figure 7.19). This would have resulted in a substantial difference in estimates of oviposition cycle duration and

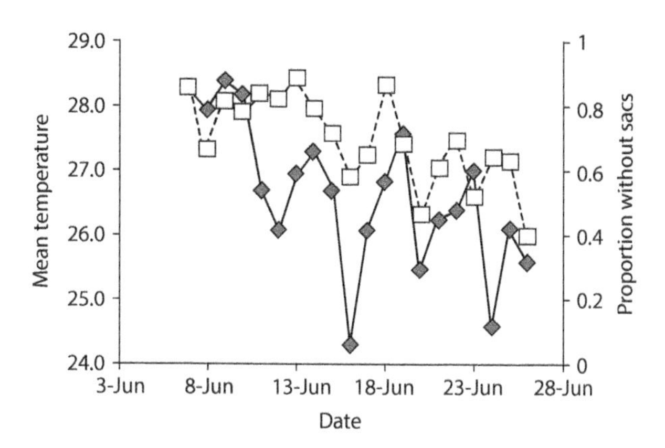

Figure 7.18 Proportion of parous *Anopheles coluzzii* without large ovariolar sacs from tent and light traps (dotted line) and mean temperature, Okyereko, Ghana.

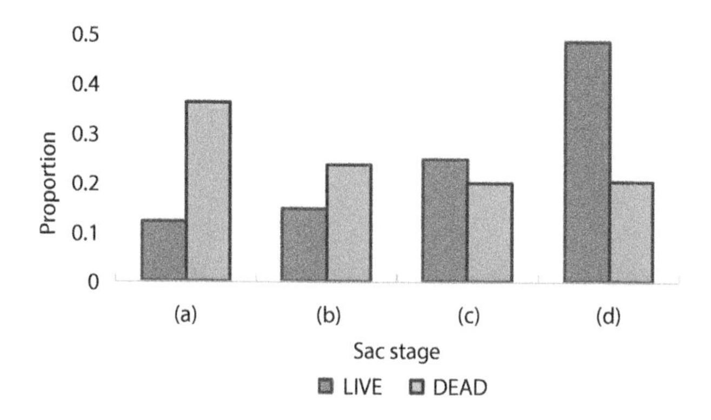

Figure 7.19 Sac stages of *Anopheles arabiensis* that had died before dissection (blue histogram) and those that were alive before dissection (red histogram), Muleba, Tanzania.

therefore of survival and vectorial capacity. It can, however, be assumed that a mosquito that has a very high proportion of large sacs in host-seeking females (such as *Chagasia bonnae* from Brazil) has a rapid oviposition cycle (Wilkes and Charlwood, 1979).

As observed in capture-recapture experiments, moon phase also affects the likelihood that some species of mosquito return to feed shortly after oviposition. This has been observed in *A. funestus* from Mozambique, *A. minimus* s.l. from Cambodia as well as *A. farauti* and *A. koliensis* from Papua New Guinea (Table 7.5).

Table 7.5 Estimated Oviposition Cycle Duration, in Days, of Mosquitoes According to Moon Phase (*Anopheles funestus* and *A. minimus* s.l. Determined by Dissection and *A. farauti* and *A. koliensis* by MRR Experiment)

	Mosquito Species			
Moon Phase	***A. funestus***	***A. minimus* s.l.**	***A. farauti***	***A. koliensis***
Waxing	2.42	2.49	2.26	2.22
Full	2.40	2.30	2.04	2.00
Waning	2.51	2.39	2.40	2.25
No moon	2.48	2.40	2.35	2.50

The complex answer to the question posed at the start of this chapter is the quantity called the basic reproductive number, R_0. It depends on knowing the following (Smith et al., 2009):

- How many times is a person bitten by vectors each day? (the biting rate m)
- How many human blood meals does a vector take over its lifetime?
- What fraction of blood meals taken by infectious mosquitoes cause infections in humans?
- How long does a person remain infectious?
- What fraction of mosquitoes feeding on infectious humans become infected?
- What fraction of mosquitoes survive sporogony?

The measurement of R_0 and the determination of the action of control measures on reproductive success are essential for the quantitative assessment of the impact of control policies on the abundance and prevalence of infection (Figure 7.20). If $R_0 > 1$, then a single infectious mosquito would tend to leave more infectious mosquitoes, and as a consequence the parasite rate would increase until it reached a steady state when new infections were balanced by cleared infections.

Exposure to mosquitoes is not uniform but rather some people are bitten more than others. Exposure depends on how bites are distributed within households, among households and among individuals over time. The factors that determine who gets bitten within a household include body size, sex, pregnancy and olfactory cues that have not yet been identified. Some households get more exposure than others, depending on their proximity to larval habitats, the house design and odours that attract mosquitoes. All of these effects combine so that a few houses harbour the great majority of the mosquitoes. It has been proposed that 20% of the people get 80% of the bites – the so-called 80/20 rule.

This ratio is found in many diseases including schistosomiasis and HIV/AIDS (Woolhouse et al., 1997). In many areas, it is seen in the distribution of mosquito numbers by houses – for example, in the numbers of *Anopheles coluzzii* collected in light traps from houses in São Tomé (Figure 7.21).

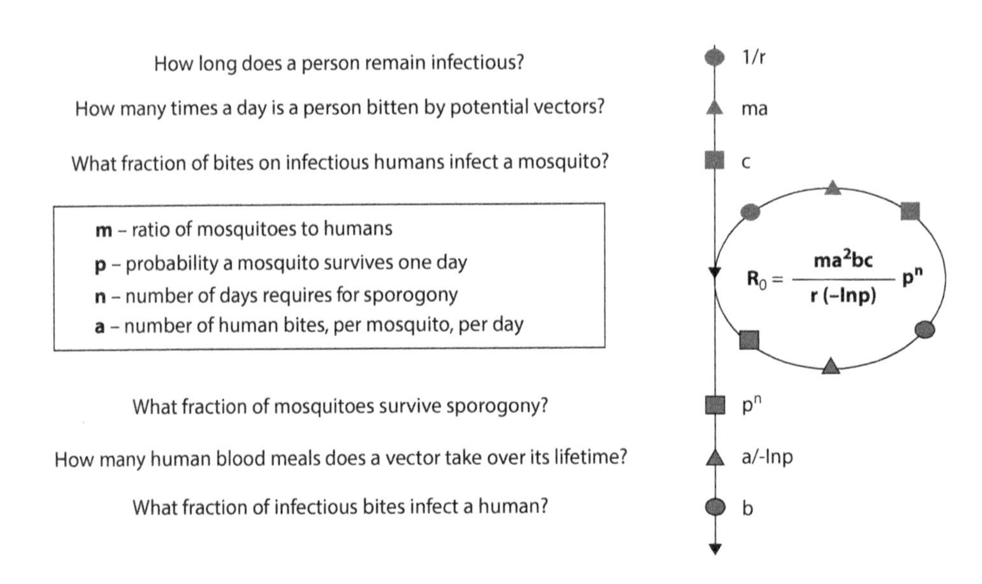

Figure 7.20 Measuring R_0. (From Smith, D.L. et al., Measuring malaria for elimination. Chapter 7 in *Shrinking the Malaria Map: A Prospectus on Malaria Elimination*, Feachem, R.G.A., Phillips, A.A. and Targett, G.A., (Eds.), The Global Health Group, San Francisco, CA, 2009.)

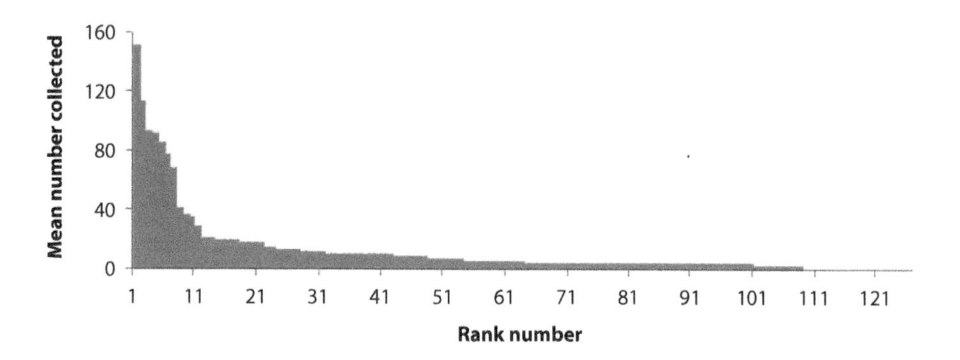

Figure 7.21 Mean number of *Anopheles coluzzii* collected in light traps by house, Riboque, São Tomé. (From Charlwood, J.D. et al., *Malar. J.*, 2, 45, 2003a.)

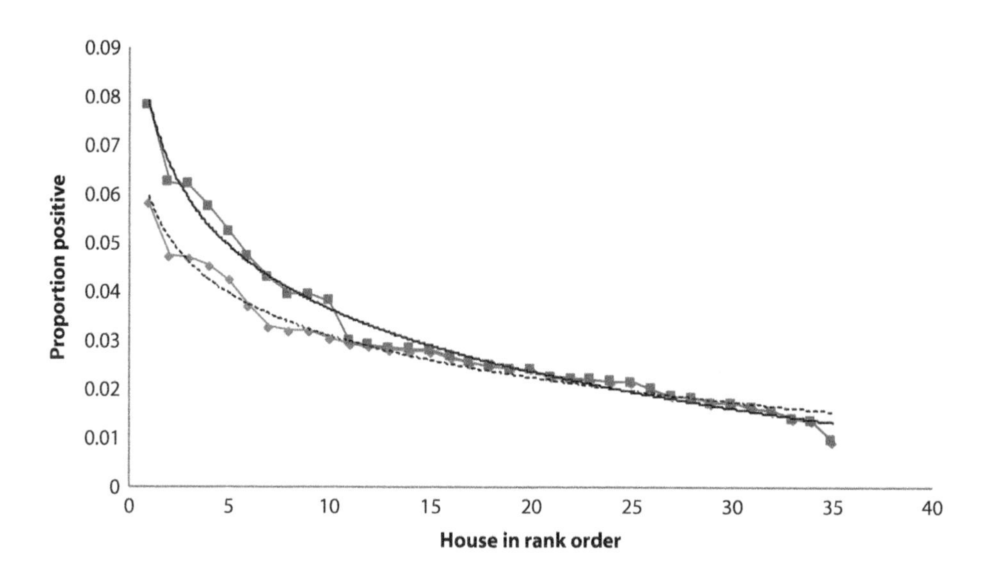

Figure 7.22 Sporozoite rates determined from light-trap collections by house from Furvela, Mozambique. The lower line (in blue) is the crude estimate of sporozoite rate, and the upper line (in red) is the corrected estimate according to pool size from Table 8.1.

It is also seen in the sporozoite rates by house in Furvela, Mozambique (Figure 7.22), and in transmission over Africa as a whole (Figure 7.23).

From the above figure, it is obvious that the most intense transmission occurs in West Africa with some further 'hot spots' in southern Tanzania and northern Mozambique. This is the 80/20 rule writ large.

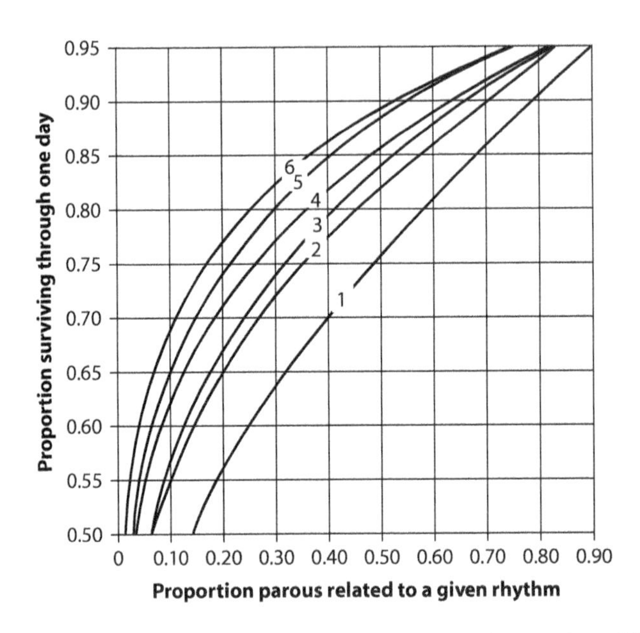

Figure 7.23 The spatial distribution of the estimated basic reproductive number of *P. falciparum* malaria at present levels of control (R_C). (From Smith, D.L. et al., Measuring malaria for elimination. Chapter 7 in *Shrinking the Malaria Map: A Prospectus on Malaria Elimination*, Feachem, R.G.A., Phillips, A.A. and Targett, G.A., (Eds.), The Global Health Group, San Francisco, CA, 2009.)

SPOROZOITE DETERMINATION

For the mosquito to transmit malaria sporozoites from the ruptured oocyst must enter the salivary glands where they will be ejected during blood feeding. The salivary glands of mosquitoes infected with Plasmodium sporozoites are impaired in their production of an apyrase, an enzyme that hydrolyses ATP and ADP to AMP and prevents platelet aggregation. This reduction in apyrase makes blood feeding more difficult and increases the probing time of the mosquito. Increased probing time results in a greater delivery of parasites and a greater number of hosts that are fed upon. Manipulating its hosts' enzyme production increases the parasites' likelihood of being distributed to more hosts.

The salivary glands of *Anopheles* and *Culex* differ. In Culicines the middle lobe of the salivary gland is about as long as the other two whilst in *Anopheles* it is shorter (Figure 7.24).

The salivary glands are located in the thorax just behind the head. They can be removed by gently pulling the head off (Figure 7.25). They are examined with a compound microscope. The technique requires a relatively high level of skill. Live specimens are required for direct observation, and the species of parasite cannot be determined. Densities can, however, be estimated.

The method most widely used today is the ELISA technique developed by Wirtz and Burkot. In the ELISA Plasmodium, circumsporozoite protein (CSP) acts as the antigen. Mosquitoes are ground up (either in pools or individually) in phosphate buffered saline (PBS) and added to ELISA plates that are treated with antibodies to the protein. On incubation, the CSP binds to the antibody. A further antibody that turns blue when a peroxidase is added is also incubated and a colour change produced in positive wells. The process is shown in Figure 7.26.

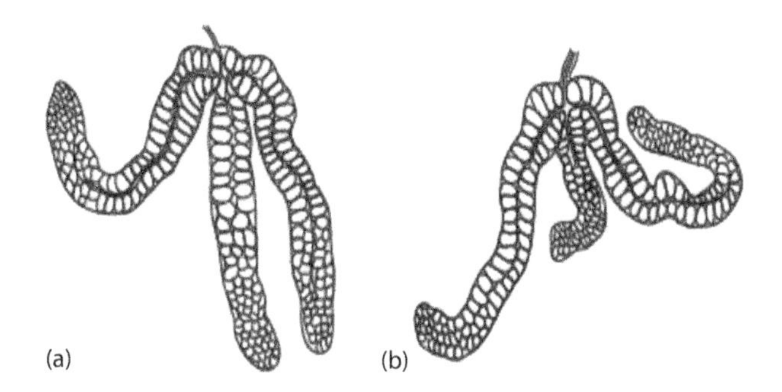

Figure 7.24 Salivary glands of (a) culicine and (b) anopheline.

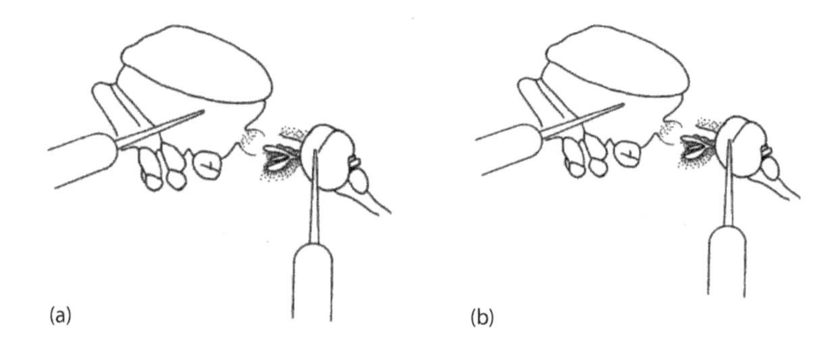

Figure 7.25 Two methods of removing the salivary glands either by (a) pulling the head of the mosquito off the thorax or (b) by squeezing the glands out once the head has been removed.

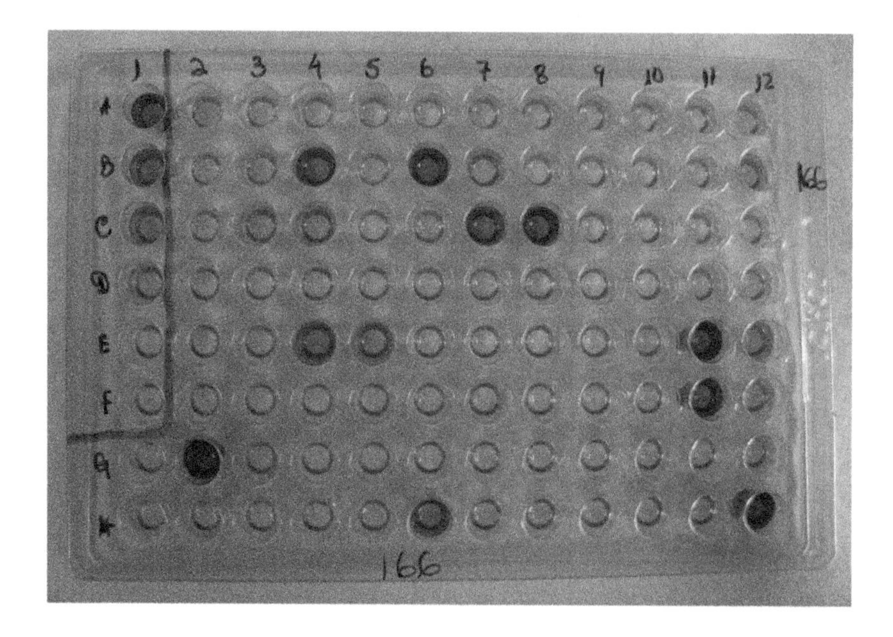

Figure 7.26 The steps involved in the process of an ELISA. (From WHO, *Global Technical Strategy for Malaria 2016–2030*, WHO, Geneva, Switzerland, 2013b.)

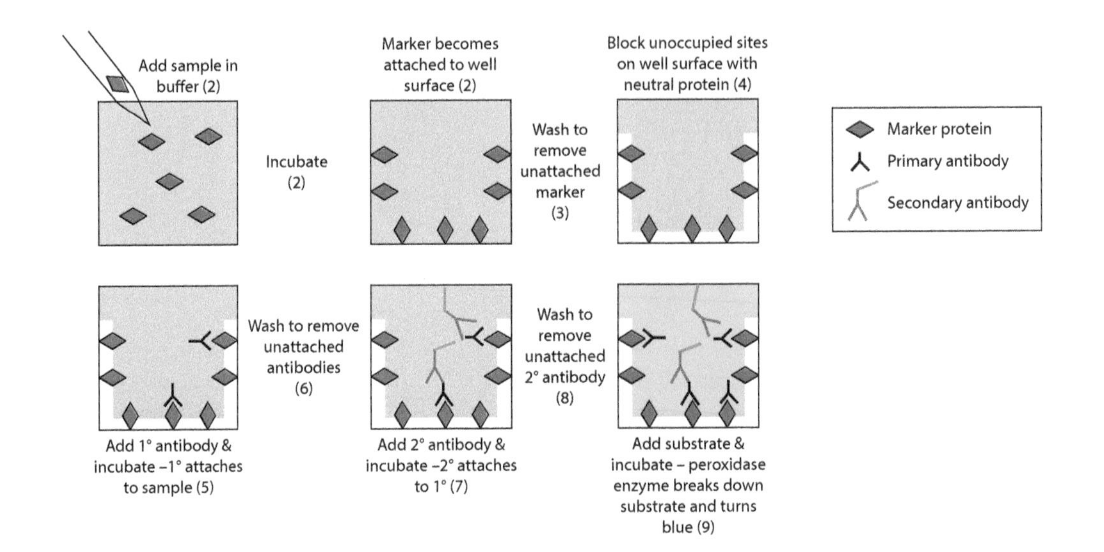

Figure 7.27 ELISA of *A. funestus* from Mozambique. A1-D1 Positive controls – different dilutions of sporozoites E1-F1 Negative controls substrate but no product Positive wells green/blue.

A sample plate is shown in Figure 7.27; note the colour difference among positive wells.

The plate can be read in a machine that gives optical densities corrected for the negative controls or it can be read visually. There was good correspondence between visual grades (separated into three classes) and machine readings when mosquitoes from Mozambique were examined by both systems (Figure 7.28).

When undertaking the ELISA it is usual to separate the head and thorax from the abdomen because there is a possibility that mosquitoes with mature oocysts are considered to be infectious. This, however, is a considerable amount of tedious work, especially with dried mosquitoes. How great the error might be was also investigated with mosquitoes from Mozambique. In this case, positivity among paired sets of heads + thoraces and abdomens was determined. Most of the mosquitoes tested were positive only in the head + thorax (blue in Figure 7.29). A smaller number were positive for both sets

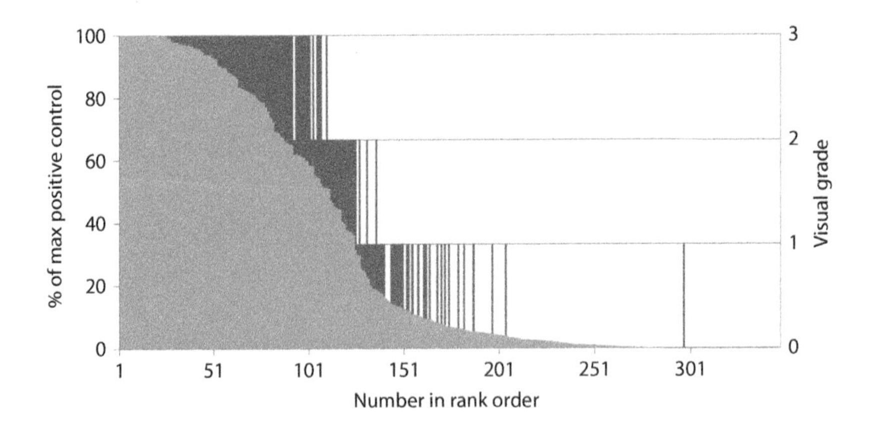

Figure 7.28 Control reading from a plate reader and visual grades estimated by eye in *A. funestus* from Mozambique analysed in pools of 10 mosquitoes per sample well.

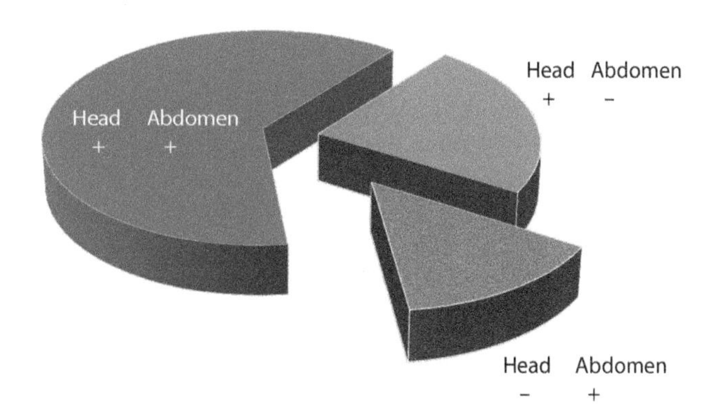

Figure 7.29 Positivity in paired head + thorax and abdomen samples of *A. funestus* from Mozambique. Blue = positive for head + thorax only, Red = positive in both head + thorax and abdomen and Green = positive in abdomen only.

of body parts (in which case the insect would be considered to be positive) (red in Figure 7.29), and only a small fraction (the green part of the pie chart) were positive in the abdomen alone. Although a pilot set of samples should, perhaps, be tested in this fashion in any large-scale study, these results indicate that for large surveys the error introduced is slight when using whole insects.

It is of interest to the entomologist to know if different samples of mosquito have different positivity rates – for example, if outdoor-biting mosquitoes that may be more likely to bite animals rather than people have different infection rates to those collected indoors. In Mozambique, there was no difference between light-trap and tent-trap infection rates implying that, as in Papua New Guinea, the indoor- and outdoor-biting population was homogeneous. There were slightly lower rates among mosquitoes resting and exiting from houses (i.e., post feeding) compared to either light-trap or tent-trap. This implies that feeding in infected mosquitoes has a mortality risk.

The effect of pool size was also investigated in these studies (Figure 7.30). The positivity rate in individual mosquitoes was higher than that determined from pools of mosquitoes. This may have been for a number of reasons, in particular the difficulty that may be associated with grinding larger numbers of mosquitoes compared to grinding a single specimen. The lower positivity rate may not be such a problem as it first appears because most studies are designed to compare an intervention to a control group and, as long as similar techniques are used in both groups, the relevant effect can be accurately assessed.

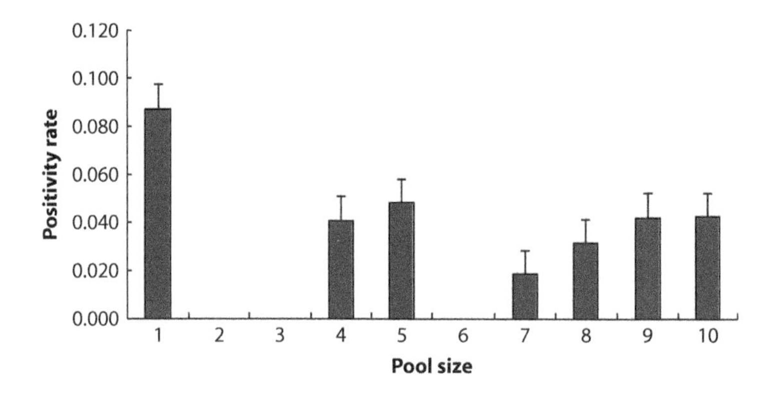

Figure 7.30 Positivity rate by pool-size *A. funestus* whole mosquitoes from Furvela, Mozambique.

Testing mosquitoes in pools (of 5 or 10 individuals) reduces the amount of work and resources needed to test large numbers of mosquitoes. As positivity rates increase, there is the possibility that two infected individuals will be present in a single well (thus reducing the estimate of positivity). This can be compensated for by determining the relative underestimate (Table 7.6). At rates above 8% it is worth doing the ELISA on individual mosquitoes.

Temperature affects the duration of the extrinsic cycle in the mosquito. If mosquito survival is not affected by higher temperatures, then the proportion of infected mosquitoes that become infectious will increase with temperature. This occurred in mosquitoes from Mozambique, where mean temperatures vary from 30°C in the hot season to 20°C in the cool season (Figure 7.31).

Sporozoite rates may change from one year to the next independent of any intervention. Thus, in Furvela, sporozoite rates decreased between the years 2002–2006 (when no intervention was in place) but the reason for this is not known (Figure 7.32).

Rates in Furvela dropped even further when an intervention of a *cordon sanitaire* of impregnated nets was installed around the main breeding site. Unfortunately, this effect was only apparent for a year or two and by 2009 rates had increased to pre-intervention levels. Malaria incidence measured at the project clinic had also increased. The reduction in effect was probably due to an increase in the resistance status of the *A. funestus*, which by 2009 were not killed or knocked down by pyrethroids on nets.

Table 7.6 Underestimates of Sporozoite Rates According to Pool Size Used in the ELISA

	Pool Size			
	5		10	
Sporozoite Rate (%)	Crude Estimate	Underestimate	Crude Estimate	Underestimate
0.5	0.5	0	0.49	0.01
1.0	0.98	0.02	0.96	0.04
2.0	1.92	0.08	1.83	0.17
4.0	3.69	0.31	3.35	0.65
8.0	6.82	1.18	5.66	2.34

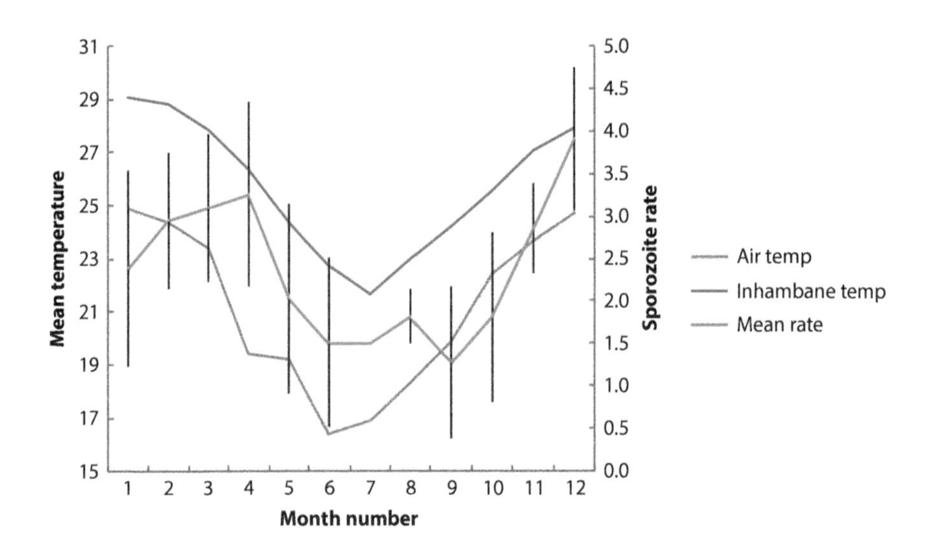

Figure 7.31 Sporozoite rates in *A. funestus* and mean air temperatures from Furvela, Mozambique.

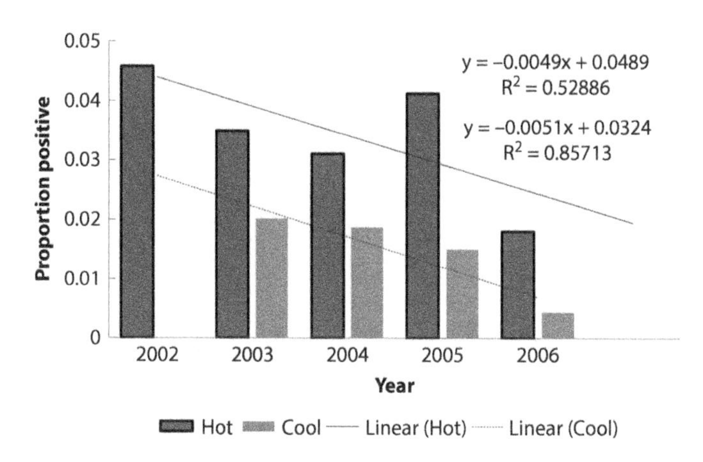

Figure 7.32 Sporozoite rates (n = 35,471) in *A. funestus* in hot and cool seasons from 2002 to 2006 in Furvela, Mozambique.

CHAPTER 8

Chemical Methods of Vector Control

The most efficacious methods of vector control depend on the reproductive potential of the species that can be classified into *r* and *K* strategists (Table 8.1).

The *r* strategists are characterised by their high rates of population increase (due to a high fecundity and short generation time), well-developed powers of dispersal (migration) and ability to locate new food sources. Because they may occur in large, but unpredictable numbers, predators may have little effect on their species. Furthermore, although *r* species are subject to disease, they are slow to take effect. Because of their high reproductive potential, *r* strategists are able to tolerate mass mortality and rapidly recover to their original density. Hence, biological control, which is a relatively slow but long-term method, is of little use against *r* strategists (such as mosquitoes). Insecticides, which can be stored for application at short notice, continue to be the most important tool in their control.

INSECTICIDES

People have used chemicals against insect pests for millenia. The Greeks used sulphur against pests almost 3000 years ago and the Romans used asphalt fumes to rid their vineyards of insect pests. Inorganic compounds such as arsenic, sulphur, borax and phosphorous were sprayed onto insect food plants. These require ingestion before working. Thus they might be used against mosquito larvae but are unsuitable for sucking insects like adult mosquitoes for which 'contact poisons', absorbed through the cuticle or tracheal system, are required.

Some plants produce such contact poisons, notably the pyrethroids produced by plants in the genus *Pyrethrum (Chrysanthemum)* (family Compositae). The plants are still grown on an industrial scale (Figure 8.1) but because natural pyrethroids break down rapidly in ultraviolet light they need to be reapplied frequently which makes them expensive. They were largely replaced by cheaper synthetic insecticides in the 1940s, an important consequence of which was that relatively few insects became resistant to them. This, in conjunction with the development of several photostable synthetic pyrethroids, has led to a resurgence in their importance. They now account for about one-third of all insecticides in use today. Not surprisingly, this has led to the development of resistance; from 10 species in 1970 to more than 80 in 2003.

Table 8.1 Appropriate Methods of Control According to the Reproduction Potential of the Pest

Method of Control	Rapid Reproducers		Slow Reproducers
	Insecticides	→	
		Biological control	
		←	Genetic control

(a) (b)

Figure 8.1 Pyrethrum flowers. (a) Drying before preparation. (b) Growing at the edge of the Volcanoes National Park in Rwanda.

The scientific development of insecticides began in 1867 with the formulation and use of the arsenical Paris green. This was used, as a larvicide, to eliminate *A. gambiae* s.l. that had invaded Brazil (by hitching a ride on the fast mail boats that operated between Dakar and Recife). In the 1920s the structures of many of the botanical insecticides, which had been used since the early 1800s, were elucidated. It was not until 1939, however, that Müller discovered the insecticidal properties of the first synthetic insecticide, DDT (dichlorodiphenyltrichloroethane), when he tested it against clothes moths.

The potential of this new insecticide was demonstrated in 1943 when an epidemic of louse-transmitted typhus was controlled in Naples. Subsequently, DDT played a major role during the Second World War in controlling outbreaks of typhus, trench fever and louse-borne relapsing fever by direct insecticidal dusting of both soldiers and civilians and their clothes. The major benefit of DDT, however, came in malaria control, where it was a principal component of the World Health Organization's global malaria eradication campaign of the 1950s and 1960s. Its continuing utility was shown in South Africa where, in addition to the introduction of ACTs for treatment, it was used to suppress an epidemic of malaria transmitted by *A. funestus* in Kwa Zulu. In this case there had been a malaria epidemic due to the combination of drug-resistant parasites and pyrethroid-resistant *A. funestus*. Thus, how much of the effect was due to the change of drug and how much to the change of insecticide is not known.

The reverse situation was encountered in Rondonia, Brazil, where changing from DDT to lambda-cyhalothrin (ICON) for Indoor Residual Spraying (IRS) (without a change in drug regime for treatment) resulted in a decline in the number of malaria cases. 55,000 houses were sprayed with ICON whilst circa 343,095 out of a possible 459,832 houses were sprayed with DDT in the rest of the state (Table 8.2).

Table 8.2 Number of Malaria Cases from Machadinho and Jaru and the Remainder of Rondonia in 1987 and 1988 Before and After Spraying with ICON in May 1988 Machadinho and June/July in Jaru

	Month	Machadinho			Jaru			Rondonia		
		1987	1988	% change	1987	1988	% change	1987	1988	% change
P. falciparum	1–4	2636	4123		3561	6984		22,490	40,417	
	5–12	12,085	2851	−76.4	6275	4519	−28.0	72,512	89,222	+23.0
P. vivax	1–4	2163	2716		4358	6049		22,426	31,656	
	5–12	6935	4807	−30.7	5444	4991	−8.3	66,933	78,230	+16.9

Source: Charlwood, J.D. et al., *Acta Trop.*, 60, 3–13, 1995a.

There was a 28%–76% reduction in *P. falciparum* following the spraying with ICON but a 23% increase in the rest of the state where DDT was used. DDT is known to act as a repellent as much, if not more, than a killing agent. Spraying the outside eaves of houses with lambda-cyhalothrin did not reduce entry rates of *A. funestus* in Mozambique so the insecticide may not have such a strong repellent effect. One interpretation of the different results from South Africa and Brazil is that a strong repellent is better than an ineffective killing agent (i.e., an insecticide to which the vector has become resistant), but that in the absence of resistance a killing agent (lambdacyhalothrin in this case) is better than a repellent.

The trial in Brazil took place before the wide-scale use of mosquito bed nets impregnated with pyrethroids as a vector control tool. Today, the use of pyrethroids for IRS is likely to be counterproductive (because it may accelerate the development of resistance in susceptible populations) and if alternatives are available they should be used. The combination of one kind of insecticide for IRS and pyrethroids on nets may reduce the likelihood of resistance developing.

Although long used as a mainstay of vector control, IRS is now considered by the WHO to only be useful if the following criteria are met:

- The majority of the vector population is endophilic.
- The vector population is susceptible to the chosen insecticides.
- A high percentage of the houses or structures in the operational area have adequate sprayable surfaces.
- Spraying is done correctly.

The considerable resource requirements, import needs, environmental concerns in their use and the potential for development of vector resistance compel highly selective targeting of IRS. This may make it only suitable for focal control of vectors.

IRS may also only be really useful in areas where there has been little development and people are poor. With an increase in wealth, in addition to affixing a tin roof to their house (in itself likely to be an effective malaria control strategy), people tend to put posters of their favourite football team or pop singers on the walls. These are removed before spraying but replaced immediately afterwards. At the same time, people have more material goods including wardrobes and other furniture. These too are not sprayed. The end result is that in even moderately well-off houses the relative surface area that is not sprayed – and therefore is a safe resting site for mosquitoes – increases substantially up to the point where the IRS is ineffective. In addition, wall often have a very rough surface, either mud or other material. This effectively increases the surface area of the wall but the amount of insecticide needed for spraying is calculated as if the surface were smooth – so that a dose per m^2 does not reflect the actual dose on the wall. This means, either that considerably more insecticide needs to be used (increasing costs), or that applied dosages are lower than recommended (which may enhance the likelihood of resistance developing).

Insecticide-based control works in a density-independent manner. In other words, a certain proportion of the population will always be killed and the effect will not be greater as the population increases.

INSECTICIDE TREATED MATERIAL

The Greek historian Herodotus in 2500 years ago described how fishermen in Egypt slept under their nets as a way to avoid the many mosquitoes (gnats) that attacked them in the delta of the Nile. He says 'south of the marshes they sleep at night on raised structures, which is of great benefit to them because the gnats are prevented by the wind from flying high; in the marsh-country itself they do not have these towers, but everyone, instead, provides himself with a net, which, during the day he uses for fishing, and at night fixes up around his bed, and creeps in under it before he goes to sleep. For anyone to sleep wrapped in a cloak or in linen would be useless, for the gnats would bite through them; but they do not even attempt to get through the net' (Herodotus, translated by A de Sélincourt, 1966). Given that a cloak would not have protected them and that the mosquitoes did not apparently attempt to feed through the nets, it would seem likely that it was the smell of the fish that was responsible. So, this may not only be the first record of people using nets to avoid mosquitoes but impregnated ones at that!

Presently the use of Long Lasting Insecticide Impregnated Nets (LLINs) is the main method of vector control used worldwide. In very early trials the effect of treated nets on mosquito behaviour was investigated in Papua New Guinea. In addition to reducing parasite prevalence in selected villages, such nets affected the vectors in a number of ways: The population size of the main vector, *A. farauti,* was estimated from capture-recapture experiments and found to have dropped from 22,451 (+/− 4073) to 13,330 (+/− 3578) after the introduction of the nets. The oviposition cycle, which had been a regular two days before the nets, became disrupted and increased to more than three days, and the biting cycle changed from a late-night one indoors to an earlier one outdoors. Numbers collected in resting collections in and around houses dropped from 6.45 per collector to 1.41, and the feeding success of the mosquito dropped – the ratio of engorged to unfed insects in resting catches dropped from 9:1 before the introduction of the nets to 1:1. Not only that, but significantly fewer had fed on man (Table 8.3).

Changes in the time of biting (from late night to earlier in the night) have been recorded from a number of other studies, although the reasons for this have rarely been investigated. In Papua New Guinea the change was largely due to mosquitoes being unable to feed on the night after oviposition. Thus, prior to the introduction of the nets, parous insects generally had large ovariolar sacs but once the nets were introduced they were collected without sacs. *Anopheles farauti* has a relatively wide host preference and there was also a shift towards more animal feeds following the introduction of the nets.

More recently, the behaviour of individual mosquitoes at nets has been investigated using high-definition cameras (Parker et al., 2017). At untreated nets, *A. arabiensis* spend much of their time on top of the net above the host's head (Figure 8.2a), whereas if the net is treated with insecticide (in this case Deltamethrin) this is much reduced and the insects tend to fly around the net rather than rest on it (Figure 8.2b).

Following IRS or the deployment of treated nets, a change in species ratios has been reported on a number of occasions (reviewed by Durnez & Coosemans, 2013). In many situations, the previously

Table 8.3 Blood Meal Source of *Anopheles farauti* Before and After the Introduction of Insecticide Impregnated Nets in Agan Village, Papua New Guinea

	Man	Pig	Dog	Mixed (man/other)
Before	235	3	95	13
After	63	6	98	6
A. koliensis				
Before	29		6	
After	2			

Source: Charlwood, J.D. and Graves, P.M., *Med. Vet. Entomol.*, 1, 319–327, 1987.

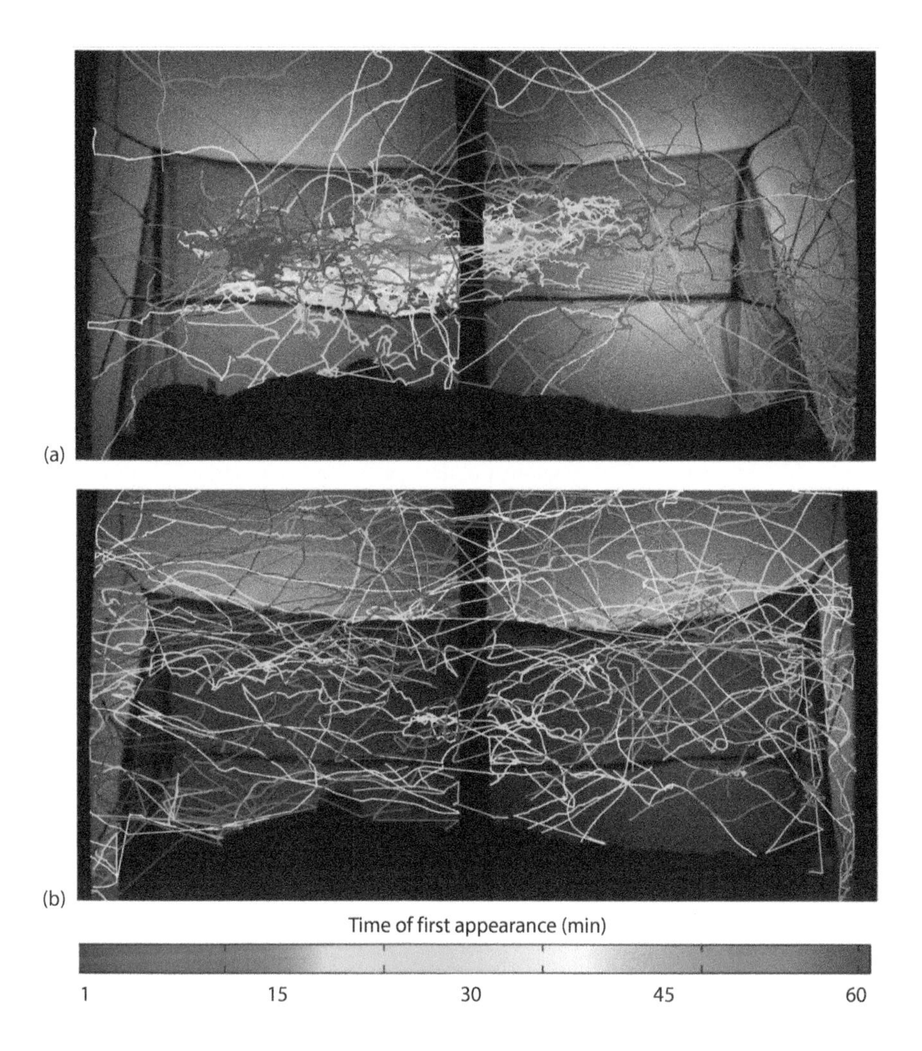

Figure 8.2 Flight activity of field-caught mosquitoes at untreated nets and LLINs. Examples of the flight tracks of *Anopheles arabiensis* in response to a human volunteer inside (a) an untreated and (b) an insecticide-treated bed net. Each image shows the total activity recorded over 60 minutes, with 10 mosquitoes released in each test. Each track is the path of an individual mosquito flight. Tracks are coded according to the time they first appear in the field of view as shown in the key below the images: blue tracks at the start through to red at the end of the test. (From Parker, J.E.A. et al., *Malar. J.*, 16, 270, 2017.)

predominant species is replaced by a close relative that may be less exposed to the insecticide. Under normal circumstances these species may be less efficient larval competitors. Apart from being less exposed to the insecticide (because of their exophilic and zoophagic tendencies), when the dominant species is largely removed they are perhaps able to dominate the larval habitat that they might share, and implies that they now occupy the larval niche previously dominated by the former species. In many areas, most recently Muleba in Tanzania, *A. gambiae* has been replaced by its sibling *A. arabiensis* (Charlwood et al., 2018) and *A. funestus* may be replaced by *A. rivulorum* or *A. leesoni*, both members of the *A. funestus* group (Wilkes et al., 1996). Outside Africa there is evidence that similar species replacement occurs among S.E. Asian anophelines, *A. dirus* s.l. being replaced by *A. minimus* s.l. (Durnez & Coosemans, 2013).

Ways of reducing the effect of metabolic resistance to the pyrethroids used on nets include the addition of synergists, which while not killing the insect require it to use the enzymes otherwise available for detoxification of the insecticide. Piperonyl butoxide (PBO) is one such chemical that acts by inhibiting enzymes involved in the natural defence mechanisms of insects, which results in pyrethroid not being detoxified and the pyrethroid on the net remaining potent despite resistance in the mosquito. A randomised control trial (RCT) was undertaken in Muleba, Tanzania (on the northern side of Lake Victoria), to determine if there was any additional benefit from such nets compared to standard nets.

The study area comprised 29,365 households and a population of 135,900. Results indicated that the additional effect of the PBO nets on malaria prevalence was evident at the end of the second year. There was 44% protective efficacy against malaria after one year and a 33% protective efficacy at the end of the second year. In areas receiving PBO long-lasting insecticidal nets compared with standard long-lasting insecticidal nets the estimated EIR was reduced by 87% during the first year and 67% during the second year.

The WHO now recommends that PBO nets be deployed for prevention of malaria where vectors are resistant to pyrethroids, provided that vector control coverage is not compromised.

INSECTICIDES FOR INDOOR RESIDUAL SPRAYING

Insecticides are delivered in a variety of ways including wettable powders (WPs), emulsifiable concentrates (ECs) and solution. There are four main classes of insecticide used in public health:

- **Organochlorines** such as DDT are inhibitors of the normal functioning of the nervous system.
- **Organophosphates** (phosphorothioates) such as pirimiphos-methyl (Actellic) act by binding the enzyme acetylcholinesterase at the nerve junction. Phosphorothioate insecticides can vapourise quickly and often have associated with them a sulphurous 'bad egg' smell.
- **Carbamates** such as bendiocarb have an identical mode of action to the organophosphates.
- **Pyrethroids** such as deltamethrin act in exactly the same way as DDT and its analogues.
- **Insect Growth Regulators** may be used against the larval stages of the mosquito.

The different classes of insecticide, their molecular target and the type of resistance that has been reported against them is shown in Table 8.4.

The WHO estimated that, on a worldwide basis, about 3 million humans are hospitalised each year due to exposure to pesticides (for both agriculture and public health) and about 220,000 persons die, almost all in developing countries. It should be noted that about two-thirds of these are suicides.

The development of new insecticides, for a number of reasons, is less of a priority today than it was earlier. Most insecticides are used for agricultural purposes and genetic engineering of the crops themselves, making them resistant to the attack of pests, which reduces the need for insecticides. Developing new insecticides is expensive. The time and cost of development and registration are thought to be US$35–45 million with an additional US$55–65 million for the cost of a production plant, which makes their development unattractive. Only 1 in 20,000 candidate chemicals ever reach the marketing stage and if they do they have a relatively short 'life expectancy' because of the development of resistance by the target organisms. Companies, which seek to maximise sales within 3–5 years of registration, are not, therefore, investing in their development.

Table 8.4 Mechanisms of Insect Resistance to the Main Insecticide Families of Public Health Interest

		Resistance Mechanisms	
Insecticides	**Molecular Target**	**Target Site**	**Enzymatic Mechanisms**
Pyrethroids, type I	Sodium channel	Kdr and super Kdr mutations	Monooxygenases + esterases
Pyrethroids, type II	Sodium channel	Kdr mutations	Monooxygenases + esterases
Organochlorates	Sodium channel	Kdr mutations	gs-transferases + monooxygenases
N-alkyl-amides	Sodium channel	No resistance reported against insect of public health importance	
Organophosphates	Acetylcholinesterase	Ace 1^R mutation	Esterases + gs-transferases + monooxygenases
Carbamates	Acetylcholinesterase	Ace1^R mutation	
Neonicotinoids	Nicotinic acetycholine receptor	Not reported	Monooxygenases
Spinosad	Nicotinic acetycholine receptor	Not reported	
Cyclodienes, Lindane, Bicyclic phosphates	GABA receptor	Rdl mutation	gs-transferases
Phenylpyrazoles	GABA receptor	Rdl mutation	gs-transferases
Avermectines	GABA receptor	Undescribed	Monooxygenases + esterases
Insect Growth Regulators	Ecdysone agonist/disruptor or inhibitor of ATP synthase, chitin biosynthesis or lipid synthesis	No resistance reported against insect of public health importance	
Bacillus thuringiensis var. *israelensis*	Microbial disruptors of insect midgut membranes	Reported against *Culex pipiens* s.l. but not described	

Source: Corbel, V. and N'Guessan, R., Distribution, mechanisms, impact and management of insecticide resistance in malaria vectors: A pragmatic review, from anopheles to spatial surveillance: A roadmap through a multidisciplinary challenge. Chapter 19, in *Anopheles Mosquitoes—New Insights into Malaria Vectors*, Manguin, S. (Ed.), InTech, Rijeka, Croatia, p. 828, 2013.

INSECTICIDE RESISTANCE

The problem of resistance of the vectors to insecticides, especially pyrethroids (the only insecticide available for use on treated fabrics such as mosquito bed nets), although long recognised, is becoming more prominent. This is due to a number of factors, primarily the wide-scale development of resistance to pyrethroids by all the main malaria vectors, by the greater understanding of resistance mechanisms (as a result of a greater understanding of the genome in vectors, particularly *A. gambiae*) and by the realisation that chemical means of control remains for the foreseeable future the mainstay of vector control and that if present control efforts do start to fail then many people will be put at risk if present control efforts do start to fail. The potential problem of resistance was recognised before the first worldwide malaria eradication campaign conducted by the WHO. At that time it was envisaged that control or elimination would be achieved by means other than insecticides following two or three rounds of IRS. That did not happen and IRS remains one of the important techniques available. But, as has been said before, 'It is tempting, if the only tool you have is a hammer, to treat everything as if it were a nail' and as Einstein famously said, 'If you always do what you always did, then you will always get what you always got' – in this case, resistance. How important resistance is operationally, however, remains open to debate because data (described below and largely taken from Hemmingway & Ranson (2000) and Corbel & N'Guessan) are conflicting. For example, despite there being a high level of knockdown resistance (kdr) in some populations of *A. gambiae*, LLINs continue to protect people. This may be because the insects are not as irritated by the insecticide as are their susceptible counterparts and so persist in their attempts to feed. In so doing they may receive a lethal dose. Resistance is the development of an ability in a strain of some organism to tolerate doses of a toxicant that would prove lethal to a majority of individuals in a normal population of the same species. It is a genetically inherited characteristic whose frequency increases in the vector population as a direct result of the selective effects of the insecticide. Resistant individuals have a higher probability of surviving insecticide treatment and, on average, will contribute more offspring than susceptible individuals to the next generation. Thus, the gene or genes conferring resistance will increase over time. Resistance arises as a result of a genetic change that alters the normal physiological, morphological or behavioural attributes of a species. There is also a distinction to be made between tolerance – in which increasing numbers of insects are killed as exposure to the insecticide increases (either by being exposed for a longer period or by being exposed to higher concentrations of the insecticide) – and resistance (when very little mortality is seen until the insects are exposed to a particular dose, which may be up to 1000 times the discriminating dose, which is twice the dose that kills susceptible insects).

Resistance is not new. In India, resistance of the main malaria vector *A. culicifacies* to dieldrin developed in 1958 and resistance to DDT in 1959, but the malaria control program continued until 1965–1966 when both DDT and HCH failed to control outbreaks of malaria. As a result, malathion was introduced in some areas in 1969 with some success but *A. culicifacies* developed resistance by 1973. Malathion resistance resulted in colossal epidemics of malaria in 1975 with 4 million cases reported as compared with 125,000 in 1965. This vector is now resistant to DDT, dieldrin, organophosphates, carbamates and pyrethroids. The number of species reported to be resistant to at least one insecticide by year is shown in Table 8.2.

A similar trend was noted in Central America and the Caribbean. *Anopheles albimanus* now exhibits multiple resistances to DDT, dieldrin, lindane and other chemicals recently used in public health. Carbamate resistance is now spreading in malaria vectors especially in West Africa. In Muleba, Tanzania, incipient resistance to bendiocarb was noted after only a single round of IRS. The way that resistance may spread in a mosquito population is shown in Figure 8.3.

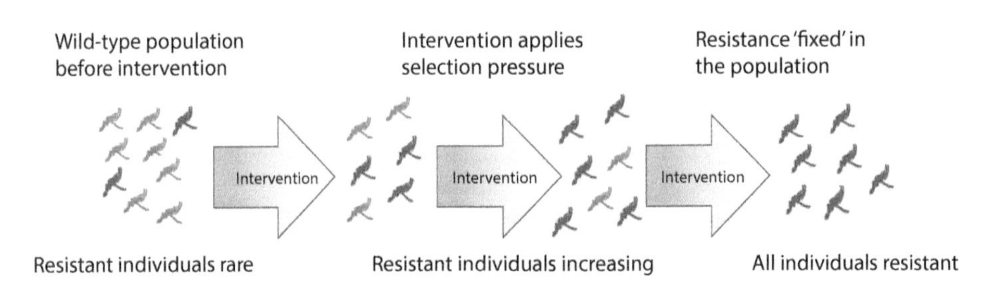

Figure 8.3 The spread of resistance in a population. Rare mutants occur in the population and are able to detoxify the insecticide. Because this generally comes with a fitness cost, they remain uncommon. Their fitness increases, however, when control with an insecticide is applied and so they increase in number (because they are more likely to survive and reproduce than susceptible insects). Over time the susceptible insects may all die and leave only resistance insects. In this case (as is the case in Muleba, Tanzania, for pyrethroid insecticides) resistance is 'fixed' in the population (susceptible insects in blue and resistant ones in red; arrows denote application of the same insecticide).

The rate that resistance will be likely to appear in a population depends on a variety of factors, some intrinsic to the insect (such as generation time) and some due to external environmental factors (such as the decay rate of the insecticide on different surfaces). Insects such as tsetse flies that have a long generation time and a slow rate of growth (i.e., K species) are less likely to develop resistance than insects, such as mosquitoes, with a shorter generation time. Factors influencing the development and spread of resistance are shown in Table 8.5.

A number of methods are available to determine if a mosquito population is resistant to insecticides. These are outlined in Table 8.6.

One way of examining the irritability of an insecticide is to use excito-repellent test boxes like those shown in Figure 8.4. Papers with the insecticide are attached to the inside walls of one box with the other acting as a control. Mosquitoes are introduced into the boxes and the numbers leaving, through the cages at the top or side, every minute are counted. If the insecticide has a repellent effect, then they will leave before they die. If it kills without repelling, then – well, they won't leave!

Nyssorhynchus darlingi and *Ny. nuneztovari* (formerly *A. nuneztovari*) from Brazil left test boxes which had DDT sprayed papers on the walls, indicating that they were irritated by the insecticide (Figure 8.5).

Table 8.5 Factors Influencing the Rate of Evolution of Insecticide Resistance

Entomological	Environmental/Chemical
Population cycle time	Insecticide selection pressure
Population size	Proportion of the population exposed
Exposure of larvae or adults	Time span of exposure
Genetic variation within the population	Insecticide dose rate
	Extent of prior exposure to insecticide
	Residual efficacy of insecticide
	Decay rate of insecticide

Source: Hemingway, J. and Ranson, H., *Annu. Rev. Entomol*., 45, 371–391, 2000.

Table 8.6 Advantages and Disadvantages of the Different Methods Available for Insecticide Resistance Determination in Mosquitoes

Method	Advantages	Disadvantages
Bioassays using WHO defined diagnostic doses of insecticide (WHO tubes or CDC bottle assays)	Standardised, simple to perform, detect resistance regardless of mechanism	Lack sensitivity and provide no information about level and type of resistance (except when using with synergists), needs live mosquitoes
Dose response bioassays	Provides data on level of resistance in population, regardless of mechanism	Requires large numbers of live mosquitoes, and data from different groups not readily comparable
Biochemical assays to detect activity of enzymes associated with insecticide resistance	Provides information on specific mechanisms responsible for resistance	Requires cold chain; not available for all resistance mechanisms; sensitivity and specificity issues for some assays (e.g., GST)
Molecular assays to detect resistant alleles	Very sensitive; can detect recessive alleles and therefore provide an 'early warning' of future resistance	Requires specialised and costly equipment; only available for a limited number of resistance mechanisms

Figure 8.4 Test boxes used to determine excito-repellency effects of insecticides, Manaus, Brazil. (From Charlwood, J.D. and Paraluppi, N.D., *Acta Amazon.*, 8, 605–611, 1978.)

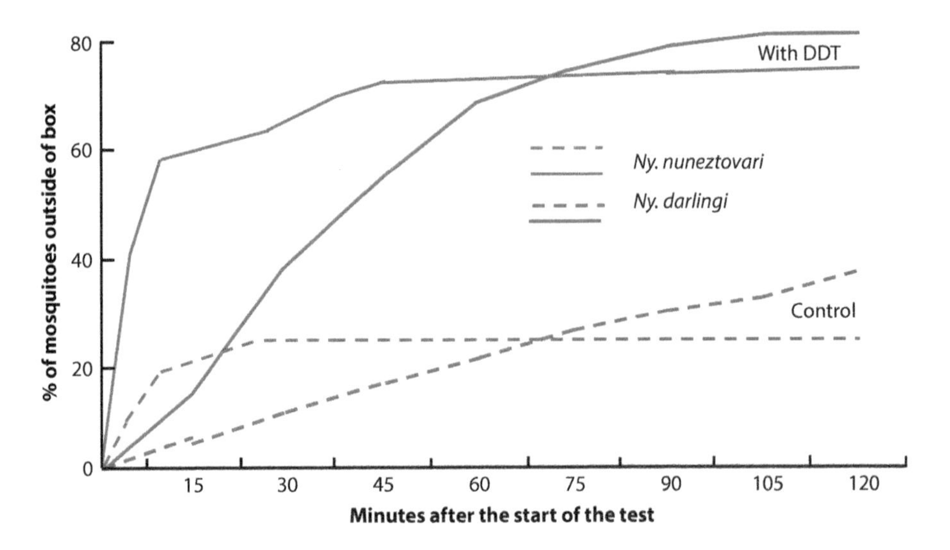

Figure 8.5 Exit rates of *Nyssorhynchus darlingi* and *Ny. nuneztovari* exposed to DDT. (From Charlwood, J.D. and Paraluppi, N.D., *Acta Amazon.*, 8, 605–611, 1978.)

A schema of potential behavioural and physiological changes associated with insecticide resistance in malaria vectors is shown in Figure 8.6.

Resistance mechanisms can be divided into four broad categories: reduced penetration, site insensitivity, metabolism and behaviour.

Target Site Resistance

Alterations in the target site that cause resistance to insecticides are often referred to as knockdown resistance (kdr) in reference to the ability of insects with these alleles to withstand prolonged exposure to insecticides without being 'knocked down'. The target site for OP and carbamate insecticides is

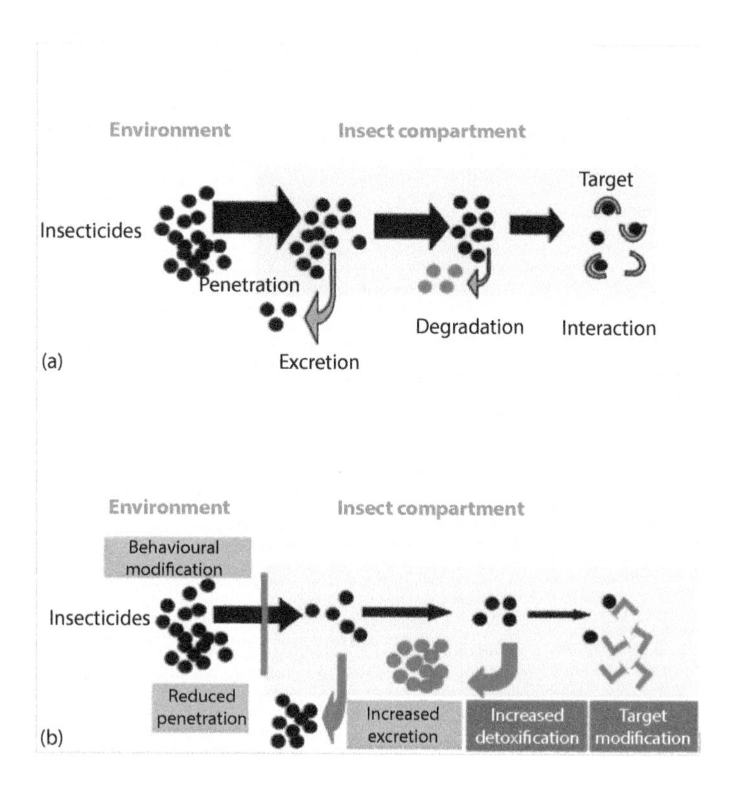

Figure 8.6 Scheme of potential behavioural and physiological changes associated with insecticide resistance in malaria vectors; (a) susceptible insect; (b) resistant insect. In a susceptible insect, the insecticide penetrates the cuticle and although some is excreted or degraded, sufficient reaches the target site that it prevents normal functions and kills the insect. In a resistant insect, less insecticide penetrates the cuticle and more is broken down before it reaches the target site, which may also be modified so that it no longer binds and so no longer kills the insect. (From Corbel, V. and N'Guessan, R., Distribution, mechanisms, impact and management of insecticide resistance in malaria vectors: A pragmatic review, from anopheles to spatial surveillance: A roadmap through a multidisciplinary challenge. Chapter 19, in *Anopheles Mosquitoes—New Insights into Malaria Vectors,* Manguin, S. (Ed.), InTech, Rijeka, Croatia, p. 828, 2013.)

acetylcholinesterase (AChE) in the nerve cell synapses. Several mutations in the gene encoding for an acetylcholinesterase have been found in insects, which result in reduced sensitivity to inhibition of the enzyme by these insecticides. There is cross-resistance between pyrethroids and DDT for kdr. Kdr resistance was first noted in areas where cotton was grown and where pyrethroids were used for pest control.

Reduced Penetration

Many formulations of insecticides are designed to enter the insect through the cuticle. Cuticular changes that reduce the rate of penetration confer resistance to a number of insecticides. Recently, measures of mean cuticle thickness in a laboratory strain of *A. funestus* using scanning electron microscopy (SEM) showed that the mean cuticle thickness was significantly greater in pyrethroid-tolerant mosquitoes than their susceptible counterparts. How widespread such a mechanism is remains largely unknown.

Metabolic Resistance

The function of >90% of metabolic genes is still unknown but basically metabolic resistance implies that enzymes responsible for the breakdown of the insecticide into harmless by-products are

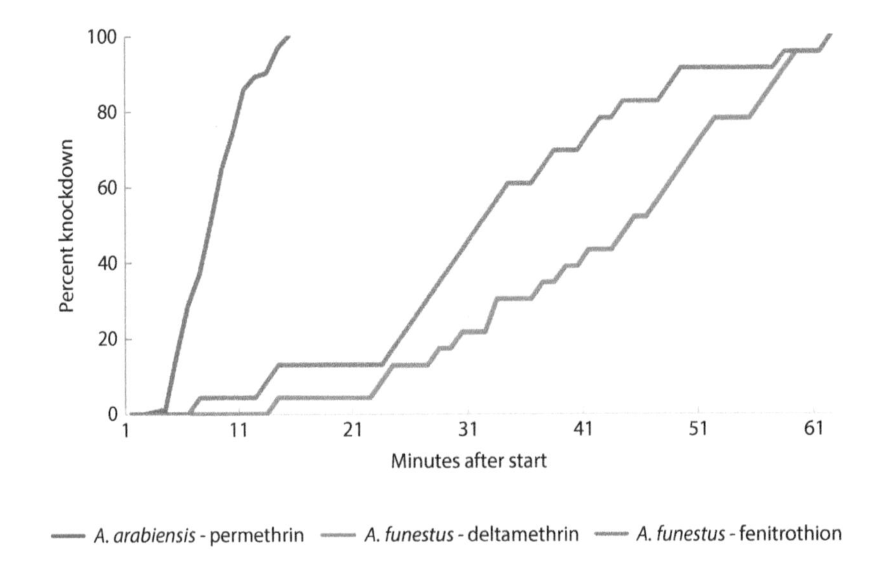

Figure 8.7 Knockdown rates of susceptible *A. arabiensis* (from Muleba, Tanzania) and resistant *A. funestus* (from Furvela, Mozambique) in exposed to LLINs in cage bioassays (all 115 of the *A. arabiensis* were killed by the nets but only 257 (59%) of 437 *A. funestus* tested were killed by the nets).

upregulated and so more insecticide is needed for it to have an effect. Thus, malathion resistance in *Anopheles* sp. was associated with an altered form of esterase that specifically metabolises the molecule at a much faster rate than that in susceptible counterparts. It is perhaps the most common form of resistance reported. It does come with a cost to the insect, however, and so, unless the selection pressure has been such that the mutations associated with resistance have become fixed in the population, it is likely that when an insecticide is withdrawn the population will revert to the wild-type susceptible population that was present prior to the intervention. Fixation has occurred in *A. gambiae* with the *kdr* gene from Muleba, Tanzania.

The difference in knockdown rates between susceptible and resistant mosquitoes can be dramatic (Figure 8.7).

Behavioural Resistance

Although insecticides induce behavioural changes in many vectors, how often these become inherited traits (the true measure of resistance) remains a matter of debate. Thus, as mentioned earlier, the biting cycle of *A. farauti* changed from biting late at night to biting earlier in the night when bed nets impregnated with permethrin were introduced into a village on the coast of Papua New Guinea (Figure 8.8), but this was because the insects that previously had been able to feed late at night were unable to do so because of the nets and so attacked in the earlier part of the night on the following evening.

Perhaps one of the few cases of behavioural resistance was that seen in the Solomon Islands where the cycle of *A. farauti* changed from late biting indoors to early biting outdoors (Figure 8.9).

Resistance is more likely to develop in insects that are exposed to the insecticide as larvae either directly by larviciding or indirectly through runoff when the insecticide is used in agriculture. Insecticides used on nets are only encountered by hungry females and then only for a short period of time. Both males and females are equally exposed to larvicides for the duration of their development. Should exposure be through runoff from agricultural use, then exposure may be to a sub-lethal dose, which may nevertheless have a selective effect on the insect.

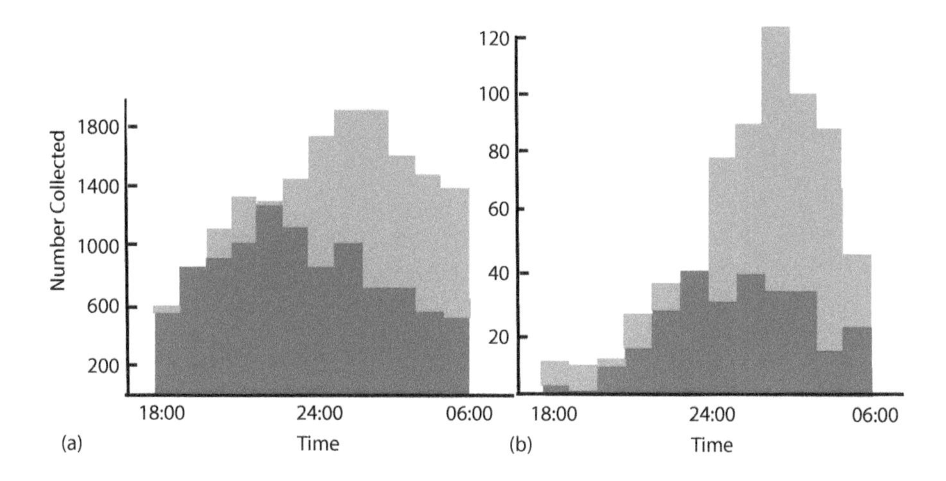

Figure 8.8 Biting cycles of *A. farauti* and *A. koliensis* from Agan village, Papua New Guinea, before (orange histogram) and after (blue histograms) the introduction of pyrethroid impregnated nets into the village. (a) *A. farauti* and (b) *A. koliensis.* (From Charlwood, J.D. and Graves, P.M., *Med. Vet. Entomol.*, 1, 319–327, 1987.)

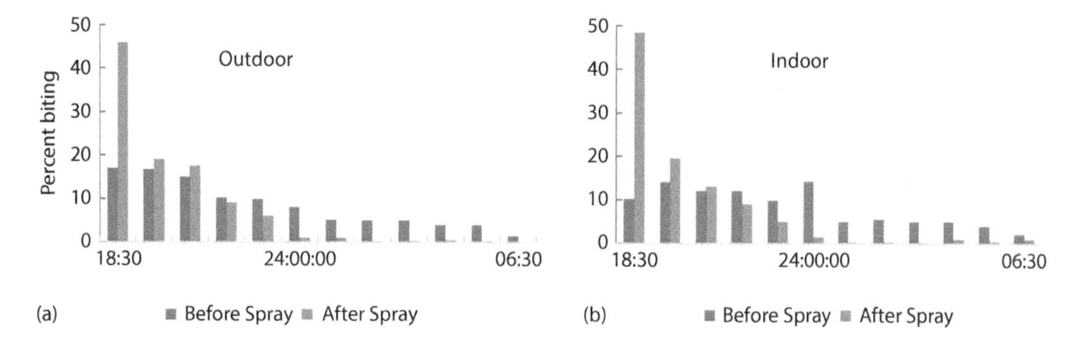

Figure 8.9 Biting cycles of *A. farauti* from the Solomon Islands before and after the long-term application of DDT for IRS. (a) Outdoor, (b) Indoor. (From Taylor, B., *Trans. R. Soc. Lond.*, 127, 277–292, 1975.)

The distribution of pyrethroid resistance among African malaria vectors is shown in Figure 8.10. The sites shown on the maps are those where testing has been undertaken. It is quite likely that resistance is more widespread than shown. Insecticides are not used only for public health; their main use is against agricultural pests. The widespread use of synthetic pyrethroids against pests of cotton has been linked to the development of kdr resistance in West Africa. What probably happens is that the water used for larval development becomes contaminated with low dosages of insecticide due to runoff from crops that have been sprayed when it rains. The low dosage encourages tolerance on the part of the mosquito and eventually leads to full-blown resistance.

A number of ways to reduce the effect of resistance have been proposed. They include (1) use of non-insecticidal control strategies (i.e., IVM); (2) reduction in the amount of insecticide applied and a more appropriate timing of application; (3) increasing the dose applied so that resistant strains are killed; (4) use of insecticide mixtures; (5) rotation of insecticides; (6) use of synergists, which depress the rate of detoxification; and (7) development of new forms of insecticide.

Despite the presence of resistant vectors (as defined by the WHO bioassays), it is likely that some extra degree of protection is provided by insecticide-treated nets compared to untreated nets.

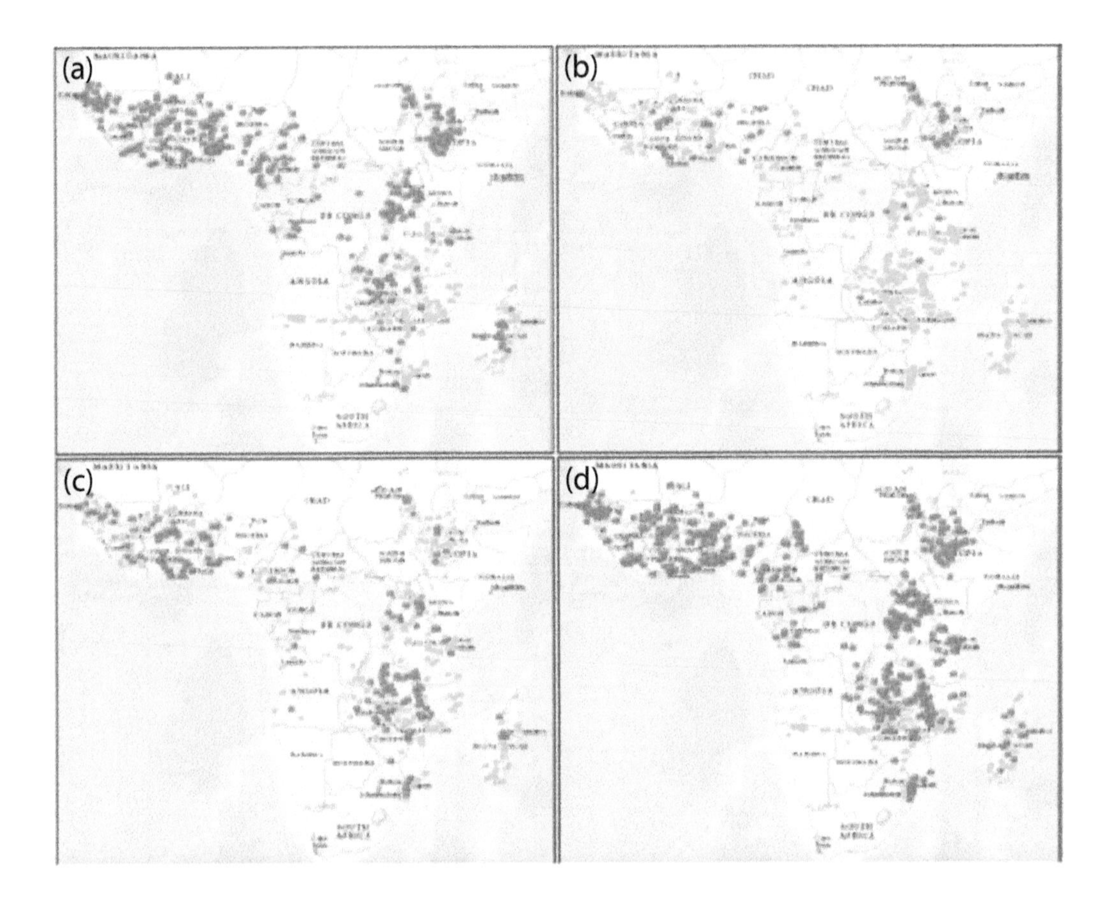

Figure 8.10 Maps showing the distribution of *Anopheles* resistant to Organochlorides, Organophosphates, Carbamates and Pyrethroids. Red dots confirmed resistance; yellow dots possible resistance; green dots susceptible. (a) Organochlorides, (b) Organophosphates, (c) Carbamates, (d) Pyrethroids. (From Huijben, S. and Paaijmans, K.P., *Evol. Appl.*, 11, 415–430, 2018.)

An analysis of several studies found that treated nets offered greater protection than did untreated nets, even when vectors were resistant to pyrethroids (Strode et al., 2014). Indeed, in a recent multicountry study (that involved 1.4 million follow-up visits) there was no evidence found that protection was reduced in areas where resistance was higher (Kleinschmidt et al., 2018). This may be because there is a delayed effect of the insecticide on mortality of the mosquito or the parasite. Oocyst development was found to be slower in deltamethrin-exposed resistant mosquitoes compared to unexposed mosquitoes (Kristan et al., 2016). On the other hand, in Senegal the emergence of Kdr resistance among *A. gambiae* was associated with a decrease in LLIN efficacy and a rebound in malaria morbidity (Trape et al., 2011).

Alternative ways of reducing the impact of resistance on the effectiveness of vector control also include the development of 'late-acting' insecticides that only kill the mosquito after several contacts. This would allow susceptible insects to lay eggs once, twice or even three times before they succumb to the insecticide, meaning that their genes will continue in the population. Because it is only old insects that transmit malaria, the effect of such an insecticide will be the same as one that kills more rapidly.

It is possible that insecticides applied inside houses act in this way against mosquitoes that are largely exophagic, as was the case in the *A. coluzzii* from São Tomé. Here, the number of positive

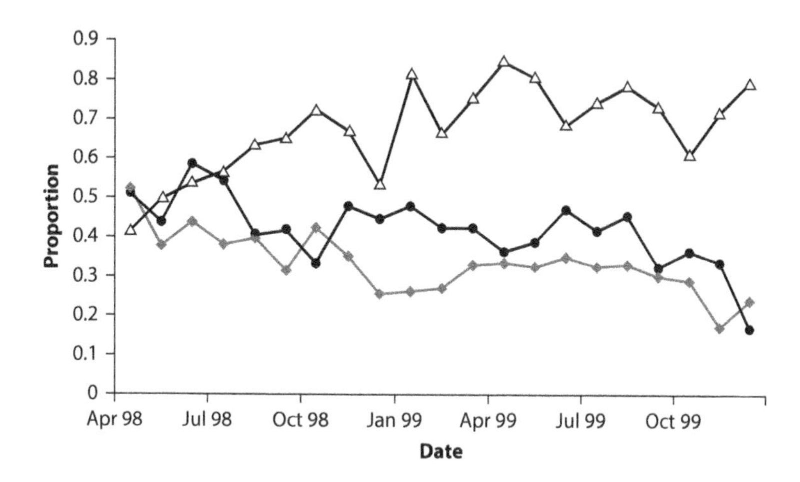

Figure 8.11 Longitudinal variation of bed net use and malaria prevalence in bed net users and nonusers in São Tomé. Key: solid circles = prevalence in non-net users, diamonds = prevalence in net-users, and triangles = proportion of attendees using bed nets. (From Charlwood, J.D. et al., *Trans. R. Soc. Trop. Med. Hyg.*, 99, 901–904, 2005.)

blood slides at a sentinel site monitoring malaria incidence decreased as bed net use increased, despite the proportion of mosquitoes feeding indoors being small (Figure 8.11). Fungi that can be sprayed on walls (or in mosquito resting sites) are another possible late-acting insecticide of this sort.

Other methods include 'mosaic' spraying in which insecticides of different classes are used in different areas or 'rotation' spraying in which insecticides are changed with different spray rounds.

In some cases, IRS may be used in conjunction with LLINS. The effectiveness of the combination of these two tools has been the subject of considerable debate because observational and randomised control trials have produced conflicting results. In Burundi, Eritrea, The Gambia and Benin there was no difference in malaria outcomes when they were deployed together or independently, whilst in Tanzania, Kenya, Equatorial Guinea and Mozambique an increased efficacy was observed in areas where they were used in combination. Results of these studies suggest that there is additional protection provided by the combination when one or other is rarely used. In the recent trial in Muleba, Tanzania, in which standard LLINs were compared to LLIN+PBO, standard LLIN plus IRS and LLIN+PBO plus IRS (with pirimiphos-methyl) against a pyrethroid-resistant *A. gambiae* population, there was little added benefit from the IRS compared to the LLIN+PBO arm, but both were more effective at reducing malaria prevalence than standard nets by themselves.

The insecticides discussed above mainly operate on the principal of rapid death of (most) members of the pest population. More recently another class of chemicals, known as insectstatics because they suppress growth and reproduction, are being developed. They include (1) substances that inhibit chitin synthesis or tanning, rendering an insect more susceptible to microbial (especially fungal) infections or by preventing normal activity because the muscles do not have a firm structural base; (2) antagonists or analogues of essential metabolites (such as essential amino acids or vitamins) whose effect is to prolong larval life and retard egg production; (3) insect growth regulators (IGRs) with juvenile hormone activity, which prevent metamorphosis, and, for certain agricultural pests, (4) sex attractants.

CHAPTER 9

Alternative Methods of Vector Control

Alternative methods for vector control include environmental management, biological control and personal protection, such as use of repellents, wearing protective clothes and sleeping under bed nets. Whereas chemical control is highly effective in the early-stage control programmes, more species-specific control, or species sanitation, will become a central component in low and moderate transmission settings when disease incidence nears elimination. Environmental management may include destruction of breeding sites by drainage, filling, impounding, or channelling streams and rivers into canals or by altering the vegetation and shade characteristics of the sites favoured by the vectors.

It is now recognised that a single method of vector control will not achieve elimination. Alternatives are needed that complement the main conventional techniques of IRS or LLIN. These can be applied together with standard, conventional control to produce 'integrated' control (Figure 9.2).

Possible methods of control of vectors can be allocated according to particular behaviours or phases in the life cycle (Figure 9.1). The effects of different interventions (in combination with LLINs at different measures of coverage) have been modelled. Their relative efficacy depends to a certain extent on the behaviour of the mosquito. Some of the techniques that show some promise are described below.

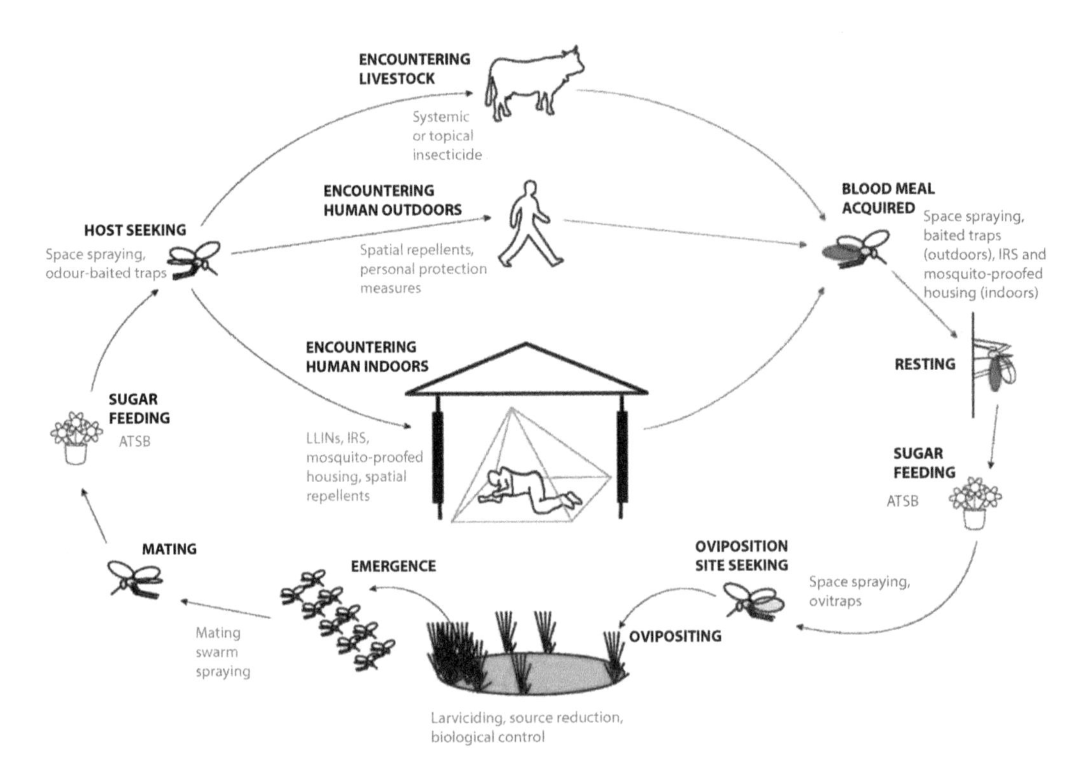

Figure 9.1 A schematic diagram of the different phases of a mosquito's life cycle and the possible methods that can be used to control it. (From Kiware, S.S. et al., *PLoS One*, 12, e0187680, 2017.)

ENVIRONMENTAL MANAGEMENT

It has long been realised that one way in which vector-borne diseases can be controlled is to manage the environment of the vectors that transmit them, i.e., to alter the habitat. As far back as the first century BC, Vitruvius, a Roman engineer and architect, designed a drainage system to flush coastal marshes and recommended that houses not be built near these marshes. The eventual draining of these marshes during the administration of Mussolini is also one of the few benefits that accrued during this time in power.

A definition of environmental management for vector control is 'the planning, organisation, carrying out, and monitoring of activities for the modification and/or manipulation of environmental factors or their interaction with humans with a view to preventing or minimising vector propagation and reducing human–vector–pathogen contact' (WHO, 1982).

Environmental management for vector control was commonly used before and during the Second World War; it was largely ignored post-war, being replaced with synthetic insecticides to control vectors (during the so-called 'spray gun war') and drugs to treat the diseases. Vector control based on environmental management has to be tailored to local conditions and requires the active involvement of local communities; environmental impact assessments during the planning stage are essential. It is also vital to have an intimate knowledge of the biology of the vector species to be controlled.

Often, environmental and biological control methods are not universally applicable, and their efficacy in reducing disease transmission depends on how well the intervention is matched to the vector's specific ecological characteristics.

One environmental control method employed successfully in several countries to control vectors in rice fields is intermittent irrigation. The method is based on the principle of supplying only as much water as is needed for optimal plant growth. Thus, on the island of Java in Indonesia, periodic drainage of rice fields and cleaning of fish ponds helped to eliminate malaria.

Biological control was carried out in China at least 2300 years ago. The Chinese collected colonies of the tree-nesting ant *Oecophylla smaragdina*, which they placed in their citrus trees to control caterpillars and wood-boring beetles. In addition, they placed bamboo runways between trees to facilitate the ants' movements. The first known example of classical biological control occurred in the eighteenth century, when mynah birds imported from India were used in the control of the red locust in Mauritius.

IMPROVEMENTS TO HOUSES

House materials correlate well with malaria prevalence and risk of infection for the occupants. Anophelines tend to bite close to the ground and 'putting your feet up' at night reduces biting rates (Charlwood et al., 1984). In São Tomé that mosquito densities were 50% lower in houses that were built on stilts compared to those built at ground level (Figure 9.2).

It is likely that the most suitable method of reducing man-vector contact is to improve houses to make them less accessible to the mosquito. Some of these improvements are relatively simple such as closing the gap between the eaves and the roofs of houses, which has been shown to reduce entry rates by approximately 43% (Figure 9.3). Indeed, using old mosquito nets for this purpose both

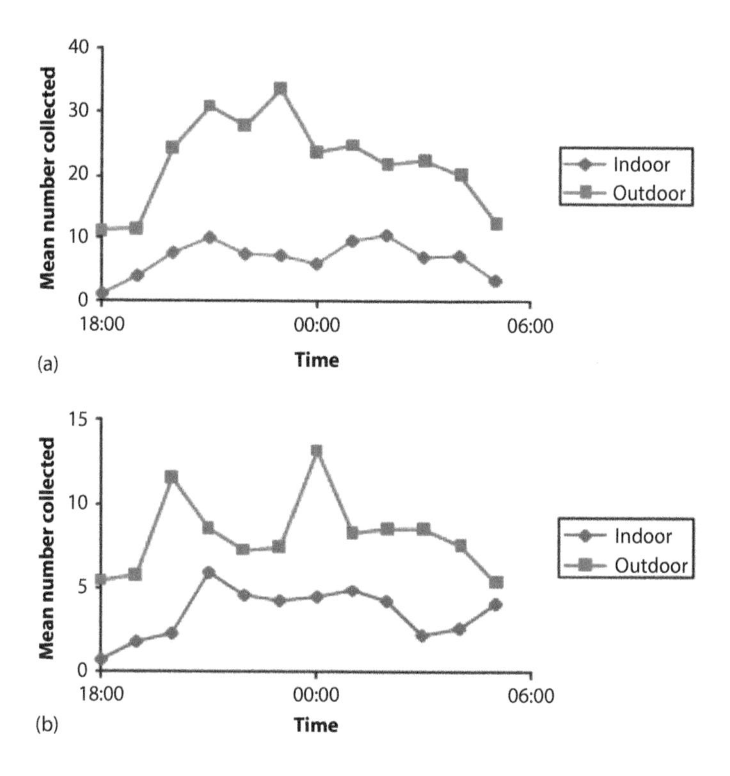

Figure 9.2 Biting cycles of *A. coluzzii* according to house type, Riboque São Tomé. (a) Ground level and (b) elevated houses. (From Charlwood, J.D. et al., *Malar. J.*, 2, 45, 2003a.)

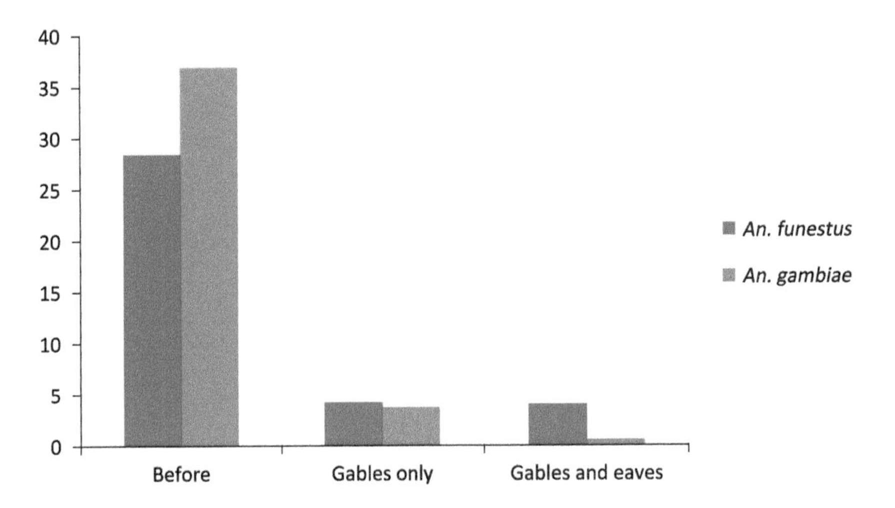

Figure 9.3 The effect of closing off the open gables and eaves of reed houses in Furvela, Mozambique, on entry rates of *A. funestus* and *A. gambiae* (determined by light-trap collection).

avoids environmental contamination with their disposal and provides a significant level of added protection to the householders. Such measures have the advantage that once in place further compliance by the householder is not required. They do not depend on an insecticide and can be done on a do-it-yourself (DIY) basis.

A further house improvement that happens, without the need for any persuasion, is the shift from thatched roofs to corrugated iron or tin roofs. Tin roofs are desired because they last for much longer than traditional roofs. They have the advantage that mosquitoes do not rest on them and are less likely to rest inside at all in houses with this kind of roof. In reality, it is quite possible that this will have a greater and more sustainable impact on malaria transmission than many other vector control methods advocated by professionals. A potential 'mosquito proof' house described by Mutt Kirby (2013) is shown in Figure 9.4. A number of possible designs for healthier houses in the tropics have been produced by the architects Knudsen and Von Seidlein (2014). They stress the need for the interior of houses to approach the 'comfort zone' for people to want to live in them.

Effective vector control by nonprofessionals has worked before. On the island of Príncipe, where sugar cane was an important crop, sleeping sickness was a major problem. One enterprising plantation owner had his employees wear sticky patches on the backs of their shirts. As a result, the tsetse flies that attacked them got stuck on their shirts and the population declined to virtually zero.

In well-built houses an alternative method of control is the use of eave tubes. These are pieces of PVC tubing, available at builders' merchants (nominally 15 cms in diameter), that are inserted into the walls to replace the other openings in the house (so the eave gap is closed and windows are screened). The tubes have insecticide-treated netting over their interior ends. Mosquitoes are attracted to the tubes due to the odour and other attractants leaving the house and so, come into contact with the insecticide and are killed. The method shows considerable promise but is only suitable for well-built houses made of brick or cement blocks. They may not be useful for houses made of reed or similar material.

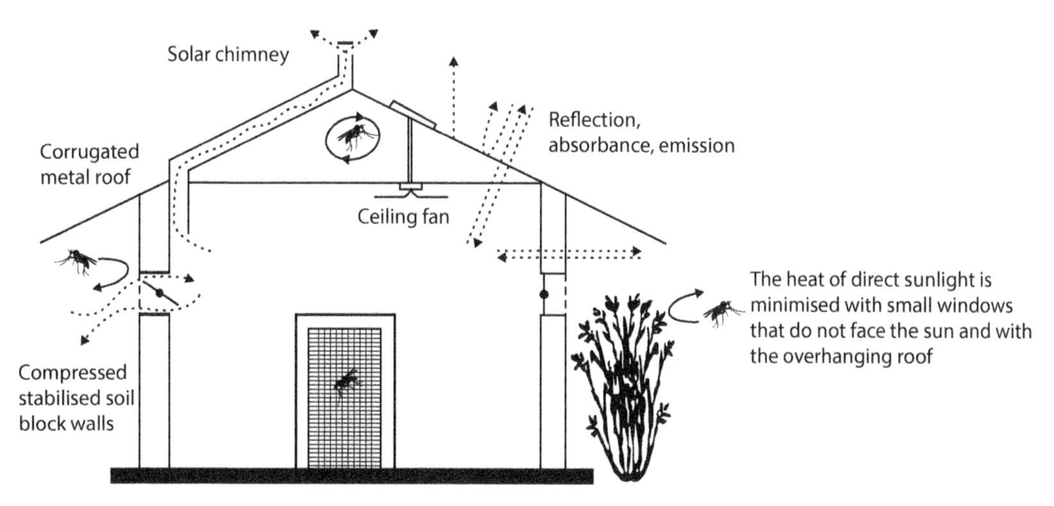

Figure 9.4 The ideal house? Mosquitoes approaching doors and windows are prevented from entering by screening material over the openings and further discouraged by repellent plants. Solar-powered fans and a solar chimney improve air movement indoors. The ceiling prevents any mosquitoes in the roof space from reaching the room and should absorb some of the heat from the metal roof. Carefully chosen construction materials improve the durability of the house. A solar-powered Suna trap placed outside close to the walls would also help. (From Kirby, M.J., House screening. Chapter 7, in *Biological and Environmental Control of Disease Vectors*, Cameron, M.M. and Lorenz, L.M. (Eds.), CAB International, Wallingford, UK, 2013.)

BIOLOGICAL CONTROL

Natural biological control is larval control by naturally occurring biotic agents. The habitat type, size permanence and water depth all determine which biological control agents can successfully establish themselves in a given location.

With *inoculative* releases, small numbers of natural enemies are introduced that are expected to reproduce in the environment and provide long-term vector suppression over successive generations. For *inundative* releases, overwhelming numbers of organisms are released to produce an immediate decline in a vector population.

The most successful biological vector control organism to date, *Bacillus* spp., produces highly specific toxins during sporulation. The living or dead spores with accompanying toxins are applied inundatively, essentially as chemical insecticides, and eaten by larvae. Inundation with zoospores of fungi can also quickly reduce a high-density host population.

The second strategy is invasion. The invading pathogens kill the host after entry into the body over a period of days or weeks. Invasive pathogens include such things as nematodes and fungi.

Two species of fish are frequently used in mosquito control: the mosquito fish, *Gambusia affinis*, and the guppy, *Poecilia reticulata*. *Gambusia* do not always reduce mosquito populations and they may eat smaller native fish. Thus, indigenous fish should be used when possible.

The effective use of arthropods for disease control falls into two strategies: to fill vacant niches and to maximise the effectiveness of local predators. The most commonly studied predators of mosquitoes include many invertebrates, such as predaceous insects, hydra and flatworms. Arthropod predators of *Anopheles* and *Culex* larvae mainly come from three orders: Hemiptera (true bugs), Coleoptera (beetles) and Odonata (dragonflies and damselflies). Mosquitoes belonging to the non-blood-feeding genera, Toxorhynchites, have predatory larvae and have been used in the control of container-breeding *Aedes*. In addition to aquatic predators, the shore fly *Ochthera chalybescens* preys on all instars of *A. gambiae* larvae by fishing them out with their sickle-shaped front legs. They have even been found to prey on adults.

ZOOPROPHYLAXSIS

Zooprophylaxis, the possible protection afforded by diverting potential vectors from humans to animals, thereby reducing transmission, has been advocated by the WHO (1982) as a possible alternative to conventional control. The effectiveness of diversion depends, among other factors, on the natural host choice of the mosquito, the biomass of alternative hosts available and the tendency of the mosquito to feed outdoors. Vectors that are generally exophagic or wide-ranging in their preference, such as *A. arabiensis*, are more likely to be diverted than those that are not. Even so, many vectors, including all other members of the *A. gambiae* complex, including the normally anthropophilic *A. coluzzii*, will feed on other hosts when they are available (Sousa et al., 2001). At the other extreme, some normally zoophilic mosquitoes can, in the absence of animals, be diverted to humans and so become occasional vectors, as occurs with some members of the *A. funestus* group (Wilkes et al., 1996).

A number of models have been developed to predict the effect of such diversion. These generally suggest that the introduction of animals, such as cattle, close to houses may actually increase mosquito densities and, in some situations, increase the frequency of mosquito bites on humans (Saul, 2003). Thus, Sota and Mogi (1989) predicted a significant reduction of malaria only when an extremely large number of easily accessible animals are introduced. Even then, however, as the number of animals increase, improved availability of blood meals may increase mosquito survival, thereby countering the impact of diversion. Other models indicate that zooprophylaxis, whilst not being sufficient in itself, may delay or prevent the development of insecticide resistance in the mosquito (Kawaguch et al., 2004). Where it has been investigated in practise, diversion to other hosts has generally had little effect on malaria transmission, although this might depend on the relative location of breeding sites, houses and alternative hosts (Charlwood, 2001). For example, entomological investigations in The Gambia indicated that the presence of cattle did not alter the risk of malaria transmission in nearby houses nor was there any difference in the prevalence of *Plasmodium falciparum* in children living close to individual cows compared to children living in households that did not own cattle (Bøgh et al., 2002). In contrast, but also in The Gambia, mosquito densities were lower in houses where horses were tethered close to houses (Kirby et al., 2008) whilst, in Pakistan, cattle ownership was actually associated with a higher prevalence of malaria in people (Bouma & Rowland, 1995).

Animal hoofprints are a potential source of stagnant water that may act as a larval habitat for vectors such as members of the *A. gambiae* complex, which potentially means an increase in mosquito numbers when cattle are present. Indeed, according to Okech et al. (2007), such sites may even influence vector competence.

Diversion of *A. gambiae*, from people to dogs and pigs, was considered to be responsible for the lower than anticipated prevalence of malaria in São Tomé (Sousa et al., 2001), whilst Hadis et al. (1997), Seyoum et al. (2002) and Mahande et al. (2007) suggested that cattle may protect humans from *A. arabiensis* and thus reduce malaria transmission. In a more recent experimental analysis, however, Tirados et al. (2011) concluded that 'the presence of large numbers of cattle does not confer effective zooprophylaxis against malaria transmitted by *A. arabiensis* or *A. pharoensis*.'

In Papua New Guinea one village (Maraga) with an exceptionally large population of pigs had a lower than expected prevalence of malaria based on the number of mosquitoes collected (Figure 9.5). This may indicate that diversion was effectively reducing transmission there.

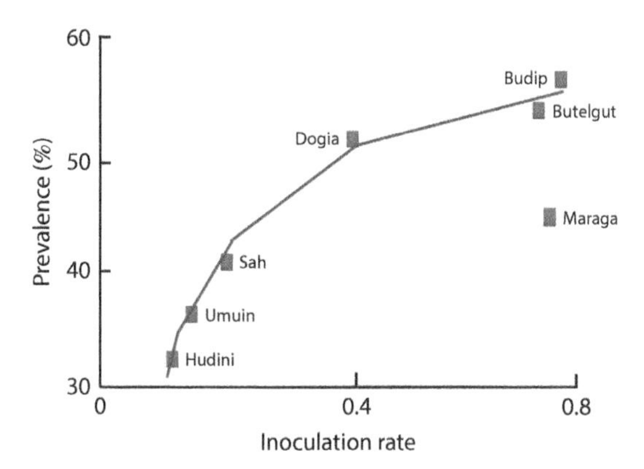

Figure 9.5 Prevalence of malaria by estimated inoculation rates among villages from Madang, Papua New Guinea. Note that Maraga, a village with a large pig population, had a lower-than-expected prevalence, possibly indicating effective zooprophylaxsis. (From Birley, M.H. and Charlwood, J.D., *Parasitol. Today*, 3, 231–232, 1987.)

TOPICAL AND SPATIAL REPELLENTS

The application of a repellent to the skin or clothing is one of the most common methods of preventing mosquito bites. Repellents have a long history. In addition to using nets and building houses on stilts to avoid mosquito bites, Herodotus describes the possible use of lamps that may have functioned as a repellent. He states, 'The Egyptians who live in the marsh-country use an oil extracted from the castor-oil plant. This plant which grows wild in Greece, they call Kiki, and the Egyptian variety is very prolific and has a disagreeable smell. Their practise is to sow it along the banks of rivers and lakes, and when the fruit is gathered it is either bruised and pressed, or else boiled down, and the liquid thus obtained is of an oily nature and quite as good as olive oil for burning in lamps, although the smell is unpleasant...' The same principal of vaporising a chemical that acts as a repellent or killing agent is used in mosquito coils and electrically heated mats. Coils are normally made of sawdust impregnated with a pyrethroid. When lit, the coil smoulders (rather like a cigarette) and the heat vaporises the insecticide immediately behind the smouldering edge. Unfortunately, the smoke from the coil contains a number of noxious chemicals (mainly microparticles) that themselves are bad for one's health.

The recent development of synthetic pyrethroid insecticides, notably metofluthrin and transfluthrin, with high vapour action at ambient temperatures has led to the development of devices that work without requiring the application of heat. There are thousands of registered products that use these active ingredients. There are sophisticated, expensive products that are used in Europe, and there are inexpensive and simpler products (e.g., mosquito coils) that are widely used throughout Africa and Asia. The primary barrier to uptake of these products is cost. They can be produced on plastic lattices or simply on plastic strips and may be effective for several weeks. They have the advantage that they protect a 'space' and so are considered to be spatial repellents (SRs). They work by disrupting the orientation of the mosquito towards the host by neural excitation, which occurs at an early stage of pyrethroid toxicity. Spatial repellents work against outdoor/day/early-evening biting mosquitoes – areas of transmission where traditional interventions are not completely effective. SRs show effect against insecticide-resistant populations and might be especially important as non-lethal effects and lowered selection pressure help prevent the emergence/spread of insecticide resistance alleles. SRs should demonstrate added benefit in areas where traditional LLINs or IRS interventions may not offer full protection or have reached their efficacy limits – especially in areas with residual transmission. Spatial repellents are an obvious area where improvements and developments will take place. A single emanator of metofluthrin reduced landing rates by 48% in Cambodia – good, but no cigar as they say.

The most commonly used contact repellent is DEET (*N*,*N*-diethyl-3-methylbenzamide). This has been used for many years and is available in a number of formulations. Alternatives to DEET include citronella and the repellent known as IG3535. Both of these are also available in either a Vaseline type of formulation or as a lotion that can be easily applied to the skin.

To a certain extent how useful a repellent may be in reducing malaria transmission depends on the feeding behaviour of the vectors, the place where they are used, and the proportion of the population using them. Those vectors, such as *A. arabiensis*, that have catholic tastes may be diverted to feed on animals (and so the repellent works) whilst vectors that primarily bite humans may only be diverted to another human host, especially if not everyone is using them. In a meta-analysis of 10 (rather heterogeneous studies) no significant protective effect of topical repellents was observed on the incidence of malaria (Wilson et al., 2014). However, the probability of avoiding infections is highly sensitive to small changes in compliance and product efficacy.

A study in Tanzania showed that placebo users living in a village where 80% of the households used 15% DEET had more than four times more mosquitoes resting in their houses in comparison to households in a village where nobody used repellent.

Many repellents are expensive and beyond the pocket of people affected by malaria. One cheaper product that can be produced where coconut palms grow is 'Mosbar'. This uses 5% Permethrin and 10% DEET incorporated into a solid bar with a coconut base. It has to be moistened and then is rubbed on the skin and is effective for several hours against mosquitoes and midges (Charlwood, 1987). It suffers from the perception that it is a 'soap' and so many people rinse it off after application, therefore defeating the purpose. Given that DEET remains the most effective repellent available, that Mosbar can be locally made and that protection against early evening outdoor biting mosquitoes is becoming more important, its further development would seem to be a good idea.

TOXIC SUGAR BAITS

Sugar feeding is another behaviour that may be susceptible to control (Foster, 1995). Attracting mosquitoes to sugar baits that contain an insecticide or even a bacterium has been tested in a number of studies with promising results. They may be most effective when used indoors, because this reduces the risk of killing nontarget organisms, and they may work best in conjunction with bed nets in that a mosquito that fails to obtain a blood meal may be energetically depleted and be more likely to take a sugar feed if one is easily available.

INTEGRATED VECTOR CONTROL

When a variety of control techniques are used in combination, then the approach is called 'Integrated Vector control'. This is obviously a sensible approach and is one recommended by the WHO (WHO, 1982). The WHO has outlined a number of the methods that can be used (Figure 9.6). The use of two synergistic methods, each of which enhances the effect of the other, is a useful approach. For example, the use of repellents around humans and at the same time the use of odour-baited traps works as a 'push-pull' system.

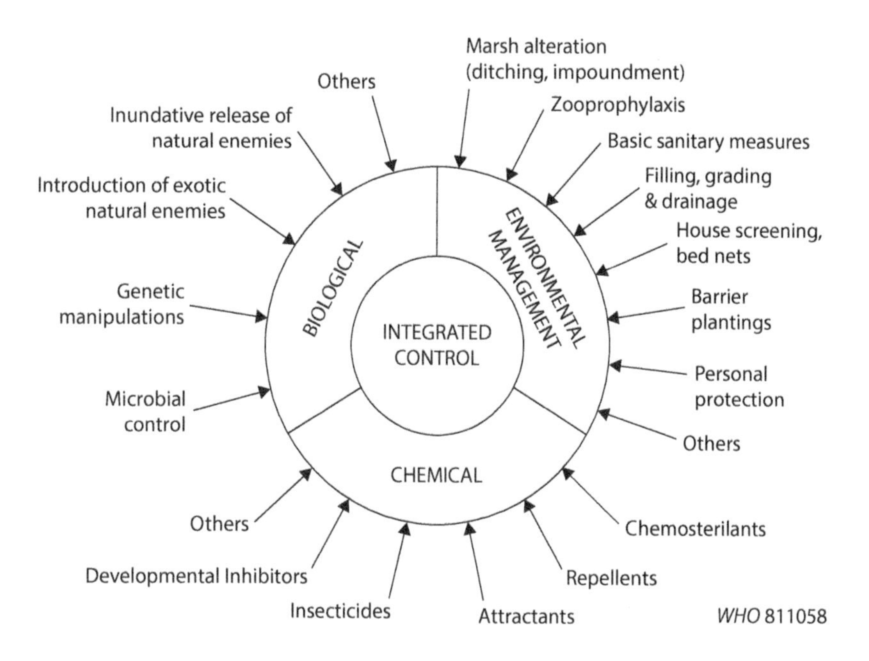

Figure 9.6 Some of the possible methods of vector control that can be used to achieve integrated control. (From Manual on environmental management for mosquito control with special emphasis on mosquito vectors. WHO offset publication No. 66. WHO, Geneva, Switzerland, 1982.)

EXPANDED POLYSTYRENE BEADS

Although not a control measure for *Anopheles*, it is perhaps worth mentioning here one alternative control measure that is effective against *Culex quinquefasciatus.* This is the use of expanded polystyrene beads applied to pit latrines and septic tanks. In the grounds of the hospital in Ifakara, Tanzania, were a number of open septic tanks with very high populations of larvae – in single dips 2213, 2405, 3002, 1982 and 2137 larvae were collected. Both nets and beads were used in an attempt to control the mosquito. Nets were introduced into the hospital wards and subsequently beads applied to the septic tanks (Figure 9.7). Mosquito numbers were monitored using light traps in the hospital (children and adult wards) and in a house 100 m away. Nets made no difference to numbers collected but, in the hospital at least, once beads had been applied to the septic tanks, numbers collected declined. In the house, numbers were unaffected indicating that the flight range of the mosquito was limited.

EDUCATION

Education and information can have an impact on the utilisation, sustainability and effectiveness of interventions. It is often the 'poor relation' in malaria control campaigns; yet, without involving local communities in the process and establishing an enabling environment, campaigns are unlikely to be sustained in the long-term. The Jesuits used to say, 'Give me a child for seven years and they will be mine for life'. They obviously knew a thing or two about the effect that early education has on shaping adult attitudes. Female and juvenile macaques on the Japanese island of Koshima learnt how to wash sweet potatoes in the sea before eating them but adult males never did. Because adult male *Homo sapiens* are as obstinate as macaques, it makes sense to target young age groups and

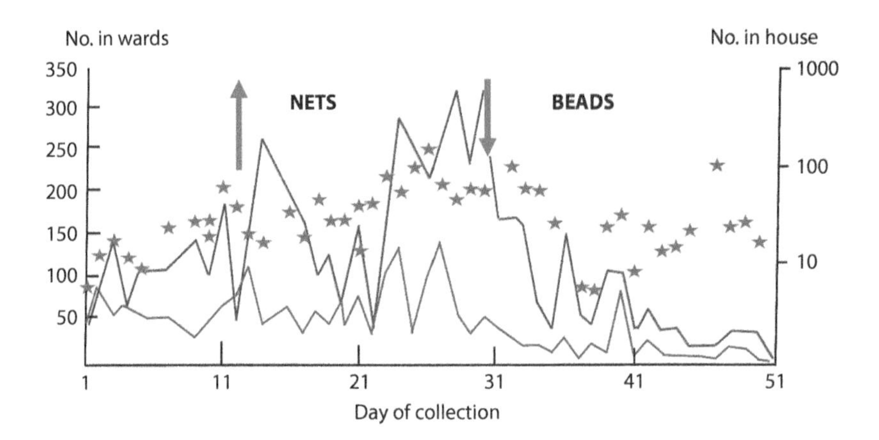

Figure 9.7 Numbers of *Culex quinquefasciatus* collected in light traps from the children's ward and adults' ward (+) of the hospital and from a nearby house (*) in Ifakara, Tanzania. Arrows indicate date of introduction of bed nets and polystyrene beads. (From Charlwood, J.D., *Trans. R. Soc. Trop. Med. Hyg.*, 88, 380, 1994.)

women in education campaigns. Womens' groups can play a useful role in applying alternative control techniques (Charlwood, 2008). Simple demonstrations of mosquito biology can have an impact. On Praia Burra, on the island of Príncipe, a flooded wheel rut, left by a motor vehicle, was the source of many of the *A. coluzzii* in the village. Larvae from this site were collected and left to emerge in a cage in the middle of the village. Once villagers saw, with their own eyes, the relationship between the larvae and the adult mosquito they eliminated the problem without any further encouragement needed (and before further mosquito samples could be collected!) (Figure 9.8).

Figure 9.8 It didn't take long for villagers to fill in breeding sites of *A. coluzzii* once they realised that these were the source of their mosquito problem.

CHAPTER 10

Surveillance and Sampling

Much of the following comes from Obsomer et al. (2013). Determining and measuring the link between species and their environment is a central research area in quantitative ecology. In the study of vector-borne diseases, maps have played an important role in orienting investigators to local conditions and in guiding epidemiological activities within a study area.

Geographic information systems (GIS) are a combination of computer technologies that integrate graphic elements with database information and enable the computation of spatial relationships. Use of GIS makes it possible to collect, manage, analyse and report spatial information about vector-borne diseases. With this information, spatial relationships associated with vectors, hosts, pathogens and the environment can be compared to learn more about the complex nature of these interactions. GIS methods are multidisciplinary, involving knowledge of physical geography, cartography, remote sensing, global positioning systems (GPS), surveillance. The key to GIS is in having one or more databases with tabular information with associated geographic position. Thus, a GIS consists of an 'organised collection of computer hardware, software and geographic data designed to efficiently capture, store, update, manipulate, analyse and display all forms of geographically referenced information'. GIS and related information technology can greatly improve survey, logistics and documentation of mosquito control operations.

GIS can be used for:

- Mapping locations of certain features
- Mapping quantities and densities
- Finding out properties of distinct areas using database queries
- Mapping change by comparing how features change over time in order to forecast future conditions (from Obsomer et al., 2013)

GIS and related geospatial tools are effective in population studies, risk mapping, predictive modelling and the analysis of both social and environmental factors related to infectious diseases. Remotely sensed data provides an important source of information to identify vector habitats and predict population abundance.

GIS methods used in the study of vector-borne diseases are aimed at identifying environmental factors responsible for pathogen transmission and survival. Thematic data layers include terrain elevation, soil type, vegetation class, hydrologic feature, meteorological condition, population size, transportation feature and the distribution of disease vectors over a geographic area.

In the case of vector data, locations are stored as pairs of XY coordinates, and each feature is linked to a row in a database table. Points represent specific locations, lines have length and direction and polygons have an area and a perimeter.

With raster data, features are represented as a matrix of cells in continuous space. Raster data have a simple structure, which is less intense computationally, and data can be easily generalised and classified for analysis purposes.

Additional qualitative or quantitative data can be added to the attribute table, either directly as new fields or through attachment to other databases. It is essential that one or more fields contain a unique identifier that distinguishes one record from another.

When a new GIS database table is created, a spatial reference is added for each record. This spatial reference may be a pair of X-Y coordinates, such as longitude-latitude, or the name of a specific geographic area.

An important early consideration in working with geographic data is determining the appropriate map scale. Map scale refers to the ratio of the distance between two points on a map and the Earth distance between the same two points. The scale of available data used in an analysis must be appropriate for the biological question under investigation.

GIS applications work best when data are used in an interactive way.

Remote sensing is the process of acquiring information about an object, area or phenomenon from a distance (i.e. from a satellite). Remote sensors record electromagnetic radiation (EMR) that is reflected or re-radiated from an object to the sensor. Once the spectral signature of a landscape object (such as a pond) is determined, this information may be used to identify similar features in an area.

1. **Spatial resolution** is the shortest distance between two objects that still permits each object to be distinguished separately.
2. **Spectral resolution** refers to the number of specific wavelength intervals in the electromagnetic spectrum to which a sensor is sensitive.
3. **Temporal resolution** is the frequency of acquiring spectral data for a particular area.
4. **Radiometric resolution** refers to differences in the sensitivity of a detector in responding to the radiant flux reflected or emitted from an area (from Obsomer et al., 2013).

The processes involved in GIS include:

1. Viewing maps that show the spatial orientation and geographic distribution of selected objects
2. Stating hypotheses regarding potential spatial relationships
3. Selecting variables for analysis
4. Extracting values for specific objects or areas from each thematic layer
5. Organising spatial data into a framework for statistical analysis
6. Applying statistical methods to the variables selected
7. Using results of statistical analyses to accept or reject hypotheses

Exploratory data analysis (EDA) is applied to geographic data to look for 'hot spots' in spatial patterns.

When coupling *Aedes* monitoring and mapping efforts, defining an optimal sampling strategy becomes of highest interest. What constitutes a sampling location? How many are needed? Where do they have to be located? How often to survey? When sample measurements are similar to each other, the sample mean is likely to be estimated with an acceptable level of precision from a few sampling locations. In contrast, when the between-location variation in the measurements is high, a larger number of sampling locations are needed. The decisions on how and where to sample do, however, remain much the same as they always have and the recommendations of earlier workers are still useful.

1. The minimum level of change that is to be detected in the analysis (for instance, 10% of change between time t and t + 1)
2. The acceptable chances of making type-1 error (i.e., concluding that change is taking place when it is not) and type-2 error (i.e., concluding that no change is taking place when it is) in hypothesis testing procedures

Designing effective sampling schemes to estimate *Anopheles* species dynamics in space and time requires decisions to be made about how to allocate sampling effort among spatial and temporal replicates.

A pilot survey is, however, required to obtain an initial approximation of the precision of the estimates linked to the variation in the field measurements. As Napoleon is attributed to have said, 'Time spent in reconnaissance is seldom wasted'.

In stratified sampling the study area is first divided into strata assumed to influence the attributes of the species measured in the field. Alternatively when random sampling is applied a number of sampling locations are selected within the strata in ratio to their relative geographical size. Stratification can occur at a variety of levels from the country as a whole to the individual village.

Much of the following comes from the WHO manual prepared in the 1960s by Gillies and collaborators (Gillies et al., 1961). Its precepts remain much the same today. The main difference being that today the development of computers, GPS units and biochemical tests for sporozoite and blood meal determination have made the job considerably easier than it was at that time. The main difficulty today lies in a shortage of manpower to carry out many of the necessary samples required for a good understanding of vector behaviour and ecology.

During an initial pre-intervention period a number of villages should be selected and catching stations established. Five houses in each settlement, or in each section in widely strung-out villages, is a convenient size of sample. An effort should be made to spread the catches out so as to include a few villages in all strata of endemicity, but most of them should be located in areas where mosquito densities are highest and transmission rates greatest. The geographical location of selected houses should be determined whenever possible using GPS units. House characteristics should also be entered into a database. Indeed, mapping of study areas is now an important aspect of all epidemiological studies.

It is impossible to lay down hard and fast rules, but two points can be made that have general application. Firstly, during the pre–intervention period, a fixed routine of catches should be followed, so that either the same or closely adjacent houses should be used at each visit, which should be on a monthly or bi-weekly basis. Secondly, once the intervention is in force, this routine can be considerably relaxed and replaced by a system of random catches ranging over the whole area of operations. Nevertheless, a proportion of villages must be retained as regular catching points right through the campaign, so as to provide a baseline for assessing the effects of such variables as rainfall and other environmental factors. The localities retained for long-term observation should, naturally, be those where the breeding potential is greatest. It is also important to carry out catches in treated houses as early as possible in the morning, especially where IRS has been used.

After spraying, a separate record should always be kept of the results from untreated dwellings. These houses will normally be few and far between, and it is quite legitimate for an entomologist to seek them out and use them as temporary catching stations. In fact, the entomologist should make a special effort to locate them, because they provide the best and most convincing evidence of the human-biting population. In so doing the entomologist is obtaining data not only of the entomological results but also on the adequacy of the overall coverage with the intervention. And in the case of IRS, provided he or she passes on the information to those in charge of spraying, especially should more than a very few scattered untreated houses be found, such scouting activities can be doubly recommended. In this way, the entomologist becomes a detective. Should it be found that transmission has not been completely broken, the entomologist becomes responsible for advising the malariologist on possible reasons for failure. This task involves conducting a number of investigations from an early stage in a campaign, many of them with negative findings, and perhaps some of them appearing to a superficial observer to be superfluous or even pointless. In order to understand clearly what is required of the entomologist it is useful to start off by defining the ways in which malaria transmission may fail to be broken and then to plan the assessment accordingly.

Table 10.1 Potential Causes of Failure and Investigations Required to Determine the Cause

Potential Causes of Apparent Failure to Break Transmission	Entomological Investigations Required
• Are the infections local or imported?	
• Has the spraying been adequately carried out both as regards dosage and cycle?	
• Has there been proper coverage of internal surfaces?	
• Have all houses, temporary as well as permanent, been sprayed?	
• Is the insecticide doing its job? • If not, is this due to (a) too long an interval between sprayings? (b) deposits of soot? (c) remudding?	Bioassay tests; spray catches; fed: gravid ratios; age grading
• Is there physiological resistance?	Susceptibility tests
• Is there outside biting by (a) main vector? (b) other species?	Night catches; Tent traps; outside resting catches
• Is there outside resting after feeding indoors? • If so, does it occur at all times or only towards the end of the period between sprayings? • Does it represent an altered pattern of behaviour?	Outside resting catches; blood meal ELISA; trap huts, trap nets, choice of host experiments
• Are cattle a complicating factor?	Bait catches; tent traps; blood meal ELISA tests on outside catch, and catches in choice of host experiments

The possible causes of failure can be set out in the form of a questionnaire, which serves to explain why one carries out the prescribed investigations and which indicates in a simple way their relative importance (Table 10.1).

INDOOR RESTING (HOUSE) CATCHES

This is one of the most basic methods for measuring the density of the human-biting section of vector populations. There are a variety of methods used for the collection of indoor resting mosquitoes including spray collections, hand catches and, most recently, the use of the prokopack sampler (as described in the collection methods). Collections from house catches should be sorted by species and by sex, and the females of vector species should be classified into unfed, fed and gravid categories; a hand lens is sufficient for this purpose.

Any members of the *A. funestus* group appearing in homes after spraying are then recorded as '*A. funestus* group' and saved for later PCR.

OUTSIDE RESTING CATCHES

Searching for mosquitoes in natural resting sites is a time-consuming and frequently unrewarding occupation. It also requires experience and local knowledge, and negative results are often of little significance. Nevertheless, the finding of mosquitoes in the daytime, away from treated surfaces, being direct evidence of the survival of vectors is of such importance that one is fully justified in devoting a proportion of the available time and staff to this problem.

The task is often made considerably easier by the use of artificial shelters. Whatever method is adopted, it is best to pick one or two of the localities where mosquitoes are often abundant and to put up a minimum of 10 shelters in each. If pit shelters are being used, a smaller number will suffice. It is possible to determine if the collection from such shelters is the equivalent of a true outdoor sample because such samples will have a high proportion of male mosquitoes in the catch. Catches can be made in these before spraying as often as convenient, say once every week or fortnight. After spraying, if they continue to yield any mosquitoes, catches can be made much more frequently and the numbers of shelter may also be increased. It may also be desirable to extend their use to other districts. Conversely, if they are consistently unproductive before or after spraying, their routine use may be abandoned. Once installed, artificial shelters can be searched in a very short time and it may be convenient to set up a number in any districts where frequent visits are being made for other purposes, and to take a quick look at them whenever one is in the district. Thus, the whole approach to outdoor catches should be much more elastic than in the case of some other types of catch. One's attitude should be that if they work, then go in for them on a considerable scale; if they do not work, abandon them. Under certain circumstances it may happen that vectors can be found resting outside in large numbers in natural sites. Obviously, under these conditions the use of artificial shelters would be superfluous.

All catches from outside should be classified according to abdominal condition. It is particularly important to distinguish unfed females, because it usually happens that the great majority of them are new emergences and their presence provides few clues as to what is occurring in relation to the intervention (unless it is larval control).

Apart from this, the importance of abdominal staging is sometimes not as fully appreciated as it might be. The basic point is that from the state of digestion and ovarian development, one can determine to which night's biting population the specimens belong. To take a simple example, when the interval between feeding and laying is two days the fed females represent those that engorged the night before, whereas the gravids are those that took a blood meal two nights (approximately 36 hours) previously. The proportions of these two groups in the whole population should be equal, apart from the slight reduction in the numbers of gravids due to natural causes of mortality. However, in many areas the ratio of fed to gravid females in unsprayed houses in the daytime is of the order of 1.5–2:1. This is a clear indication that many more gravid than fed females are resting outside and that the former group is more exophilic. Similarly, in untreated trap huts, it is the capture of fed females in window traps that is of most significance, because this indicates the existence of a group showing relatively transient contact with the interior of houses.

In houses treated with insecticide, the position is radically different. With DDT and perhaps pyrethroids, mosquitoes are either killed rapidly or irritated and leave, so that few survive to be caught indoors in the daytime until the insecticidal effect has fallen to a very low level. But with a slow-acting agent, freshly fed females are commonly found in houses even though a high proportion of them may subsequently die. In such situations, the ratio of fed:gravid females may show an almost complete absence of the latter group. As the insecticide wears off, gravid females begin to reappear in house catches until the insecticidal effect has almost disappeared, when normal fed: gravid ratios are found again.

NIGHT CATCHES

Because it is unusual to find mosquitoes alive in well-treated houses in the daytime, night catches often provide the only evidence of the attack rate on humans. Catches should be made both inside houses and outside, the relative emphasis on each depending on the co-operation of householders, the extent to which people sleep or make merry outside and the prevalence of wild animals. To be carried out with any frequency, night catches demand a fair number of assistants as well as adequate supervision. If the staff is available, collections may be made several nights a week. If not, one night a week should be devoted to this work, the catchers working in rotation and sleeping in between spells of duty.

Light traps for indoor-biting mosquitoes and tent traps for outdoor-biting mosquitoes can be very useful, with great saving of time and labour in areas where they have been shown to give results comparable with hand-catching on bait. It is sometimes convenient to select four villages where *Anopheles* are likely to be most abundant and to visit them each on a monthly basis. In areas where mosquito prevalence is markedly seasonal, night work may be confined to the wet periods of the year. Alternatively, short periods of intensive night work may be called for at any time in response to the general parasitological or entomological picture.

SPOROZOITE RATE DETERMINATION

Routine estimation of sporozoite rates in known vectors is essential under the following conditions:

1. During the pre-intervention period
2. During the initial period of the campaign, say for the first two years
3. In the case of local or temporary foci of transmission
4. In marginal areas or round other potential sources of introduced infection

It is obvious that the pre-intervention data on transmission must be as complete as possible, and the overall level and seasonal fluctuations in the sporozoite rate form an integral part of this picture. But in the day-to-day assessment of a campaign the policy adopted will depend on the circumstances. In the early stages, when parasite levels may still be expected to be high, during the first two, or perhaps, three years, it is important to maintain a watch on transmission by entomological methods. The development of the ELISA for testing for sporozoites means that it is now feasible to test many thousands of mosquitoes and obtain meaningful results even if rates fall to very low levels.

BLOOD MEAL ELISA

During pre-control surveys smears should be made from mosquitoes collected in a wide variety of resting sites and localities, so as to provide a general picture of host preference in the area. From then on it may be taken as a guiding rule that smears should be tested from freshly fed females from all situations in which there is any doubt as to the source of blood. Thus, in villages where cattle or other domestic animals are present in any numbers, smears should be made as a routine. They should also be taken from mosquitoes caught in outside shelters anywhere.

SUSCEPTIBILITY TESTS

Tests to confirm the presence of normal levels of susceptibility to the insecticide in local populations of vectors are carried out in the course of the pre-intervention period. The mosquitoes to be tested should be collected from as wide a variety of localities as possible and tested by the standard WHO method or by CDC bottle assay.

If serious doubts have arisen as to the uniform susceptibility of the population, rather large numbers may have to be tested, say of the order of 1000–2000. When working on such a scale with the standard kit, be aware that the occasional specimen may survive through accidental avoidance of contact with the test papers. Consequently, it is essential to try and obtain eggs from any individuals that survive, and to test the progeny for true inherited resistance. Samples of both surviving and dying insects should be kept, individually, for subsequent biochemical assays that may shed further light on the nature of the resistance mechanism involved.

LARVAL SURVEYS

Larval surveys form an integral part of any general malariological investigation and as such will be of most importance in the pre-intervention period. Once the campaign has started, larval searching will normally only be conducted in the following circumstances:

1. Where the available vector breeding sites are very restricted or localised
2. For collecting additional evidence on the persistence of vectors, when catches of adults have fallen to vanishing points
3. For obtaining wild-caught material for adult or larval susceptibility tests

The search for localised breeding sites will be demanded particularly where it is known that little suitable water is available for breeding. Such situations exist very generally in built-up areas and also, very strikingly, in semidesert regions, where dry season breeding sites may be restricted to a few human-made wells or storage containers. The same also holds true for some areas with well-marked dry seasons, where dams or tanks may be the only source of water. Paradoxically, the same situation may arise under high rainfall conditions in forested areas. The entomologist has a duty to seek out and define these foci because the curtailment of spraying, based on carefully conducted larval surveys, could lead to considerable savings during the later stages of a campaign. In the same way, he or she may also conclude that localised larvicidal measures should be introduced at an earlier stage in the campaign.

Preliminary surveys are of particular value when it comes to assessing the status of the *A. funestus* population after spraying. Because the members of this species complex are most easily identified in the larval stages, the presence of nonvector species in larval collections indicates the need for caution in identifying adults caught biting or resting outside. Conversely, the consistent absence of any other forms during preliminary surveys adds significance to the finding of *A. funestus*-like adults later in the campaign. Although the larvae of *A. funestus* are notoriously difficult to find when densities are low, the continued absence of this species from all types of possible breeding sites is useful supplementary evidence of near eradication.

BIOASSAY TESTS

It is desirable to carry out a certain number of observations to ascertain that the insecticide is continuing to give an adequate kill of mosquitoes coming into contact with treated surfaces.

In allocating his or her time the entomologist must primarily concentrate on measuring and checking the behaviour of mosquito populations. Nevertheless, in the early stages of a campaign, a certain amount of effort must be put into the task of ensuring that the dosage and cycle are suited to local conditions. To this extent, bioassays represent part of the routine assessment. Bioassays do, however, require that a known susceptible mosquito be used in the test. This may mean the establishment of a colony, which is in itself a considerable undertaking.

It should be emphasised that there is no basic difference between assessment and surveillance regardless of the parasite picture.

It is as well to remember the 10 rules of ecological sampling as described by Krebs (1999):

1. Not everything that can be measured should be.
2. Find a problem and state your objectives clearly.
3. Collect data that will meet your objectives.
4. Some questions are impossible to answer.
5. Decide on the number of significant figures needed before you start.
6. Never report an ecological estimate without some measure of its possible error.
7. Be sceptical about the results of statistical tests of significance.
8. Never confuse statistical significance with biological significance.
9. Code your data and enter it in some machine-readable format
10. GIGO – garbage in, garbage out.

If your data is rubbish, then so will your conclusions be rubbish.

Surveillance seeks to determine:

- What are the major mosquito vectors in the country and what is the specific contribution made to malaria transmission by each of the known or suspected vector species in each ecological zone? What are the long-term trends in vector species composition?
- What is the insecticide susceptibility status of the known or suspected vectors?
- Are the vector control interventions being implemented in the country effective against the vectors responsible for transmitting the malaria parasites in the different epidemiological situations encountered? Are those interventions still effective in the face of insecticide resistance and/or changes in mosquito behaviour?
- What are the specific behaviours of the mosquito vectors that may impact on the effectiveness of insecticide-based control interventions? For example, do they bite indoors or outdoors? Do they feed preferentially on humans, domestic animals or both? Do they rest inside houses or other structures, or do they rest outdoors? At what time of night do the different vector species bite?
- How do human behaviour and changes in human behaviour affect the effectiveness of interventions?
- What are the characteristics of the preferred larval habitats of each of the known or suspected vectors, and what is the geographic distribution of those sites?

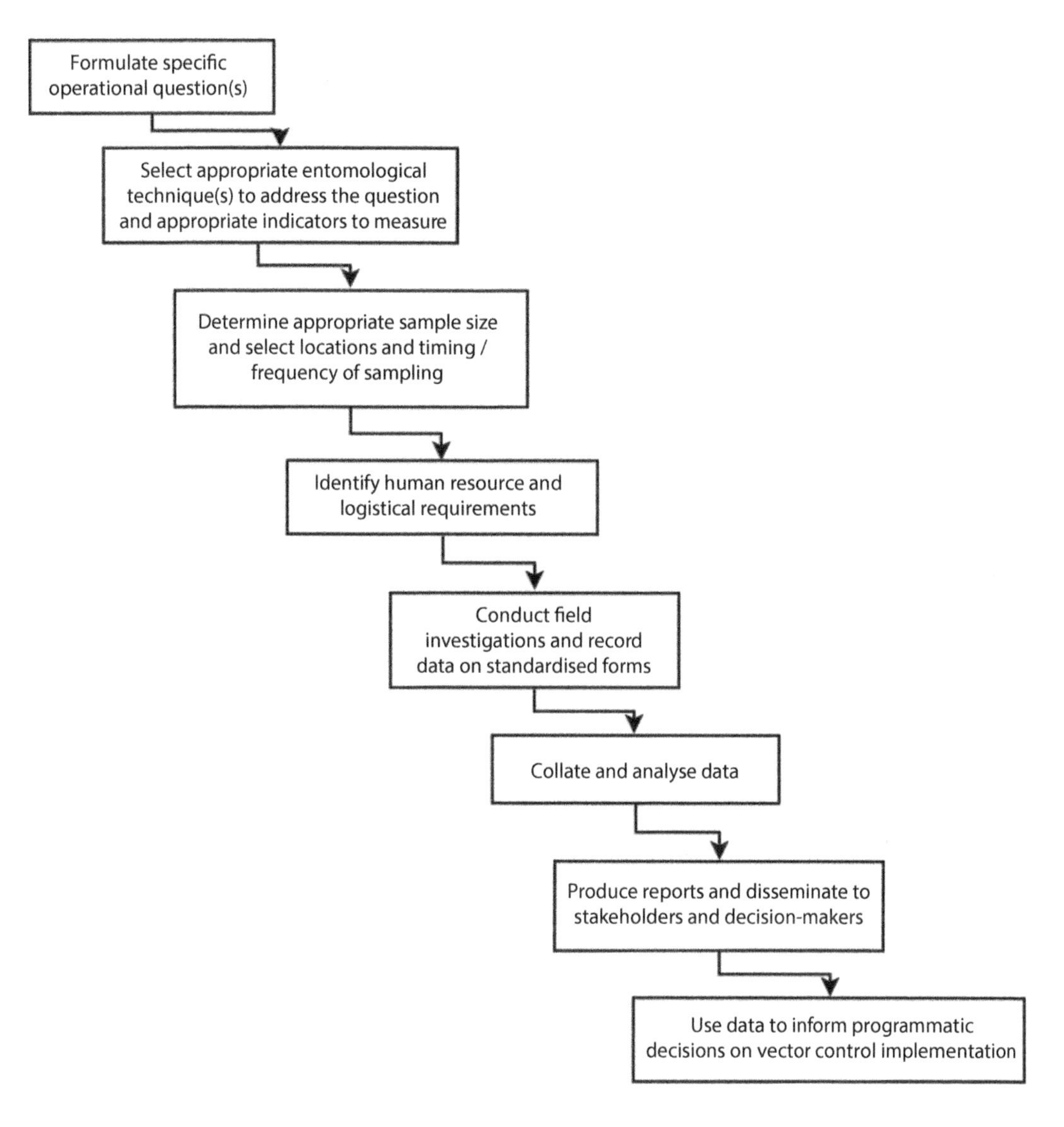

Figure 10.1 The surveillance flowchart. The main aim is for the entomological data available to inform policy in an appropriate manner.

A flowchart describing the various steps to be undertaken during surveillance is shown in Figure 10.1. The point of such a systematic approach is to eventually produce concrete recommendations for suitable decisions to be made by the operational managers of a control program based on suitably collected data designed to answer research questions.

CHAPTER 11

Epidemics

According to the WHO definition, epidemics are the occurrence of a number of cases of disease highly in excess of that expected in a given place and time.

Man-made and environmental changes frequently result in epidemics of parasitic disease. Several reviews have identified the primary drivers of these changes (Molyneux, 1997, 2003; Patz et al., 2004).

These include:

1. Movements of non-immune populations in areas where transmission occurs: Such movements may be of an organised nature – e.g., mobilisation of the workforce in Brazil to exploit forest resources has resulted in malaria epidemics. Alternatively, they may occur without formal organisation, e.g., movements of workers involved in mining for gold or gems in the Amazon and Southeast Asia.
2. Climatic changes: For example, temperature increase is considered to be a cause of highland malaria in Kenya and Ethiopia. Unusual levels of rainfall following periods of drought also result in epidemics of malaria in East and South Africa.
3. Urbanisation results in populations being exposed to new organisms, vectors being established in new habitats and peridomestic reservoirs acting as the source of infection. The dengue and zika outbreaks in Latin America were due to the establishment of *Aedes aegypti*.
4. Change in vegetation such as the growth of thickets of the plant lantana in Uganda, which provided a habitat for *Glossina fuscipes*, provoking epidemics of Rhodesian sleeping sickness. Another example is deforestation, which has resulted in exposure to leishmaniasis in the Amazon and malaria in Southeast Asia (Walsh et al., 1993).
5. Conflict, civil unrest and associated population disruption have profound impacts on parasitic diseases. Epidemics are frequently associated with such events; those diseases that are associated with generalist vectors are more prone to create health problems in conflict environments (Molyneux, 2003). Table 10.1 lists recently documented conflict-related changes in parasitic disease epidemiology.
6. Agricultural development projects, particularly those associated with irrigated agriculture and development of monocultures, are associated with changed patterns of insect-borne infections, particularly malaria (Ijumba and Lindsay, 2001), leishmaniasis and schistosomiasis (Patz et al., 2004).

Factors affecting the likelihood of an epidemic occurring include:

1. A recent increase in amount or virulence of the agent
2. The recent introduction of the agent into a setting where it has not been before
3. An enhanced mode of transmission so that more susceptible persons are exposed
4. An environment conducive to interaction between the host and the agent
5. A change in the susceptibility of the host response to the agent and/or factors that increase host exposure or involve introduction through new portals of entry

CLASSIFICATION OF EPIDEMIC TYPES

As described by Molyneux (1997) epidemics and outbreaks can be classified according to their manner of spread through a population into: common source, point source, continuous source or intermittent source. Their propagation can be due to a combination of different factors or by a more defined route.

A *common-source* outbreak is one in which a group of people are all exposed to an infectious agent or toxin from the same source. If the group is exposed over a relatively brief period, so that everyone who becomes ill develops disease at the end of one incubation period, then the common-source outbreak is further classified as a *point-source* outbreak. The initiation of the plague in Europe from a ship that had traveled to China.

In some common-source outbreaks, exposure can occur over a period of days, weeks or longer. A *continuous common-source* outbreak is one in which the exposure and hence the dates of onset occur over a prolonged period of time, and as might be expected an *intermittent common-source* outbreak is one in which exposure, and hence cases of disease, occur intermittently.

A *propagated* outbreak results from transmission from one person to others. Usually, transmission is by direct person-to-person contact, as with syphilis. Transmission can also be vehicle-borne (e.g., transmission of HIV by sharing needles) or vector-borne (e.g., transmission of malaria by mosquitoes). In propagated outbreaks, such as was the case with Ebola, cases occur over more than one incubation period. The epidemic usually wanes after a few generations, either because the number of susceptible persons falls below some critical level required to sustain transmission (as happens with measles on islands) or because public health action is taken that interrupts transmission.

Some epidemics have features of both common-source epidemics and propagated epidemics. So called mixed epidemics may have a common source followed by secondary spread, which is not uncommon.

Some epidemics are classified as *other* in that they are neither 'common source' in its usual sense nor propagated from person to person. Outbreaks of zoonotic or vector-borne disease can result from sufficient prevalence of infection in host species, sufficient presence of vectors and sufficient human–vector interaction. For example, the epidemic of malaria that occurred in São Tomé after the introduction to the island of a chloroquine-resistant parasite from Angola in the early 1980s.

With regard to vector-borne epidemics (including malaria), there are a number of factors associated with the vector that make them more likely to arise (Table 11.1).

Table 11.2 gives some examples of changes in *Plasmodium falciparum*/*P. vivax* ratios associated with anthropogenic change (conflict, irrigation, mining) that might influence the likelihood of epidemics.

In order to understand epidemics accurate surveillance is required.

Surveillance provides:

- An accurate assessment of the status of health in a given population
- An early warning of disease problems to guide immediate control measures
- A quantitative base to define objectives for action
- Measures to define specific priorities
- Information to design and plan public health programmes
- Measures to evaluate interventions and programmes
- Information to plan and conduct research

In short, surveillance data provide a scientific and factual basis for appropriate policy and disease-control decisions in public health practise, as well as an evaluation of public health efforts and allocation of resources.

Table 11.1 Common Themes Associated with Changing Vector-Borne Parasitic Diseases

- Epidemics are often associated with generalist vectors.
- Animal reservoirs or mixing vessels are associated as food sources for such vectors.
- Animal reservoirs may be domestic, or wild animals or intensively reared species.
- Reduced biodiversity (often associated with 7 and 8 below) encourages expansion of adaptable generalist vectors and reservoirs.
- Ratios of *P. vivax:P. falciparum* change with increasing *P. falciparum*
- Extractive activities (such as logging) generate the development of anti-malaria resistance.
- Water resource development (dams, microdams, irrigation, aquaculture) generates change in vector-borne disease patterns over variable time frames.
- Malaria and Japanese encephalitis – Acute/rapid
- Schistosomiasis/dracunculaisis – Medium
- Filariasis – Chronic/long term/slow
- Deforestation impacts on vector-borne infections via behaviour of human reservoirs and vectors

Source: Molyneux, D., *Trans. R. Soc. Trop. Med.*, 97, 129–132, 2003.

Table 11.2 Examples of Changes in *Plasmodium falciparum*, *P. vivax* Ratios Associated with Anthropogenic Change (Conflict, Irrigation, Mining)

	Projected Changes	References
Tajikistan	Health systems disruption, conflict migration, chloroquine resistance in *P. falciparum*	
Afghanistan/Pakistan	Chloroquine resistance in *P. falciparum*	Rowland and Nosten (2001)
Sri Lanka, Mahaweli	Irrigation on large scale	Amerasinghe and Indrajith (1994)
Thar Desert, Rajasthan, India	Irrigation on large scale Establishment of *Anopheles culcifacies*, efficient *P. falciparum* vector and dominance over *A. Stephensi*, a poor vector	
Amazonia, Brazil	New breeding sites for efficient vectors *A. Darlingi* through mining, deforestation, road building	De Castro et al. (2006)

Source: Molyneux, D., *Trans. R. Soc. Trop. Med.*, 97, 129–132, 2003.

HOW TO MEASURE EPIDEMIC THRESHOLDS

Determination of epidemic thresholds using the median and third quartile method:

1.1 Look at the number of malaria cases at a specific health facility or district by month for the past five years, if available. Exclude epidemic years.
1.2 Determine the median for each month (for example, each January for the last five years). Rank the monthly data for each month for the five years in ascending order. Identify the number in the middle of each month's series. This is the median. Repeat this process for each month in the five years.
1.3 Determine the third quartile for the monthly series by identifying the fourth highest number from the bottom in each data series (because this is ranked in ascending order). This is roughly the third quartile representing the upper limit of the expected normal number of cases.
1.4 Plot the median for each data series by month for the five-year period and join the points with a line (Figure 11.1). This line represents the lowest limit of expected number of cases.
1.5 The area between the two lines (the median and the third quartile) represents the 'normal channel'. Anything above the third quartile means that the epidemic threshold has been exceeded.

The pattern of transmission in Eritrea is similar to that in Sudan (Figure 11.2).

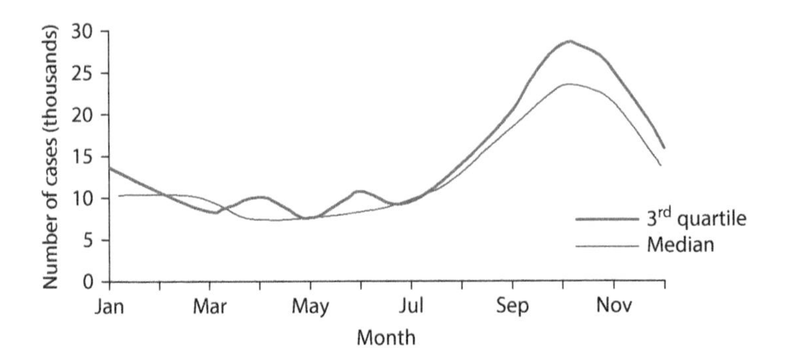

Figure 11.1 Epidemic thresholds based on median and third quartile methods.

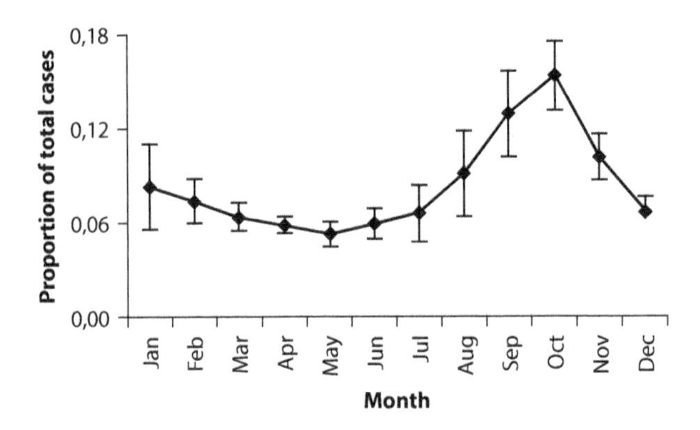

Figure 11.2 Mean proportion (and S.D.) of the total annual number of malaria cases for the years 1986, 1987, 1993–1996, by month, for all age groups, from all refugee camps, Eastern Sudan. (From Charlwood J.D. et al., *Acta Trop.*, 80, 1–8, 2001b.)

INDICATORS

Epidemic thresholds: Once the epidemic thresholds have been determined (Step 1.3), the data manager will be able to monitor the health reports coming to the health facility/sentinel site on a weekly basis.

Although the epidemic threshold is calculated by month (for ease of data manipulation), the data manager can use the cumulative number of cases reported for the month as a guide against the epidemic threshold. Alternatively, the epidemic thresholds by month can be crudely converted to a weekly epidemic threshold chart by dividing the cumulative monthly totals by four weeks. This will roughly indicate the expected number of cases for each week during the month.

Consider, the epidemic threshold for the month of January was determined to be 120 cases for a particular catchment area of a health facility/sentinel site. The expected number of cases for each week of January (dividing by 4) would be approximately 30 cases per week. For example, if in the third week of January of this year the cumulative number of cases were 100 cases, then this would have exceeded the expected 90 cases for the three weeks.

If the epidemic thresholds have been exceeded, this should signal an alarm. The head of the health facility/sentinel site should be made aware of this, and the malaria coordinators should also be notified as soon as possible for further investigation and action.

A spatial version of vectorial capacity (VC) called VCAP has been developed to propose a spatial version of the formula, allowing assessment of vectorial capacity for each pixel in a given area. To be able to do so, the VCAP is only driven by minimum air temperature (*Ta*) and rainfall. Rainfall and temperature are used as inputs to the model because they have an impact on vectorial capacity. It is possible to use minimum *Ta* derived from MODIS for monitoring risks of malaria transmission in highland regions including Eritrea and Ethiopia where a high proportion of the population lives at risk of epidemic malaria. Currently, the USGS EROS Center uses this temperature derived from MODIS night soil temperature (*Ts*) on an 8-day basis jointly with rainfall data derived from the Tropical Rainfall Measuring Mission (TRMM) downscaled to 1 km spatial resolution to produce a 1 km VCAP map every 8 days, specifically for the epidemic regions of sub-Saharan Africa.

CHAPTER 12

The Diseases – Malaria, Filariasis and Dengue

MALARIA

Much of the information below comes from Professor Sir Nick White's chapter in Manson's *Tropical Diseases* (White, 2003).

Malaria is a protozoan disease of the red blood cells transmitted by the bite of an infected mosquito (Figure 12.1). It has been described since ancient times. In his *Epidemics*, Hippocrates describes the regular attacks of intermittent fever associated with the different parasites. Fever is one of its classical symptoms. In Europe, seasonal fevers were associated with marshes. In Latin, *palus* means marshy ground and hence the French term Paludisme. In Italian, it was also associated with often pungent marshes *mal* (bad) air, hence malaria. It has perhaps been responsible for half of all human deaths since *Homo sapiens* first walked the Earth and despite recent gains in control remains a leading cause of death in Africa, where 89% of the world's malaria deaths occur. In Africa, it is also responsible for 25%–35% of all outpatient visits and 20%–45% all hospital admissions. It is especially severe in pregnant women and causes an estimated 400,000 severe pregnancy-related cases and 10,000 maternal deaths every year. It is an economic problem as well as a health problem and costs an estimated $12 billion a year and in Africa is considered to slow economic growth by 1.3% a year.

There are five species of human malaria: *Plasmodium falciparum*, *P. vivax*, *P. malariae*, *P. ovale* and *P. knowlesi*.

The life cycle of the parasite, first described by Ronald Ross, involves sexual reproduction in the mosquito and clonal reproduction in the human (Figure 12.2).

Frequent crossing occurs between different clones of parasite during sexual reproduction in the mosquito (Babiker et al., 1995).

Only a few (maybe 6–10) sporozoites are injected by the mosquito. Within 45 minutes they have either found refuge in a hepatocyte in the liver or have been killed by erythrocytes. In the liver, a phase of asexual reproduction takes place. (All reproduction in the human is asexual.) After 5.5 days for *P. falciparum* or 15 days for *P. malariae*, merozoites are released into the bloodstream. There they very quickly invade erythrocytes.

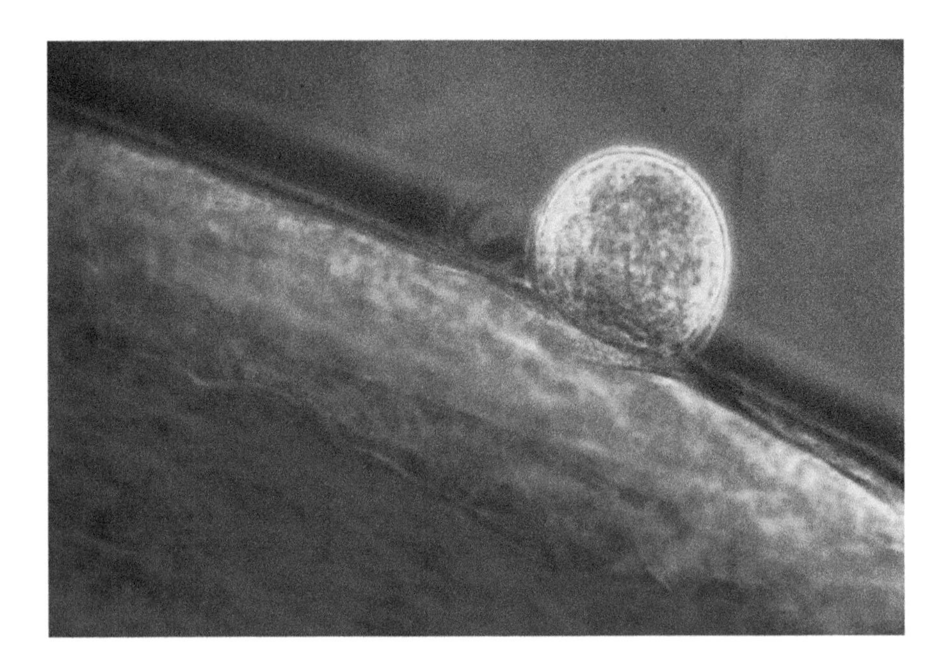

Figure 12.1 A mature oocyst (full of sporozoites and ready to burst) on the stomach wall of a recently blood-fed *A. farauti* from Papua New Guinea.

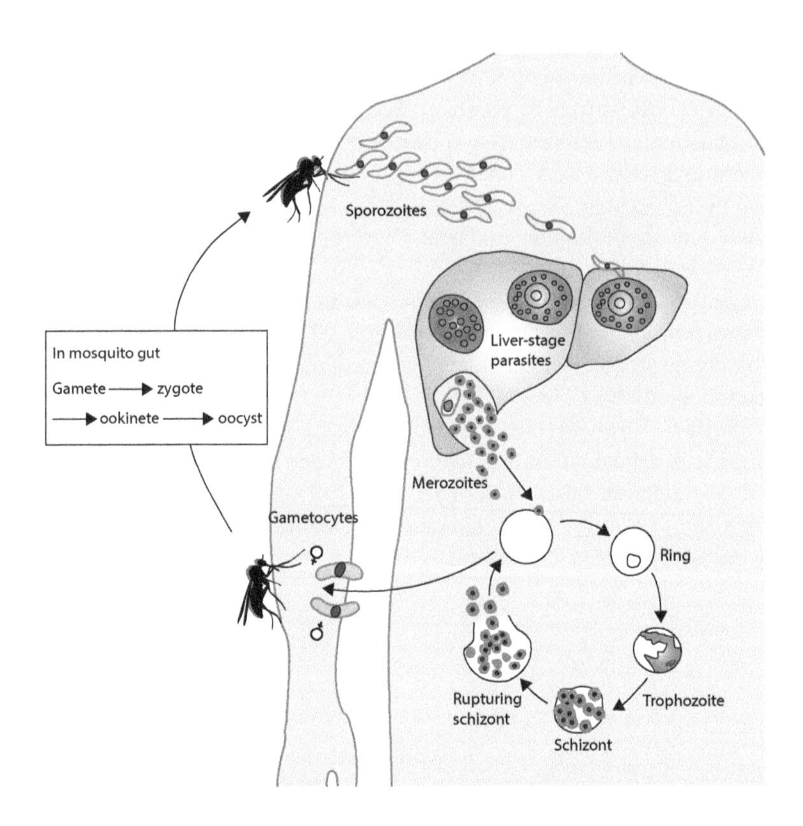

Figure 12.2 Life cycle of the malaria parasite between mosquito vector and human host. (From Greenwood, B.M. et al., *Lancet*, 365, 1487–1498, 2005.)

In some instances, the primary incubation period can be much longer. In *P. vivax* and *P. ovale* infections, a proportion of the intrahepatic parasites do not develop but instead remain inert as sleeping forms or 'hypnozoites', to awaken weeks or months later and cause the relapses that characterise infections with these two species. A variety of forms of *P. vivax* exist including one (*P. vivax multinucleatum*) in which multiple parasites invade a single red cell (Rajan and Charlwood, 1984). During the hepatic phase of development considerable asexual multiplication takes place and many thousands of merozoites are released from each ruptured infected hepatocyte. However, as only a few liver cells are infected, this phase is asymptomatic for the human host.

The species differ in many of their biological properties, in particular the duration of the extrinsic cycle in the mosquito (which is temperature dependent – Table 10.1) and the prepatent period – after the sporozoites are injected into a susceptible human host before they are detectable – and the incubation period (the time between sporozoite injection and symptoms) (Table 12.1).

There are a number of other differences in the parasites that affect their transmissibility. In particular, *P. vivax* gametocytes take 48 h to mature so they are present in the peripheral blood from the start of an infection whilst *P. falciparum* gametocytes take 7–10 days to mature (the initial stages being sequestered in the bone marrow). There are a number of factors that may induce gametocyte production. Drugs, such as chloroquine, may enhance production and so may have augmented the rise in cases that occurred at the end of the last century. Anaemia may also boost production but this may just be the effect of long-term infection. Other factors, such as the production of antibodies to mosquito saliva, may also induce gametocyte production – which would help explain the rapid rise in cases seen in places where transmission is seasonal. Once mature, gametocytes of *P. falciparum* may also be found at highest densities in capillaries close to the skin rather than in the veins. This also increases the likelihood that they will be acquired by feeding mosquitoes.

Once a mosquito has acquired an infection then there are also some possible side effects on its behaviour that increase the likelihood of transmission.

Mosquito survival is linked to activities performed in each oviposition cycle, some of which are more hazardous than others. For example, there may be significant mortality associated with feeding but not with subsequent resting. Parasite development in the mosquito is largely temperature dependent. It is in the parasite's best interest to reduce the number of hazardous events experienced by the mosquito during its development by extending the oviposition cycle. In a study in Mozambique, infected *A. funestus* were more likely to return to feed without sacs compared with uninfected females (Charlwood and Tómas, 2011). Thus, their cycle was longer than that of uninfected insects. Most mosquitoes have just one or two oocysts. Nevertheless, these draw nutrients from the mosquito, which may then be more likely to take a sugar meal after oviposition. Sugar feeding is likely to depress host-seeking with the result that the cycle is extended.

Once the incubation period is over and the mosquito has sporozoites in the salivary glands (and so is infectious rather than just being infected), then the situation is reversed and the more probing

Table 12.1 Duration of the Extrinsic Cycle (Sporogony) of the Different Species of Malaria at Different Temperatures

	30°C	24°C	20°C
P. falciparum	9 days	11 days	20 days
P. vivax	7 days	9 days	16 days
P. malariae	15 days	21 days	30 days
P. ovale	15 days at 26°C		

and feeding that the mosquito undertakes the greater the amount of transmission. A number of factors affect this. A simple mechanical blockage of the hypopharynx may reduce the mosquito's ability to salivate so it may make several attempts to feed on more than a single host. The enzyme apyrase, involved in feeding efficiency, is also reduced in sporozoite-infected females which may also result in increased probing. The mosquitoes may also take smaller (and so more frequent) blood meals. All of these factors may enhance transmission.

The duration of an untreated infection also differs. In *P. falciparum*, it can last 2 years and in *P. malariae* 40! In addition, *P. vivax* is well known for its relapses, which occur at different times following an initial infection as a result of the release of parasites that remain for long periods as hypnozoites in the liver. The usual interval between relapses in the Chesson strain of the parasite is around one month but others may relapse after a year (Table 12.2).

The different species and different stages of the parasite can be distinguished microscopically (Figure 12.3).

In general, only ring stages, trophozoites or gametocytes are seen in *P. falciparum* infections because shizogony often occurs in the sequestered parasites in the deeper organs.

Other differences between the different species of malaria are outlined in Table 12.3.

Transmission intensity of malaria differs considerably between different areas. In some places people may be exposed to one infective bite every 10 years and in others to two or three infective bites every night. For convenience, intensity of transmission is divided into four strata from hypoendemic to holoendemic, although these are not necessarily hard and fast differences and many places have malaria that is intermediate between them. The characteristics of the different classifications are shown in Table 12.4. In particular, as malaria transmission is reduced, the possibility of epidemics increases. This is due to the fact that immunity in the population becomes low or nonexistent.

The characteristics of severe malarial infections also differ according to transmission intensity. In areas of high transmission, anaemia is the most common life-threatening symptom. In areas of mesoendemic and hyperendemic transmission both anaemia and cerebral malaria are the symptoms associated with severe malaria, whilst in hypoendemic areas cerebral malaria and, in older age groups, renal failure occurs most frequently. The age distribution of these

Table 12.2 Prepatent and Incubation Period in the Different Species of Malaria

	Prepatent Period (Days)	Incubation Period (Days)
P. falciparum	11.0 (2.4)	13.1 (2.8)
P. vivax	12.2 (2.3)	13.4 (2.7)
P. malariae	32.7	34.7
P. ovale	12.0	14.1

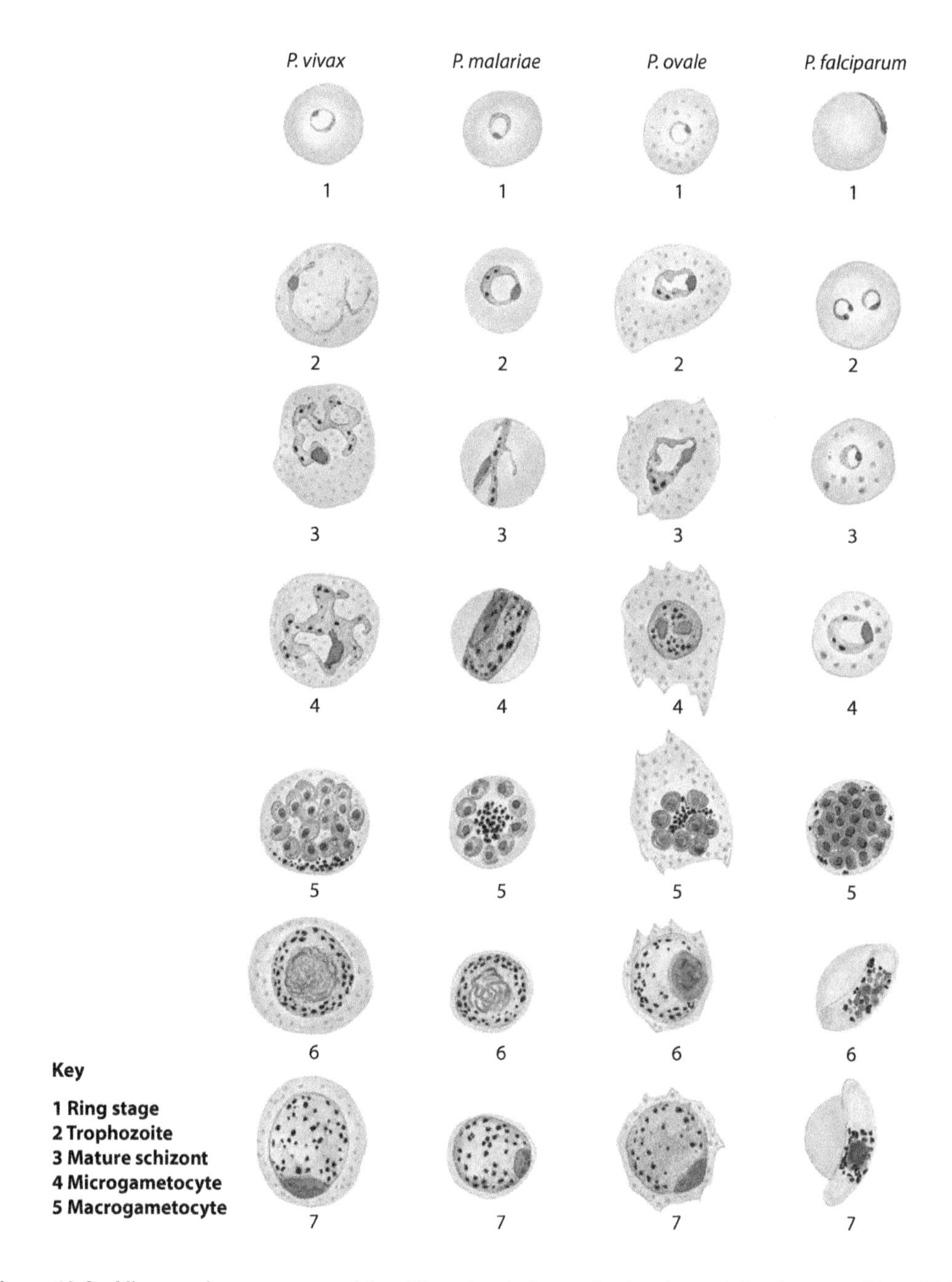

Figure 12.3 Microscopic appearance of the different malaria species by stage of development. (From WHO, *Basic Malaria Microscopy (part I and II)*, World Health Organization, Geneva, Switzerland, 1991.)

Table 12.3 Phases of Development and Duration of Infection in the Different Species of *Plasmodium* Attacking Humans

	P. falciparum	*P. vivax*	*P. ovale*	*P. malariae*	*P. knowlesi*
Exoerythrocytic (hepatic) phase of development (days)	5.5	8	9	15	5–7
Erythrocytic cycle (days)	2	2	2	3	1
Hypnozoites (relapses)	No	Yes	Yes	No	No
Number of merozoites per hepatic schizont	30,000	10,000	15,000	2000	?
Erythrocyte preference	Young RBCs but can invade all ages[a]	Reticulocytes	Reticulocytes	Old RBCs	None identified
Maximum duration of untreated infection (years)	2	4	4	40	?

[a] Parasites causing severe malaria are not selective in red cell invasion.

Table 12.4 Characteristics of Different Endemic Levels of Malaria

Criterion	Hypoendemicity	Mesoendemicity	Hyperendemicity	Holoendemicity
Spleen rate 2- to 9-yr-olds	0%–10%	11%–50%	Constantly >50%	Constantly >75%
Parasite rate 2 to 9-yr-olds	0%–10%	11%–50%	Constantly >50%	Constantly >75% in infants 0–11 months
Stability	Unstable	Unstable	Stable	Stable
EIR	<0.25	0.25–10	11–140	>140

symptoms according to transmission intensity is shown in Figure 12.4. The parasite infected cells are removed from the blood when they pass through the spleen. As a result the spleen enlarges mainly as aresult of cellular multiplication and structural change. The size of the spleen is a classic indicator of the level of malaria transmission. In some situation the spleen may become sufficiently enlarged for it by itself to be a health problem, a condition known as Tropical splenomegaly (Crane et al., 1985).

Plasmodium falciparum is the most dangerous parasite due to its ability to invade all types of blood cells (whereas *P. vivax* attacks only young cells and *P. malariae* attacks old ones) and because of the phenomenon of sequestration, in which cells stay in the deep blood vessels. Sequestration in *P. falciparum* hides the parasites from the spleen, the organ which effectively eliminates the parasite. The effect of sequestration on the assessment of parasite densities means that patients with the same densities of parasite in the peripheral blood (who are therefore considered to have the same parasitaemia) can actually have very different numbers of parasites invading their blood. In a patient without sequestration most of the parasites are in the circulating blood but in a patient with sequestered parasites only 20% are in the circulating blood, giving rise to 60 times more parasites in such a patient.

In part sequestration occurs because the infected cell has characteristic knobs on its surface, which make the parasites 'sticky' (Figure 12.5).

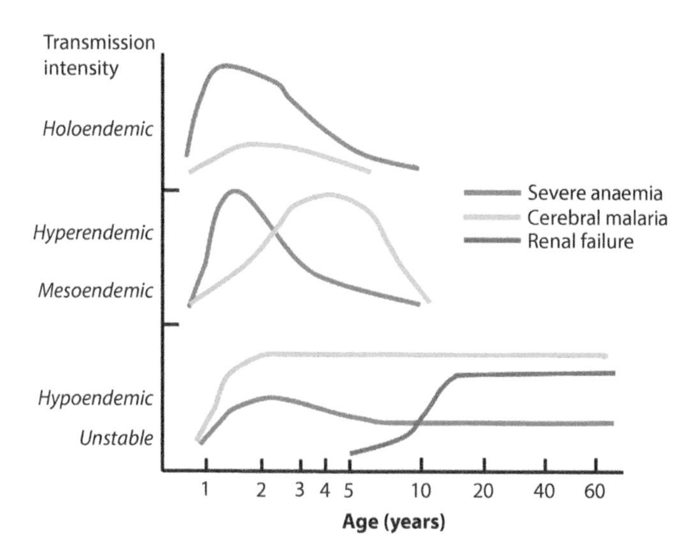

Figure 12.4 The relationship between age and the clinical presentations of severe malaria at different levels of malaria transmission. (Reprinted from *Manson's Tropical Diseases*, (22nd ed.), White, N.J., Malaria, Chapter 73, Copyright 2003, with permission from Elsevier.)

Figure 12.5 Scanning electron micrographs of (a) an infected blood cell surrounded by several uninfected cells and (b) the interaction between an infected red blood cell and an uninfected cell that appears to be mediated by a protrusion of regularly spaced 'knobs'. (From Garcia, L.S., *Diagnostic Medical Parasitology*, ASM Press, Washington, DC, 2007.)

Groups of infected cells will often stick together around a central cell so that they look like flowers. Such rosettes are likely to block small capillaries, making passage of uninfected cells through them difficult. When this happens in vessels in the head, the brain may be starved of oxygen, resulting in cerebral malaria.

The geographic distribution of the different malaria species varies. In much of Africa *P. vivax* is absent.

Combinations of drugs are now used to treat malaria. The rationale behind the use of a combination of drugs is that the probability of the parasite developing resistance to both drugs is the product of the individual probabilities. The drugs need to be always used in combination for this to work. Unfortunately this is not always the case.

The diversity of genes encoding two merozoite surface proteins (*msp*-1 and *msp*-2) of *P. falciparum* was examined in parasites infecting four households in a village in the Kilombero Valley, Tanzania. The polymerase chain reaction (PCR) was used to characterise allelic variants of these genes. In each household, extensive polymorphism was detected among parasites in the inhabitants and in infected mosquitoes caught in their houses. Similar frequencies of these alleles were observed in all households. Hence, cross-mating and gene flow occur extensively among parasites. Figure 12.6 shows that the *msp*-2 alleles of parasites of finger-prick samples of two inhabitants of a single hut, in the blood meals of fed mosquitoes from the hut, and in oocysts in the mosquito, were very diverse. The alleles in the mosquitoes could only have been derived from gametocytes taken up by the mosquitoes, but most were not present in the blood samples examined.

In Idete, another village in the Kilombero Valley, the relationship of the incidence of *P. falciparum* infection to entomological inoculation rates (EIRs) was studied in 163 children, less than one year of age, to determine likely effects of transmission-reducing interventions on infection incidence. A total of 66,727 *A. gambiae* s.l. and 17,620 *A. funestus* were caught in 1056 light-trap collections from 139 houses over a period of more than two years. Time-period specific human biting rates were estimated for 11 village neighbourhoods (Figure 12.7).

Sporozoites were detected by ELISA in 4.4% of the *A. funestus* and 2.5% of the *A. gambiae.* 817 pairs of blood slides with approximately two-week intervals between slides were used to estimate incidence of parasitaemia by fitting reversible catalytic models to parasite positivity data. Estimated EIRs during the four weeks preceding each inter-survey interval averaged 1.6 (s.d. = 2.1) per adult per night. Parasites were present at the end of 31% of the 443 intervals that commenced with a parasite-negative slide. The proportion of bites resulting in human infection was strongly dependent on mosquito density, which varied by neighbourhood and season (Figure 12.8).

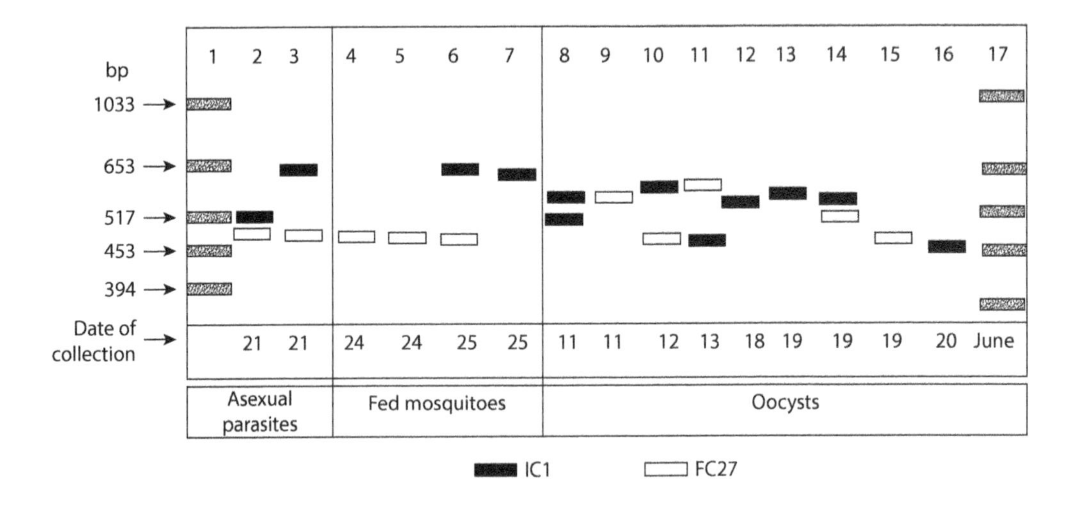

Figure 12.6 Schematic illustration of the MSP-2 alleles of parasites in fingerprick blood samples (lanes 2 and 3), blood meals from fed mosquitoes (lanes 4–7) and oocysts from mosquitoes (lanes 8–16) collected from a single house in Ifakara. Alleles are classified by size and sequence of PCR-amplified fragments. Lanes 1 and 17 are size markers. (From Babiker, H.A. et al., *Parasitology*, 111, 433–442, 1995.)

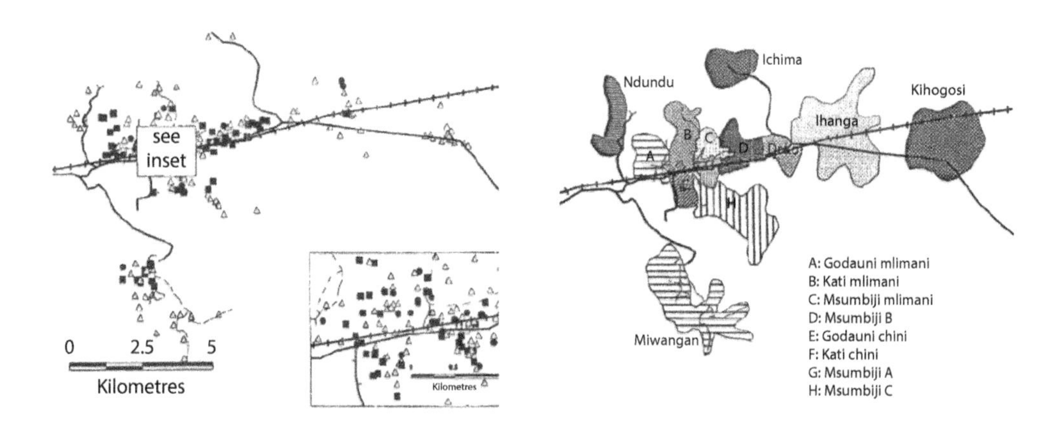

Figure 12.7 Maps of Idete village, Tanzania. Left location of houses: (open triangles) from which parasitological data were obtained; filled squares houses from which entomological data were obtained; filled circles houses from which both entomological and parasitological data were obtained. Right: neighbourhoods of Idete. (From Charlwood, J.D. et al., *Am. J. Trop. Med. Hyg.*, 59, 243–251, 1998b.)

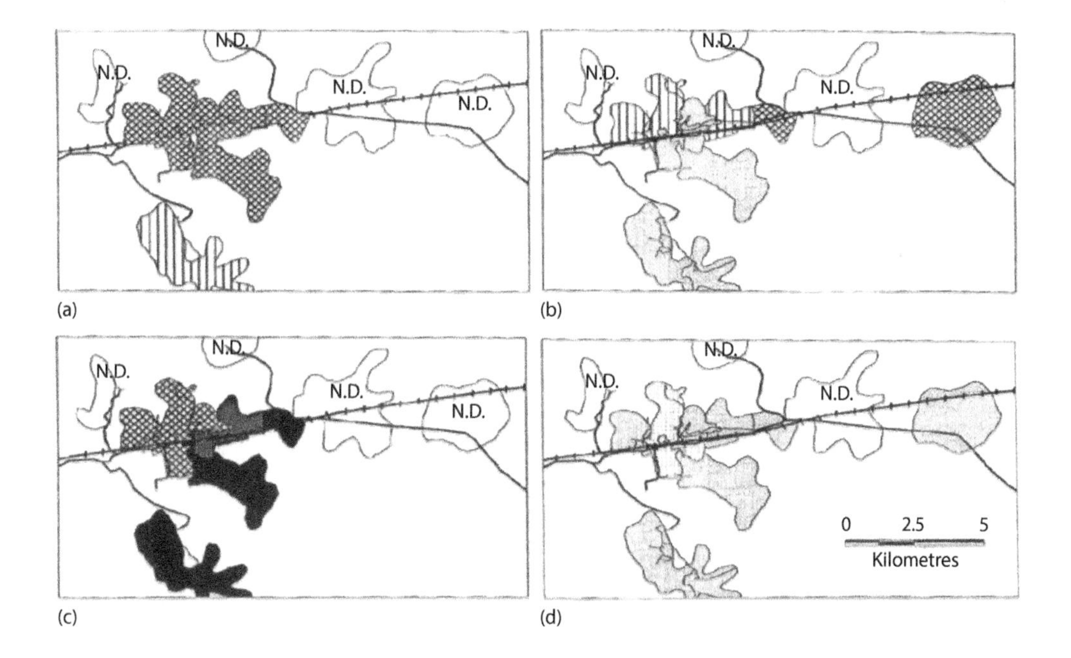

Figure 12.8 Estimated average human biting rates in Idete by neighbourhood (a) *A. funestus* wet season (January to July), (b) *A. funestus* dry season (August to December), (c) *A. gambiae* s.l. wet season, (d) *A. gambiae* s.l. dry season. (From Charlwood, J.D. et al., *Am. J. Trop. Med. Hyg.*, 59, 243–251, 1998b.)

Incidence of malaria increased with EIR up to one infectious bite per adult per night (Figure 12.9). However, higher levels of transmission did not result in a correspondingly higher incidence. The proportion of sporozoite-infected mosquitoes that resulted in patent blood-stage infections consequently showed a very marked decrease as EIR increased. This saturation in the force of infection was evident in the seasonal pattern, which indicated that there was no increase in the force of infection during the main part of the wet season when anophelines were much more abundant. This saturation is important because it implies that substantial reductions at high levels of exposure will

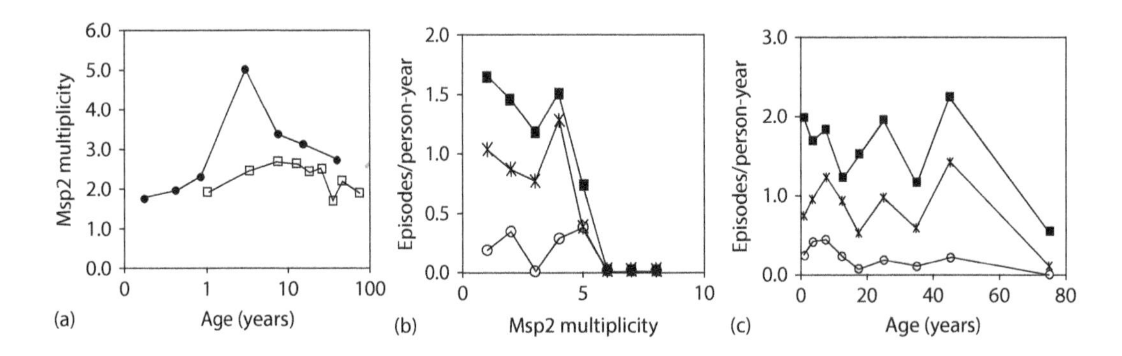

Figure 12.9 Incidence of clinical malaria by exposure -*-= not adjusted for age, open squares = age adjusted using a Poisson regression model. Different subfigures correspond to different measures of exposure. Error bars correspond to 95% intervals for the age-adjusted estimates. (From Smith, T.A. et al., *Am. J. Trop. Med. Hyg.*, 59, 252–257, 1998b.)

not necessarily reduce either prevalence or incidence. Each tenfold increase in EIR corresponded to a 1.6 increase in fever plus parasitaemia (95% confidence interval = 1.4–2.0). Therefore, reduction of human vector contact will probably reduce morbidity even at very high exposures. Incidence showed little relationship to estimated cumulative numbers of inoculations since birth but decreased steeply with cumulative time infected with trophozoites.

Even when the estimated EIR was more than five per person per night, 30 of 75 microscopy slides at the end of the fortnightly intervals were negative. Thus, there must be a considerable decrease in the proportion of bites that result in patent infections at high EIRs. To estimate this proportion, the crude infection rates were first converted to estimates of the equivalent force of infection (instantaneous infection rate *h*), which allows for the possibility that a child both gained an infection and recovered from it during the same interval.

The estimates of *h* are higher than the crude infection rates at all values of *mas* (Figure 12.10). However, the proportion of sporozoite inoculations resulting in an infection, obtained as the ratio *h/(mas)*, is much less than unity at all values of *mas.* This ratio decreases steeply from a value of almost 0.07 at the lowest exposures to 0.006 at an EIR of four or more per person per night. The observed saturation in incidence implies that distinct inoculations of the same host interfere with each other, although why this might be so remains unknown.

The high recovery rates in infants are consistent with many infections being cleared before they become patent, but we did not observe an increase in recovery rate. At very high exposure the relationships between morbidity and mortality risks to EIR remain unclear.

The absence of a protective effect associated with cumulative exposure to inoculations is evidence against an important role for long-term genotype-specific responses in clinical immunity against *P. falciparum*. The negative correlation of morbidity with cumulative time exposed to erythrocytic stages suggests that long-term clinical immunity is induced by exposure to blood stages irrespective of the number of different genotypes (clones) inoculated. Protection would therefore be associated with a gradual buildup of the immune responses to asexual stage parasites, irrespective of their genotypes, not to genetic variation in the inocula. If the frequency of inoculations does not account for long-term immunity then, in contrast to chemoprophylaxis, the protection provided by bed nets is unlikely to interfere very much with natural protection because the reduction of mosquito bites has little effect on parasite prevalence. Thus, personal

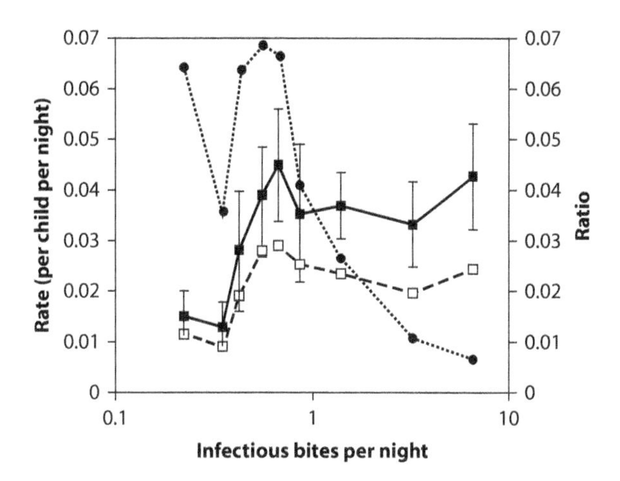

Figure 12.10 Infection rates and force of infection by overall entomological inoculation rate. Open squares = infection rate (transitions/total days at risk); filled squares = estimated force of infection (infections/person/day); filled circles = ratio of force of infection to infectious bites per night. Error bars indicate 95% confidence intervals. (From Smith, T.A. et al., *Am. J. Trop. Med. Hyg.*, 59, 252–257, 1998b.)

protection against mosquito bites will be effective in reducing the incidence of clinical malaria in children less than 18 months of age in areas of highest transmission. Protection in this age range is also unlikely to result in any substantial increase in the overall incidence of clinical malaria when growing older.

FILARIASIS

Lymphatic filariasis is caused by infection with parasitic worms belonging to three species: *Wucheria bancrofti*, *Brugia malayi* and *B. timori*. They differ from malaria and present a different set of 'trade-offs' in order to maintain transmission. For example, humans are the definitive hosts (so this means that two infective inocula are needed for transmission to humans – one a female worm and one with a male). This makes transmission less efficient because the worms must find each other inside the human host before they can mate and produce the infective stage, the microfilariae. The cycle inside the mosquito is shown in Figure 12.11.

Inside the mosquito there is no multiplication of the parasite as there is with malaria, but even a single worm can cause considerable damage to the mosquito. The microfilariae migrate through the gut and eventually enter the flight muscles in the thorax where they moult and develop into L2 (sausage stage) and finally L3 larvae. L3 larvae migrate from the thorax to the head and proboscis, a process that may damage the mosquito and reduce its ability to fly. The L3 larvae leave the mosquito when it feeds and are deposited on the skin of the host, which they must penetrate in order to gain access to the human host. The L3 develop to fourth-stage larvae (L4) as they migrate through the human body to the lymphatic vessels and lymph nodes, where they develop into adult worms. There, once they have a mate, the female worms produce large numbers of microfilariae (the infective stage) that circulate in the bloodstream whence they are ingested by feeding mosquitoes to complete the cycle. Microfilariae of *W. bancrofti*, the most common parasite, are only present in

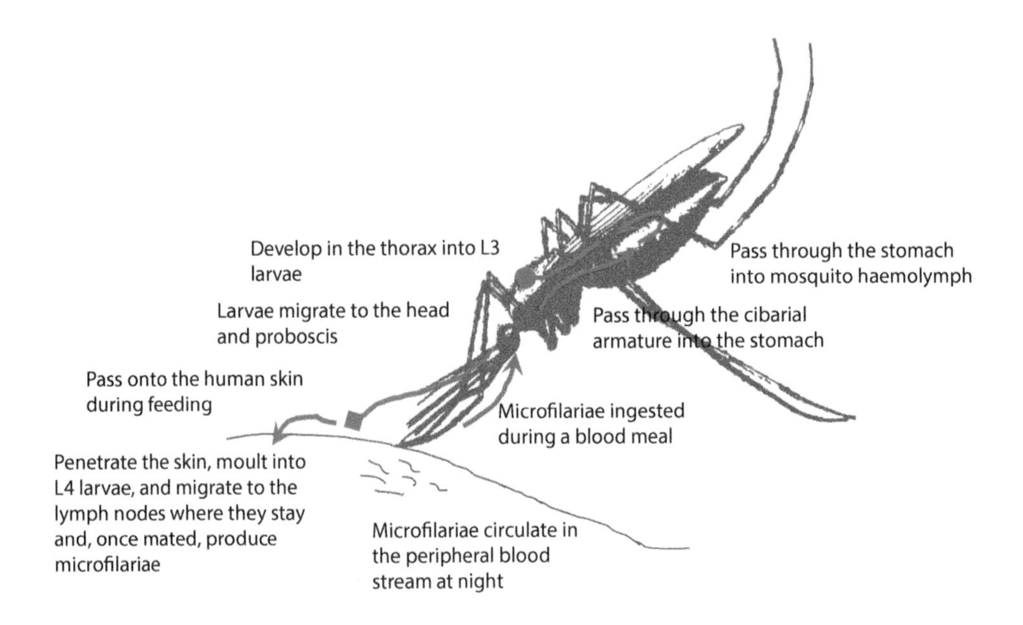

Figure 12.11 Transmission cycle of *Wucheria bancrofti* in the mosquito.

the peripheral blood during the night, whilst those of. *B. malayi* are most abundant during the day, the time when the respective vectors are biting. This means that to determine infection rates of *W. bancrofti* by microscopy, surveys need to be performed at night.

Transmission to the mosquito is also inefficient. Microfilariae that are ingested have to pass through the cibarial armature of the mosquito, which act as a first line of defence against them. In anopheline mosquitoes, the pharyngeal armature is well developed, so that microfilariae are damaged when they are ingested. In *Anopheles* mosquitoes, the proportion of microfilariae that reaches the L3 stage increases as the number of ingested microfilariae increases (facilitation). Low densities of microfilariae are associated with a much lower rate of development to L3. In contrast, in *Aedes* vectors of filariasis, low densities of ingested microfilariae have a high likelihood of survival but, the proportion of ingested microfilariae that survive to become L3 larvae decreases as more microfilariae are ingested, a process known as limitation. Only a limited number of microfilariae develop in the mosquito. As with malaria, the mosquito has to survive through the developmental time of the parasite, which may be 10–12 days depending on temperature. The larvae leave the mosquito during feeding and are deposited on the skin of the host, rather than inside the host like malaria parasites. This also reduces the likelihood of successful transmission, and once a mosquito has shed its microfilariae it reverts to being uninfected and uninfectious.

All of these factors reduce the efficiency of transmission. However, unlike malaria, filariasis can be transmitted by a wide variety of mosquitoes from *Anopheles*, *Culex*, *Mansonia* and *Aedes*. This increases the chance of transmission. In addition, adult worms may live for more than six years and produce microfilariae every day of their lives. This also enhances transmission.

Control of filariasis can be obtained by mass drug administration (MDA) against the worms combined with the usual vector control tools (bed nets for night-biting mosquitoes and repellents for those that bite in the day).

DENGUE AND ITS VECTORS

The vectors of dengue, yellow fever and the newly minted Zika viruses differ in their ecologies from malaria vectors to such an extent that, despite feeding on the same host (human), they can be used as an example of ways to avoid competition.

Aedes (Stegomyia) aegypti (also known as *Stegomyia aegypti*) and *A. albopictus* (also known as *Stegomyia albopicta*) differ in that *A. aegypti* is a domestic mosquito living close to and around houses that feeds almost exclusively on humans, whilst *A. albopictus* (the Asian tiger mosquito) is a more sylvatic insect that will feed on humans if they are available but will also feed on a range of other hosts. This makes *A. aegypti* a more efficient vector and the more dangerous of the two species.

Aedes aegypti has a lyre-shaped pattern of silver scales on its thorax, whilst *A. albopictus* has a line of scales down the middle of the thorax. Unfortunately, these scales are easily rubbed off when they are collected. Both have black and silver bands of scales on their legs (Figure 12.12).

Both use small containers for their oviposition sites (*A. aegypti* will lay eggs in almost any suitable container, whilst *A. albopictus* tends to use tree holes and cut bamboo). Unlike the eggs of anophelines, the eggs of these mosquitoes can survive desiccation for several months. This means that even if there were an effective control program in place against the adult mosquito, once the eggs were submerged in water (as happens to dry containers after the rain) the population would rebound to pre-intervention levels within a couple of weeks.

The ability of the eggs to survive in this way has also helped in the spread of the mosquitoes from their original Southeast Asian (*albopictus*) and African (*aegypti*) origins. Because of human-aided transport in general, especially of such things as used car tyres, the mosquitoes now occur anywhere the temperature is sufficiently warm in the summer to sustain them (Figure 12.13). They pose an increased threat as global warming continues. In addition to the mosquito, dengue has also spread from a relatively confined area to become a global pandemic. In the Americas it was controlled as part of the campaign against yellow fever in the 1930s such that, by the 1970s, it had been removed from large areas. Control measures were discontinued ('Where is the problem? There are no mosquitoes anymore'), and there was a subsequent resurgence of mosquitoes along with a resurgent dengue epidemic. There was a time, not so long ago, when going to Brazil was an open invitation to get dengue.

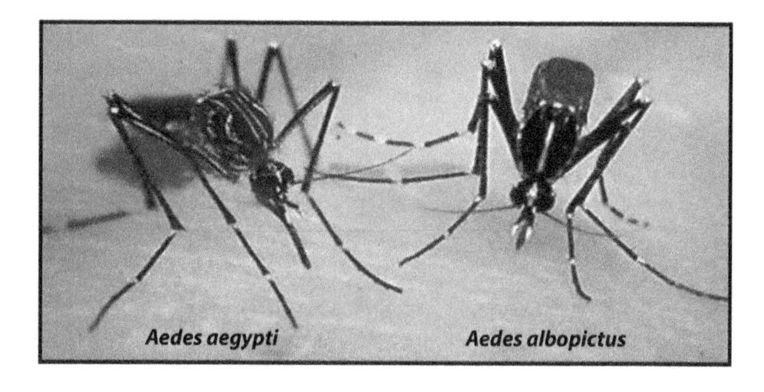

Figure 12.12 *Aedes aegypti* on the left and *Aedes albopictus* on the right. (Courtesy of the Medical Entomology Laboratory, University of Florida, FL.)

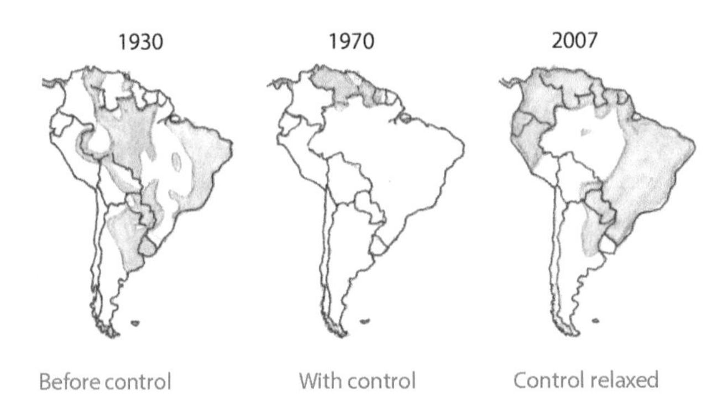

Figure 12.13 Distribution of *Aedes aegypti* in South America: 1930, 1970 and 2007. (Courtesy of Division of Vector-Borne Infectious Diseases, CDC, Fort Collins, CO; Adapted from Gubler, D.J., *Clin. Microbiol. Rev.*, 11, 480–496, 1998.)

The mosquitoes bite during the day (as opposed to anophelines that attack at night) and, at least as far as *A. aegypti* is concerned, will take small blood meals on a daily basis, which lead to the development of just a few eggs rather than a full egg batch (as is the case for anophelines). Thus, feeding and oviposition are effectively decoupled even in uninfected insects. Indeed, in Thailand it has been calculated that the mosquito requires five blood feeds before it will develop eggs – each feed leading to possible transmission of the virus. This makes estimates of mosquito age by dissection more difficult than it is with anophelines. *Aedes albopictus* differs in this respect in that it is a more direct feeder and is more likely to be gonotrophically concordant than *A. aegypti*. While *A. aegypti* is a very stealthy and easily disturbed feeder, *A. albopictus* is much more direct. This makes the former very difficult to catch in landing collections. The kind of collection setup used in Cambodia (Figure 12.14) has not been tried with *A. aegypti*.

Originally a disease of the forest, where it was presumably transmitted by *A. albopictus*, it has been people who have tended to transport both the vector and the virus into towns and cities. The African mosquito, *A. aegypti*, was probably an unwanted stowaway on slaving ships during the eighteenth and nineteenth centuries. Ever increasing trade, especially after the Second World War and the relaxation of control measures in the 1970s, has meant that both vector and disease have

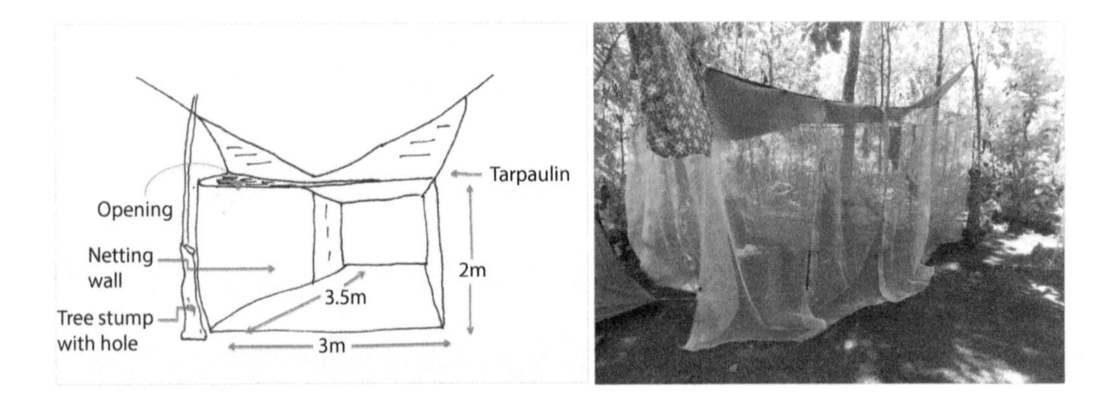

Figure 12.14 Experimental setup for the collection of *Aedes* sp. in Cambodia. Two hosts sat in the netting cage and collected *A. albopictus* that entered through the opening in one corner. Mosquitoes were collected from each other and off the walls of the cage (often engorged specimens that had managed to feed before collection). (From Charlwood, J.D. et al., *Parasit. Vectors*, 7, 324, 2014.)

spread over the past few years. There are more than three billion people living in endemic areas with an estimated annual incidence of more than 100 million cases every year. The mosquito was present in South America in the 1930s but due to a vigorous control programme had largely disappeared by the 1970s. The programme was scaled down in the 1970s and soon 'defeat was snatched from the jaws of victory' such that by the 1990s the mosquito had regained all of its former territory.

Given the difficulty of collecting adults, and the fact that the habitat for eggs and larvae are containers, it is these that are usually used to sample the mosquitoes.

Ovitraps (Figure 12.15a) are easy to make and easy to sample. They can be used to determine both the horizontal and vertical distribution of oviposition (Figure 12.15b).

Breeding of *Aedes* in septic tanks has recently been reported. This puts them in direct competition (as far as the immature population is concerned) with the classic septic tank breeder *Culex quinquefasciatus. Culex quinquefasciatus* is a much less efficient vector of disease (in this case filariasis) than *A. aegypti.*

'Water poison', as it was known in the Chinese *Encyclopaedia of diseases and remedies*, produced during the Chin dynasty (AD 265–420), may have been dengue. Similarly, the epidemics of disease that occurred in Africa, Asia and North America in the late eighteenth century may also have been dengue. Indeed, the term dengue may derive from the Kiswahili term *dinga* or *denga*, 'ki denga pepo', which means 'seizure caused by an evil spirit'.

The disease is very difficult to diagnose accurately. Symptoms can be mild to excruciating – not for nothing has it also been called 'break-bone fever'. Extreme cases result in dengue haemorrhagic fever (DHF) or dengue shock syndrome (DSS), both of which are often fatal.

Dengue consists of four different serotypes (DEN1, 2, 3 and 4) that are all very similar, having common group epitopes on the protein envelope around the virus. Before the rapid spread of the vector in the 1970s the serotypes were limited in their distribution, but 30 years later they had spread throughout the tropics and subtropics (Figure 12.16), the spread often being accompanied by major epidemics such as those that occurred in the Americas. Like the yellow fever virus, they belong to the genus *Flavivirus.* Three to 14 days after being bitten by an infected mosquito a person

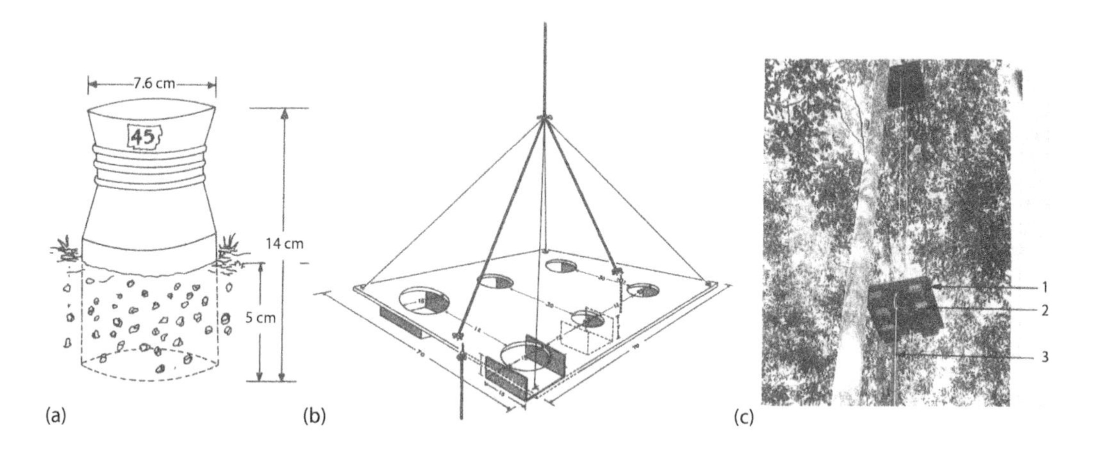

Figure 12.15 Ovitraps such as this are useful for the collection of container-breeding mosquitoes. The container in this case was a cut-off plastic bottle. Water and leaf litter were added. For *Aedes* mosquitoes the bottle is painted black on its exterior surface. (a) Paddle of hardboard wrapped, perhaps with a waterproof filter paper (such as Benchcote), can be suspended into and above the water. Eggs that have been laid on the paddle can be easily sampled. (b) Several such containers can be used and sampled sequentially in order to obtain data on temporal dynamics of the mosquitoes. (c) The system was used to study the vertical distribution of mosquito eggs and larvae in forest close to the town of Manaus, in Brazil. (From Lopes, J. et al., *Ciencia e Cultura*, 37, 1299–1311, 1985.)

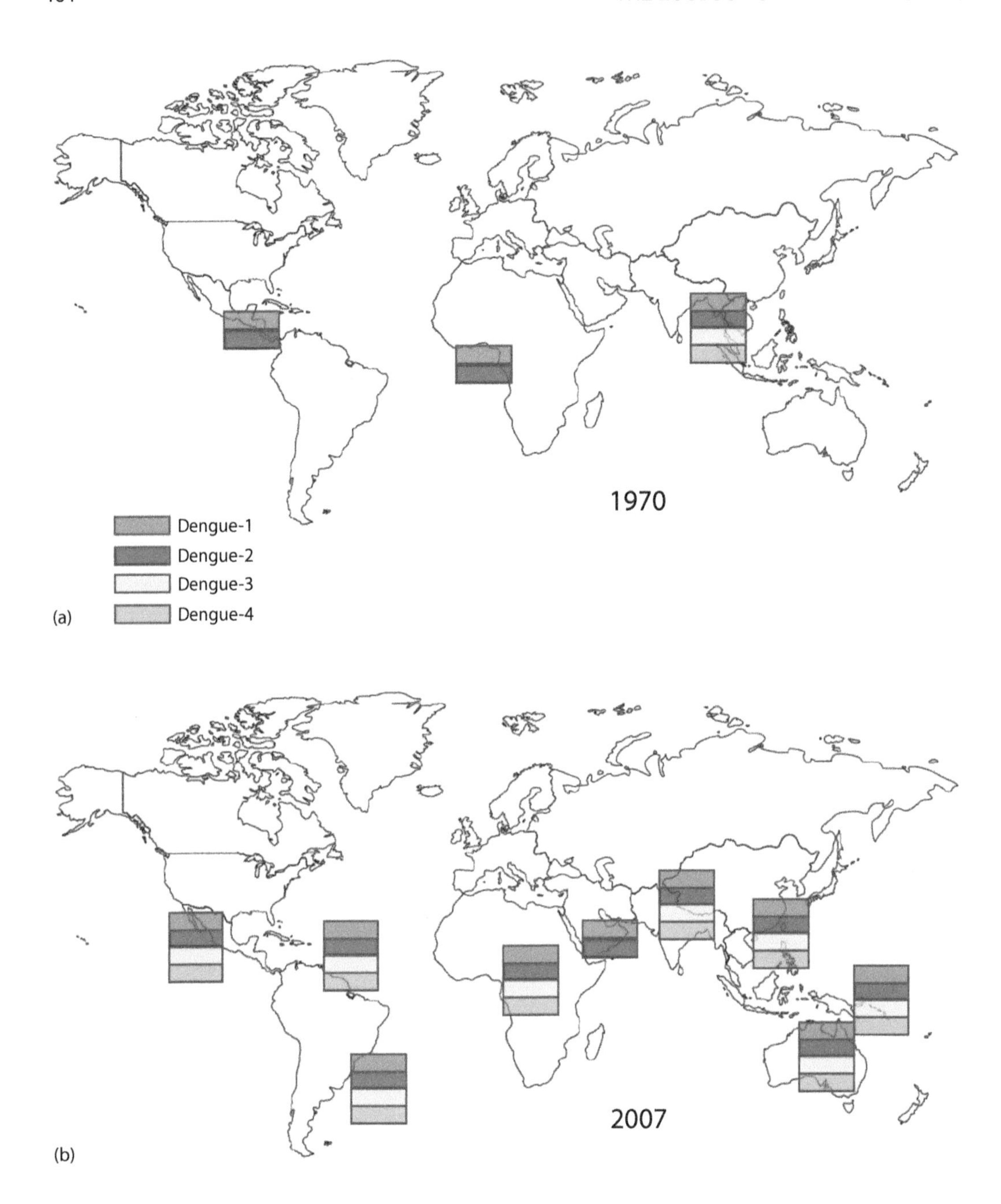

Figure 12.16 The global distribution of dengue virus serotypes, (a) 1970 and (b) 2007. (Adapted from Mackenzie, J.S. et al., *Nat. Med.*, S98–109, 2004.)

comes down with a high fever, often accompanied by vomiting, headache, general malaise and aching joints. Symptoms can last for days or even weeks. Infection provides lifelong protection against that particular serotype, but infection with a second or even third serotype is actually worse than the primary infection. This is perhaps because the antibodies to the earlier infection bind to the virus but do not kill it (as they presumably do to the initial infection if it invades the body a second time). The virus antibody complex is taken up by macrophages (that is their job), but the virus is able to replicate inside them and so infection is enhanced.

Personal protection is perhaps the best way to avoid dengue, although insecticide-treated curtains may be useful. Spatial repellents may be another possibility. These remain in development because although they reduced biting (by both *Aedes* during the day and anophelines at night), this was not sufficient to justify their use on a wide scale. In Cambodia, an emanator of metofluthrin reduced biting by approximately one-third, whilst four emanators surrounding a collector still only reduced landing rates by two-thirds (Table 12.5).

Other methods of control are the use of a strain of mosquito with a *Wolbachia* infection and the release of insects with a dominant lethal gene.

Wolbachia is a bacterium that affects a large number of arthropods. It has been estimated that between 20% and 70% of insect species are infected. A number of mechanisms have evolved to enhance its transmission. In some insects, the number of males produced is reduced (because it is only eggs that transfer the bacterium). In mosquitoes, due to a phenomenon known as cytoplasmic incompatibility infected males are unable to fertilise uninfected females, which then lay sterile eggs (Figure 12.17) because the cycle of mitotic division of the parental chromosomes in the first few divisions are out of sync and so division fails and the egg dies.

Table 12.5 The Effect of One or Four Emanators of Metofluthrin on Landing Rates of *Anopheles minimus* from Pailin, Cambodia

Intervention	Total	Collected	Median (IQR)	Mean/Hour	Rate Ratio 95% CI	*p*-value
Control	3693	4 (7)	1.72	1	–	–
Metofluthrin 1	1292	2 (5)	0.63	0.488	0.453–0.527	–
Metofluthrin 4	864	1 (3)	1.33	0.307	0.277–0.342	<0.001

Source: Charlwood, J.D. et al., *Med. Vet. Entomol.*, 30, 229–234, 2016a.

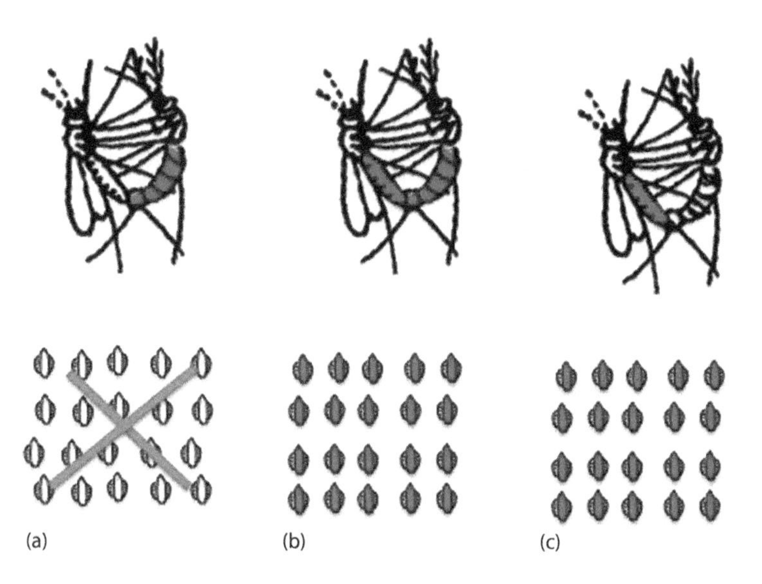

Figure 12.17 How *Wolbachia* spreads in a mosquito population. (a) When an infected male mates with an uninfected female, any eggs she lays are sterile and do not hatch. (b) Mating between a pair of infected individuals results in fertile infected eggs. (c) When an uninfected male mates with an infected female, her eggs are fertile and infected. This is an example of cytoplasmic incompatibility and demonstrates why the bacterium spreads rapidly through the population.

Infected females, on the other hand, that mate with uninfected males produce healthy, infected offspring, as do infected males mating with infected females (thus, this increases the proportion of infected eggs laid in the population). This can be used against the mosquito in that large numbers of *Wolbachia* infected males can be released into an uninfected population (effectively mimicking a sterile insect technique [SIT]) and the population is suppressed. Such an approach was used in Brazil as a response to the Zika virus outbreak.

Infected female mosquitoes are also less efficient vectors, and some groups (such as eliminatedengue.com) recommend that infected females are released for this to take effect. Such releases would, however, undermine any efforts to make use of cytoplasmic incompatibility because females in the wild need to be uninfected for the technique to work.

Release of Insects carrying a Dominant Lethal (RIDL) is a control strategy using genetically engineered insects that carry a lethal gene in their genome. Lethal genes cause death in an organism, and RIDL genes only kill young insects, usually larvae or pupae. Similar to how inheritance of brown eyes is dominant to blue eyes, this lethal gene is dominant so that all offspring of the RIDL insect will also inherit the lethal gene. This lethal gene has a molecular on/off switch, allowing these RIDL insects to be reared. The lethal gene is turned off when the RIDL insects are mass reared in an insectary with tetracycline in the larval water, and it is turned on when they are released into the environment. RIDL males are released to mate with wild mosquitoes, and their offspring die when they reach the larval or pupal stage because of the lethal gene. This causes the population of insects to crash. This technique is being developed for some insects and for other insects has been tested in the field.

There are, however, concerns about using tetracycline on a routine basis for controlling the expression of lethal genes. There are plausible routes for resistance genes to develop in the bacteria within the guts of GM-insects fed on tetracycline and from there, to circulate widely in the environment.

CHAPTER 13

Sampling Techniques

A number of books have been devoted to methods for sampling mosquitoes in the field, notably Service's *Mosquito Ecology: Field sampling methods*, recently updated by John Silver (2008). This 1400-page book describes a huge variety of sampling techniques, some of which, such as the collection of emerging insects using mosquito nets, have been used for many years (Figure 13.1).

There are, however, a few techniques that have proved useful that are not included, some of which are described below.

THE FURVELA TENT TRAP

Outdoor exposure is best measured in human landing collections (HLC), in which mosquitoes are caught attempting to bite the exposed lower legs of collectors sitting outside, because sitting outside is what people are doing anyway. They have been used extensively in the past, largely just to sample malaria vectors. Landing collections impose risks to the collectors, however, and risk-free collection methods are required. Indeed, the WHO recently recognised the necessity for novel sampling tools to conduct entomological surveillance outdoors (WHO, 2017). The objective of such surveillance would be to assess vector species composition, abundance and the time and place of biting. The primary requisites of such methods are that they catch mosquitoes that would normally bite outside and that the numbers caught should be comparable with landing collections. Something simple and inexpensive is also desirable.

The Furvela tent trap (Figure 13.2) fulfils these conditions. Mosquitoes are collected by a battery-operated CDC light trap (without the light) placed horizontally on the outside of a tent circa 4 cms from an opening approximately the size of the diameter of the trap body in the tent door.

NET FOR MEASURING THE OUTPUT OF MOSQUITOS FROM A MARSH—CLAIRFOND, MAURITIUS.

Figure 13.1 Ronald Ross in Mauritius using a mosquito net to sample emerging mosquitoes.

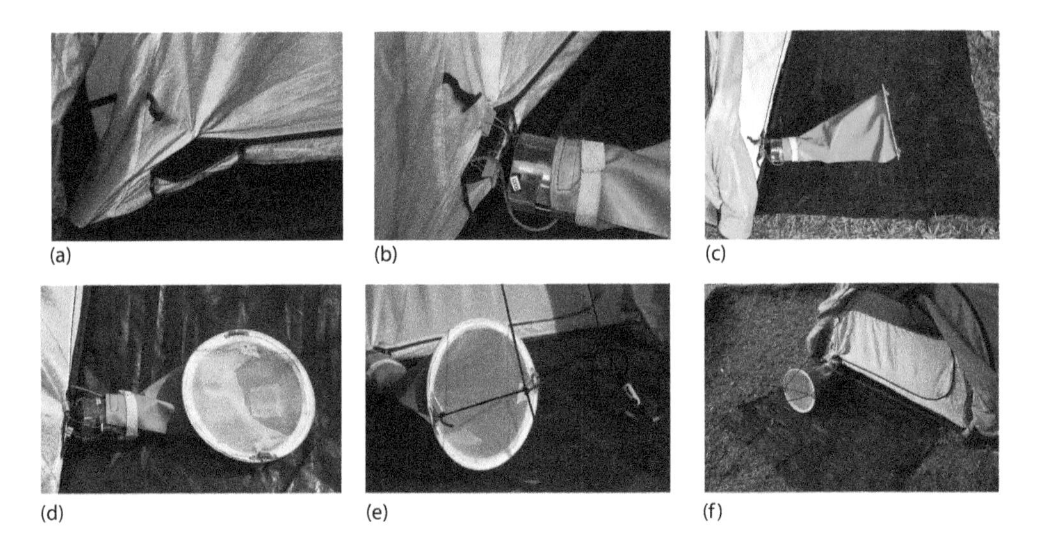

Figure 13.2 The Furvela tent trap. Installation procedure: (a) sew open the door of the tent to be approximately the diameter of the standard CDC light trap; (b) attach a CDC trap (without light or lid) to the tent approximately 4 cms from the opening, pass the lead to the trap into the tent where it can be switched on by the sleeper; (c) attach a conical netting collection bag (with optional rain cover) to the trap; (d) eyelets are sewn into the end of the bag; (e) these attach to guy ropes to keep the trap suspended horizontally in front of the opening; (f) the complete trap is ready for use. (From Charlwood, J.D. et al., *Peer J.*, 3848, 2017b.)

The trap is, therefore, made using available 'off-the-shelf' components, including the tent itself. Collections in the Furvela tent trap are closely correlated to CDC light-trap collections used to monitor indoor-biting mosquitoes (Figure 13.3). This makes it especially useful for the measurement of changes in indoor/outdoor ratios following the application of methods to control indoor-biting mosquitoes. In Massavasse, Mozambique, these ratios changed following the application of bendiocarb to the walls of a house (Figure 13.4).

Recently, the tent trap has been used as the monitoring tool in the evaluation of an intervention that targets outdoor-biting mosquitoes in Cambodia (Charlwood et al., 2016a).

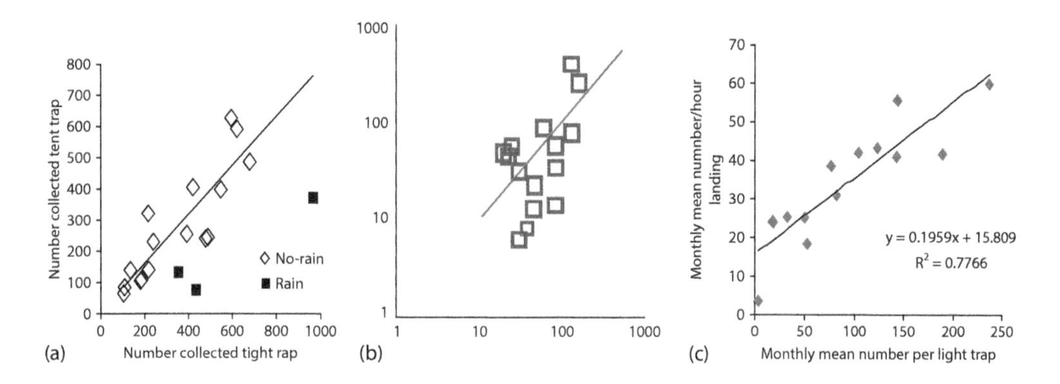

Figure 13.3 (a) Relationship between numbers of *A. coluzzii* collected in a Furvela tent trap and a CDC light trap in Okyereko, Ghana – the black squares represent nights when it rained; (b) relationship between numbers of *A. gambiae* in Furvela tent traps and light traps in Tanzania; (c) relationship between numbers of *A. coluzzii* in light traps and landing collections from São Tomé.

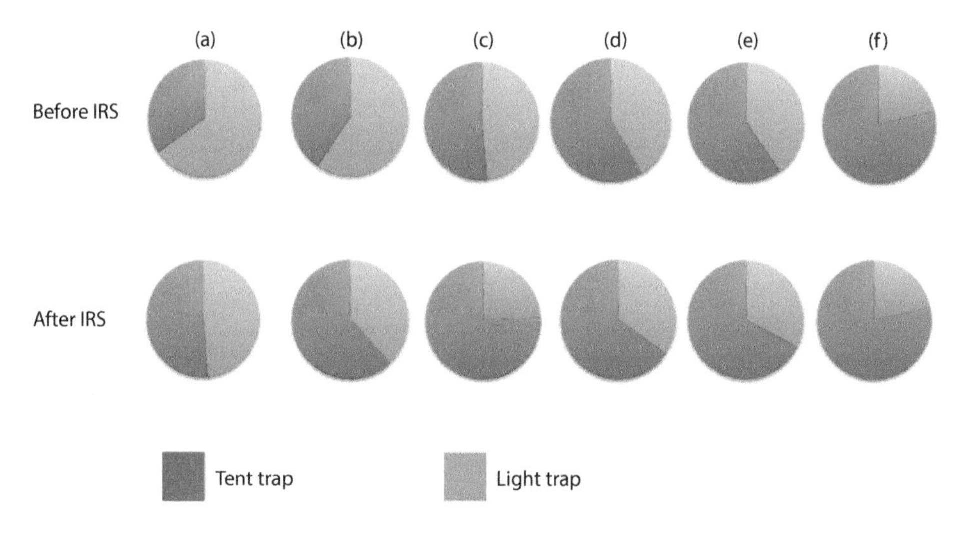

Figure 13.4 Indoor and outdoor ratios of the principal mosquitoes collected in Massavasse, Chockwe District, Gaza Province, Mozambique, before and after the walls were sprayed with bendiocarb at 0.4 gm ai m^2: (a) *A. funestus* (*n* before spray = 3754, *n* after spray = 1742), (b) *Culex* spp. – mainly *Cx. quinquefasciatus* (*n* before = 4267, *n* after spray = 2208), (c) *A. pharoensis* (*n* before = 2642 *n* after spray = 2078), (d) *A. arabiensis* (*n* before = 5406, *n* after spray = 512), (e): *Aedes muscides* (*n* before = 1085 *n* after spray = 395), (f) *Mansonia africana* (*n* before = 38173, *n* after spray = 12,054). (From Charlwood, J.D. et al., *Geospatial Health*, 7, 309–320, 2013b.)

THE SUNA TRAP

The Suna trap (Biogents, Germany) is a conical counterflow trap (52 cm high × 39 cm diameter) with a synthetic lure that mimics human odour to attract mosquitoes (Figure 13.5).

It has been used as a control tool against *A. funestus* on Rusinga Island, western Kenya, where it reduced transmission of malaria transmitted by *A. funestus* but was less effective against *A. gambiae* or *A. arabiensis* (Homan et al., 2016). A 12-volt battery drives a fan that sucks mosquitoes up through a tube with a 10 cm diameter opening into a collection bag. Netting can be placed between the tube and the fan so that collected mosquitoes are not damaged when caught. A non-return gate, that is activated when the fan is switched off, or when the tube is removed, means that under these circumstances the tube acts as a collection cage. The remainder of the base of the trap is perforated with numerous small holes through which the attractant is blown. There is also the possibility of adding carbon dioxide to the trap via an external source connected with a tube to an outlet in the trap. In the absence of gaseous carbon dioxide or dry ice, carbon dioxide can be generated using a sugar and yeast mixture in 2l of water in a 5l plastic bottle (Smallegange et al., 2010). In the trial in Rusinga Island, 2-butanone was included as a substitute for carbon dioxide. The trap is placed approximately 30 cms off the ground in places where mosquitoes might be expected to be present (Figure 13.6). In the trial in Rusinga Island the traps were suspended close to houses, and the battery was recharged by a solar panel supplied to householders (Homan et al., 2016).

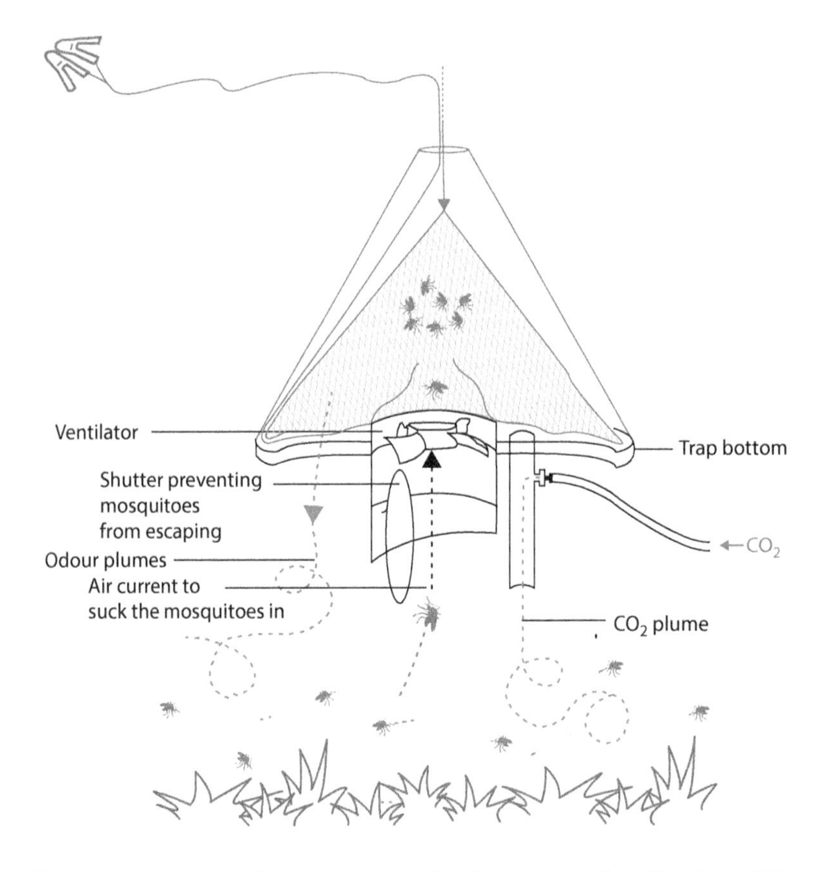

Figure 13.5 The Suna trap showing its components and method of operation. (Courtesy of Biogents.)

Figure 13.6 The Suna trap in operation in Eritrea with carbon dioxide generated by a yeast sugar mixture in a 5 L bottle.

The trap was less successful in Laos (Tangena et al., 2015) and when used in Eritrea (Figure 13.6) it failed to collect *A. arabiensis* when compared to either landing collections or the Furvela tent trap.

Nevertheless, traps like this show promise as a future control tool, especially in 'push-pull' systems. The need for carbon dioxide for it to work (unless that source comes from a nearby house) may, however, limit its use as a routine monitoring tool (Killeen, 2016).

SOME COMMENTS ON LIGHT TRAPS

In a recent meta-analysis from a multisite comparison, relationships between human landing collections (HLC) and light-trap collections were quite variable (Briët et al., 2015), although in São Tomé there was a good relationship between the two methods. Nevertheless, it was concluded that light traps provide a low-cost way of achieving high sampling intensities and provide valuable entomological measures of the impact of vector control on human exposure.

One light trap that was developed in Papua New Guinea combines a double bed net with an updraft light trap (Figure 13.7). This collected more mosquitoes than either a bed net trap on its own or a double net with a light trap suspended inside the net but collected fewer than comparable landing collections. The trap suffers from the fact that it needs a considerable space in which to be set up. (It is included here because the picture looks so good!)

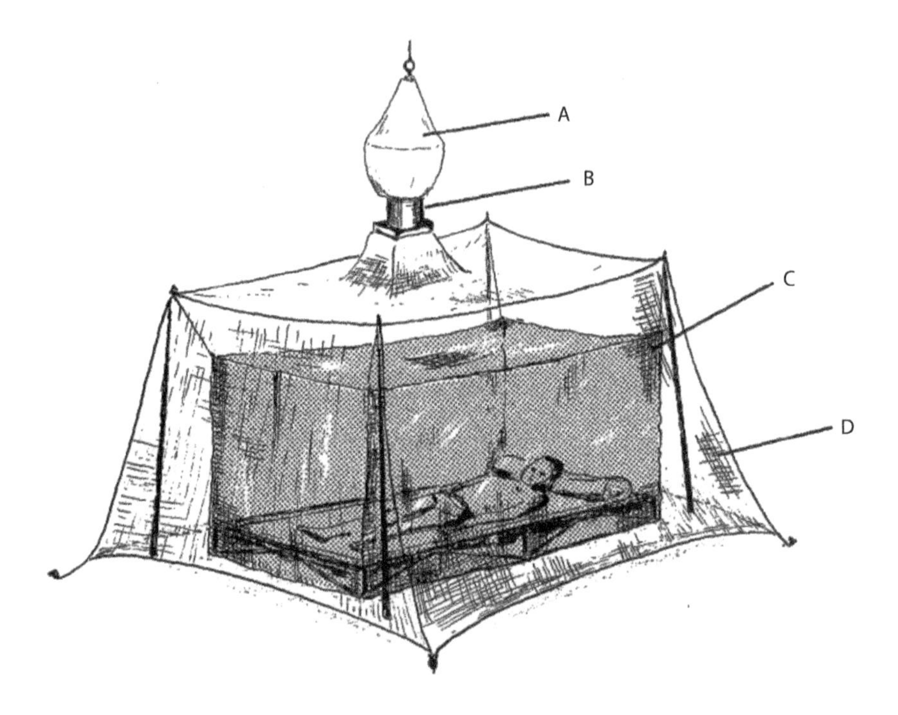

Figure 13.7 Combined bed net and updraft light trap. (From Charlwood, J.D. et al., *J. Vector Ecol.*, 11, 281–283, 1986c.)

EXIT COLLECTIONS FROM HOUSES

This is a simple and risk-free collection method for endophilic mosquitoes that does not require an energy source. Many endophilic mosquitoes leave houses at dusk. They are positively phototropic at this time and so leave via the lightest part of the house. This is the doorway if it is left open. By covering the door with netting, mosquitoes can easily be aspirated as they attempt to leave by a collector inside the house. The collection can be done without the need for a torch, because exit activity has generally ceased before it gets dark. The technique can be seen in a YouTube video: https://www.youtube.com/watch?v=SL8FeIuY1GM.

RESTING COLLECTIONS

Information on the blood-meal source and resting site of malaria vectors can inform models of transmission and potential control measures. In order to undertake assessments of feeding and resting behaviour, suitable collection methods for sampling insects are required. A lightweight, motorised collector, the Prokopack, has recently been developed by Vazquez-Prokopec (2009). It can be made with readily available materials (Figure 13.8) and is cheaper than the CDC backpack sampler.

The Prokopack has been shown to work well for the collection of indoor, peridomicilliary and outdoor resting mosquitoes. In Tanzania, it collected approximately 1.5 times as many indoor resting *A. gambiae* as did experienced collectors performing manual collections (Table 13.1). It is now probably the method of choice for sampling resting mosquitoes.

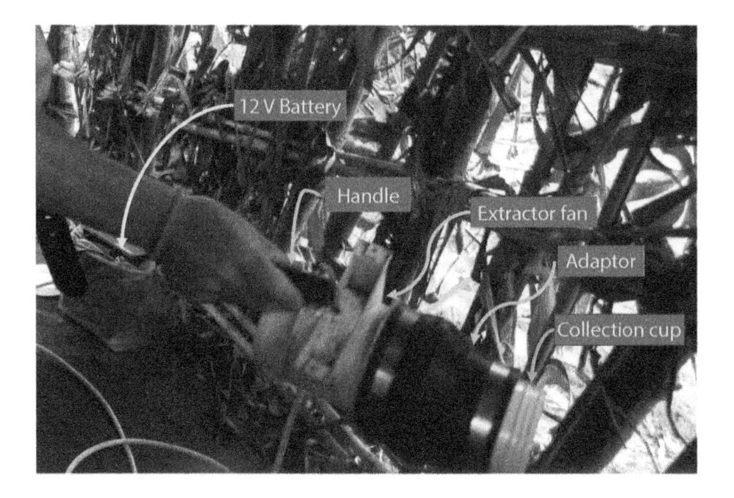

Figure 13.8 The Prokopack collector, showing the components, in use in Tanzania.

Table 13.1 Mean (s.d.) Number of Female and Male *A. gambiae* Collected by Manual Aspiration and Prokopak Collection According to the Order in Which the Collection Was Undertaken, Kyamyorwa, Muleba, Tanzania

	Aspirator		Prokopak	
Order of Collection	1	2	1	2
		Females		
Roof	2.36 (1.5)	1.09 (0.9)	3.49 (3.3)	3.33 (3.0)
Walls	4.76 (2.2)	0.94 (0.6)	6.71 (4.0)	3.04 (1.9)
		Males		
Roof	0.93 (1.3)	0.50 (0.5)	1.70 (1.4)	0.95 (1.1)
Walls	1.97 (2.6)	0.67 (1.2)	4.89 (5.0)	1.17 (1.0)

Source: Charlwood, J.D. et al., *Peer J.*, 5155, 2018.

It is often of interest to estimate the total number of resting mosquitoes in a particular site (such as a house or animal shelter). There are a number of possible ways to do this; one might be to calculate the Lincoln Index following the release of marked individuals, but this has the disadvantage that it is not certain that the marked insects distribute themselves homogeneously within the resting site or that they are not affected (in the short term at least) by the marking procedure. Another might be to undertake pyrethrum spray collections, but not all sites are suitable for this procedure. An alternative is to undertake removal sampling as described by Southwood (1978). The principle of removal sampling is that a known number of animals are removed from the resting site in sequential samples, thus affecting subsequent catches. The rate at which collections decline is directly related to the size of the total population and the number removed. For removal sampling to function adequately a number of assumptions must be met: the catching procedure must not affect the probability of an animal being caught; the population must remain stable during the catching period; and, most importantly, the chance of being caught must be equal for all animals. It is also important that the sampling effort in each round of collection is the same. In other words, the time spent performing each sample round must be the same and no extra effort must be involved in any particular sample. A relatively large proportion of the population must also be caught to obtain reasonably precise estimates. Numbers collected on each trapping interval must decline for estimates to be meaningful (Charlwood et al., 1995d). Zippin's

(1956, 1958) method, based on maximum likelihood, which provides an estimate of the standard error, can be used to estimate the total population. As described by Southwood (1978), the total catch T = *n1* + *n2* + *n3* + *n4* + *n5* where *n1*...*n5* are the number of insects caught on each respective round of sampling. Then the value of $\sum_{i=1}^{k}(i-1)y$ is found, where k = the number of samples and i = 1 and y_i = the catch on the *i*th occasion. Thus, following the example given by Southwood (1978), supposing sampling that 65, 43, 34, 18 and 12 insects (a total of 172 insects) are caught in five rounds of sampling, then $\sum_{i=1}^{k}(i-1)y$ is

$$(1-1)65+(2-1)43+(3-1)34+(4-1)18+(5-1)12$$

$$=0+43+68+54+48$$

$$=213$$

Following this, the ratio R is determined where

$$R=\sum_{i=1}^{k}(i-1)\,yi$$

Thus, $R = 213/172 = 1.238$
and

$$R=\frac{q}{p}-kq^{k}/(1-q^{k})$$

where p = the probability of capture on a single occasion and $q = 1-p$ and the estimated size of the total population is

$$\tilde{N}=T/(1-q^{k})$$

Zippin (1956) provided a set of graphs for k = 3,4,5 or 7 (Figures 13.9 and 13.10) which avoids the mathematics of these last steps.

For k = 5 and R = 1.24 the value of $(1-qk)$ is 0.85 derived from Figure 13.9C so that, in our example, the total estimated population $\tilde{N}$ is 172 ÷ 0.85 = 202.

The standard error of $\tilde{N}$ is given by:

$$\text{S.E. of } \tilde{N}=\sqrt{\frac{\tilde{N}\left(\tilde{N}-T\right)T}{T^{2}-\tilde{N}\left(\tilde{N}-T\right)\left[(kp)^{2}/(1-p)\right]}}$$

where p is read from Figure 13.9. In the present example $p = 0.33$.

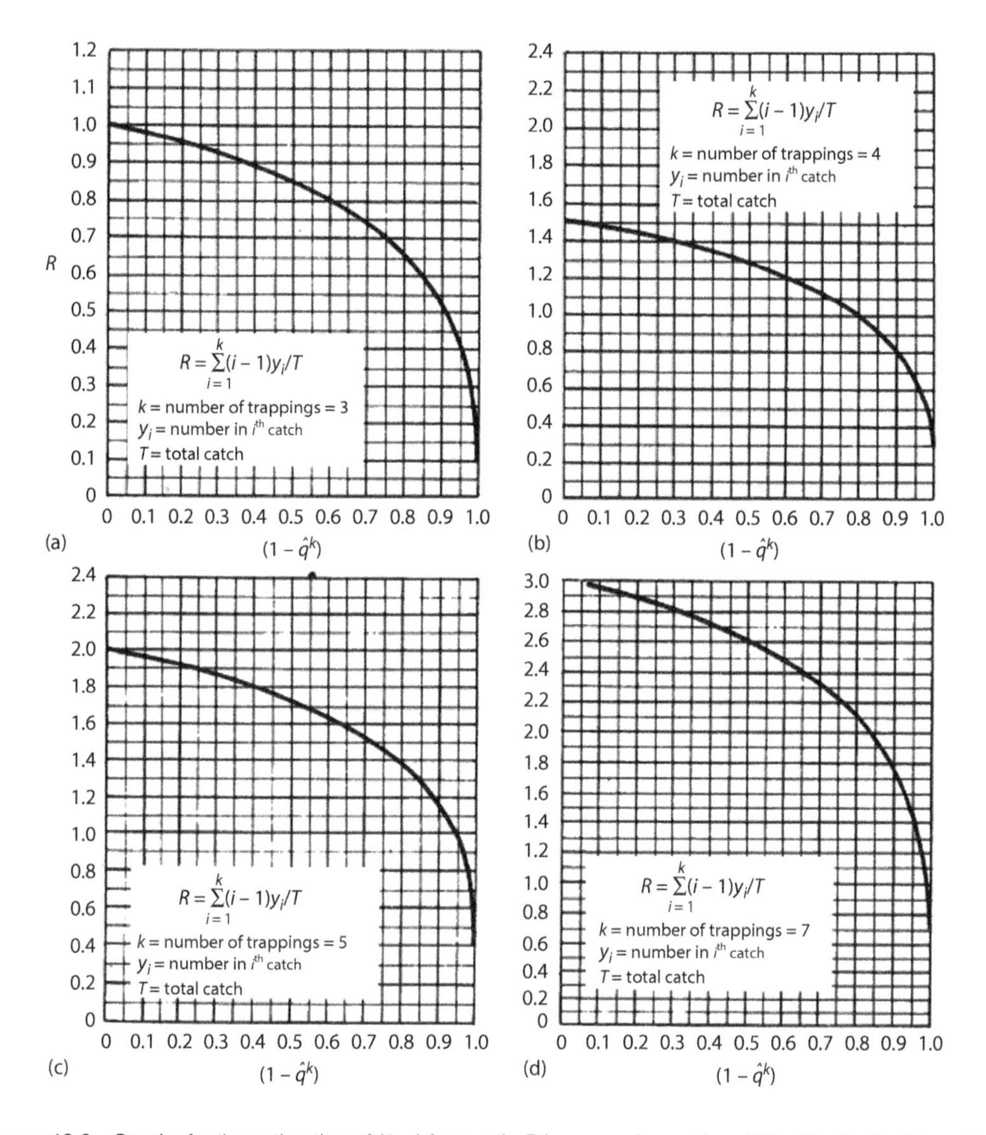

Figure 13.9 Graphs for the estimation of (1-*q*) from ratio *R* in removal sampling. (After Zippin, C., *Biometrics*, 12, 163–189, 1956.)

$$\text{Thus, in the example above S.E. of } \tilde{N} = \sqrt{\frac{202(202-172)172}{172^2 - 202(202-172)\left[(5\times 0.33)^2/(1-0.33)\right]}}$$

$$\text{S.E. of } \tilde{N} = 14.46$$

And 95% confidence intervals of the estimate is $202 \pm 2 \times 14.46 = 202 \pm 28.9$.

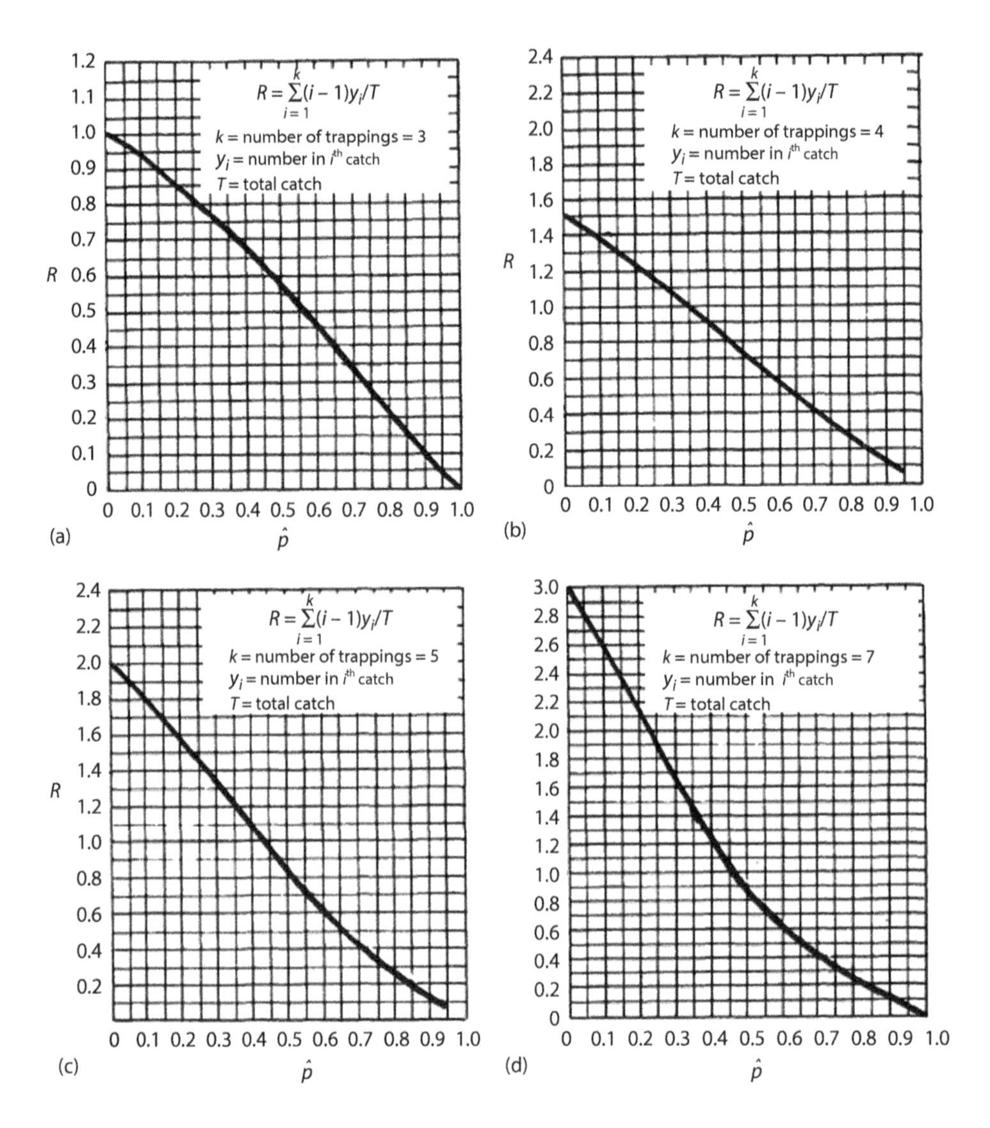

Figure 13.10 Graphs for the estimation of p from ratio R in removal sampling. (After Zippin, C., *Biometrics*, 12, 163–189, 1956.)

WINDOW TRAPS

In many areas electricity for recharging batteries used in light or tent traps may be a problem. As their name implies, window traps are passive traps that catch mosquitoes attempting to leave houses through the window (in other words via the lightest part of the room). Traps are basically cages (that can be made to fit the window), one side of which is converted into an entrance funnel or slit. A sleeve is sewn into one side, allowing the mosquitoes to be removed with an aspirator (Figure 13.11). Many of the mosquitoes that attempt to leave the house (either at dawn or during the night) are attracted to the light from the window and consequently are caught as they try to escape.

They provide information on the endophagic but exophilic fraction of the population when anti-mosquito measures are not in use or information on the number entering and leaving when such measures (such as nets) are in use.

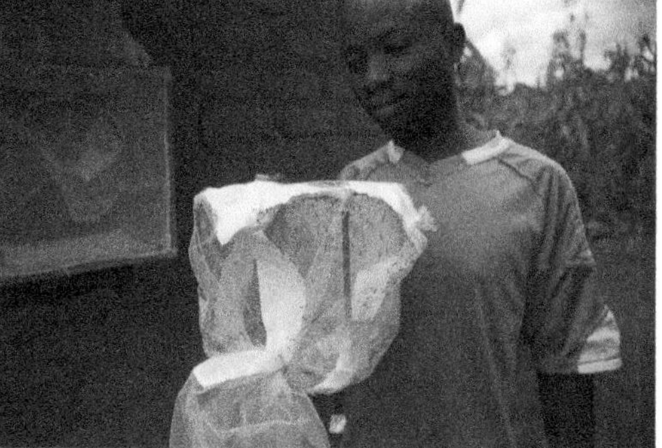

Figure 13.11 Window trap as used in Kyamyorwa. Note the inverted cone 'lobster pot' design that allows mosquitoes to enter but not leave. Entomological assistant, Kulwa, holds the collection from a single morning.

When nets are used, mosquitoes that have entered a house but been unable to feed will attempt to leave during the night. The efficiency of window traps then varies according to moon phase. In periods when there is no moonlight the cue to leave provided by the trap is absent.

From a series of collections made between 23 October 2014 (ISO week #43) and 14 March 2015, 70 window-trap and 17 Furvela tent-trap collections were made. Changes from the two collection methods were similar (r = 0.93, $p < 0.001$) (Figure 13.12). Numbers in both collections declined during this period (from a peak of 1018 in the window trap and 243 in the tent trap). The house was sprayed with primiphos-methyl on 23 February 2015, and numbers in the window trap fell to zero but numbers in the tent trap persisted, albeit at a very low density.

Anopheles gambiae was caught in approximately equal numbers from window trap and tent trap whilst other species were collected in greater numbers in the tent trap compared to the window trap, (Figure 13.13) indicating that the tent trap was, indeed, collecting exophagic mosquitoes.

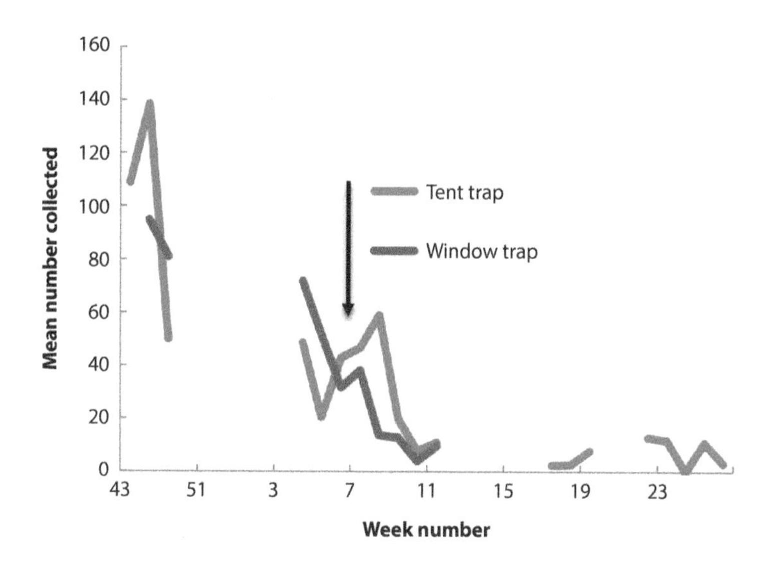

Figure 13.12 Exit window-trap and Furvela tent-trap collections of *A. gambiae* s.l. from a bedroom in the village of Kyamyorwa, Muleba District, Kagera Region, Tanzania. The arrow marks the time when the interior walls of village houses were sprayed with pirimiphos-methyl (Actellic) at 1g ai per m² (prior to the spray cross-correlation between mean weekly numbers in the window trap and numbers in the tent trap r = 0.93, $p = >0.001$). (From Charlwood, J.D. et al., *Peer J.*, 3848, 2017b.)

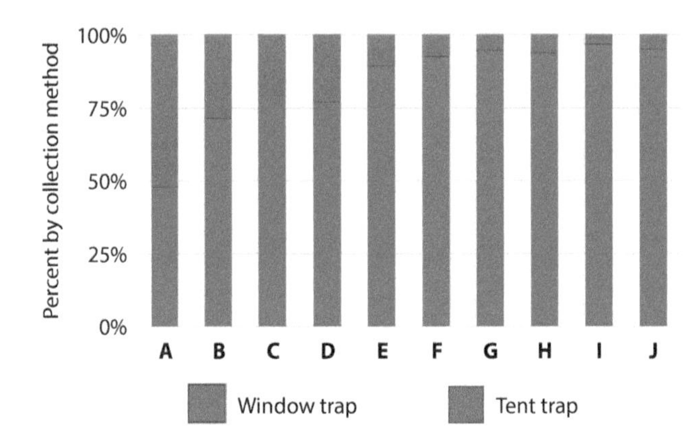

Figure 13.13 Window-trap/tent-trap ratios of mosquitoes from Kyamyorwa, Muleba District, Tanzania: (a) *Anopheles gambiae* sl (n = 10512) (b): *A. funestus* (n = 81), (c) *A. coustani* (n = 27), (d) *A. zeimanni* (n = 282), (e) *Cx. quinquefasciatus* (n = 471), (f) *Cx. tritaeniorhynchus* (n = 21), (g) *Coquelettidia fuscopennata* (n = 130), (h) *Mansonia* spp. (n = 737), (i) *A. squamosus* (n = 81), (j) *A. pharoensis* (n = 49). (From Charlwood, J.D. et al., *Peer J.*, 3848, 2017b.)

EXPERIMENTAL HUTS

There are a number of reasonably well-known designs of experimental huts available in a number of books. One hut that has yet to make it to these is the Furvela experimental hut used to measure exit direction of endophilic mosquitoes (Figure 13.14).

The hut is an octagonal roundhouse or *rondável* 3.2 m in diameter and 3.15 m high (Figure 13.14). Each side consists of a hinged 80 × 160 cm panel made with reed, a local building material much used in this part of Africa. Above each of the panels there is an 80 × 35 cm opening that can be

Figure 13.14 Experimental hut used for the assessment of exit behaviour of *A. funestus* and *A. gambiae* from Mozambique. (From Charlwood, J.D. and Kampango, A., *Malar. World J.*, 3, 3, 2012.)

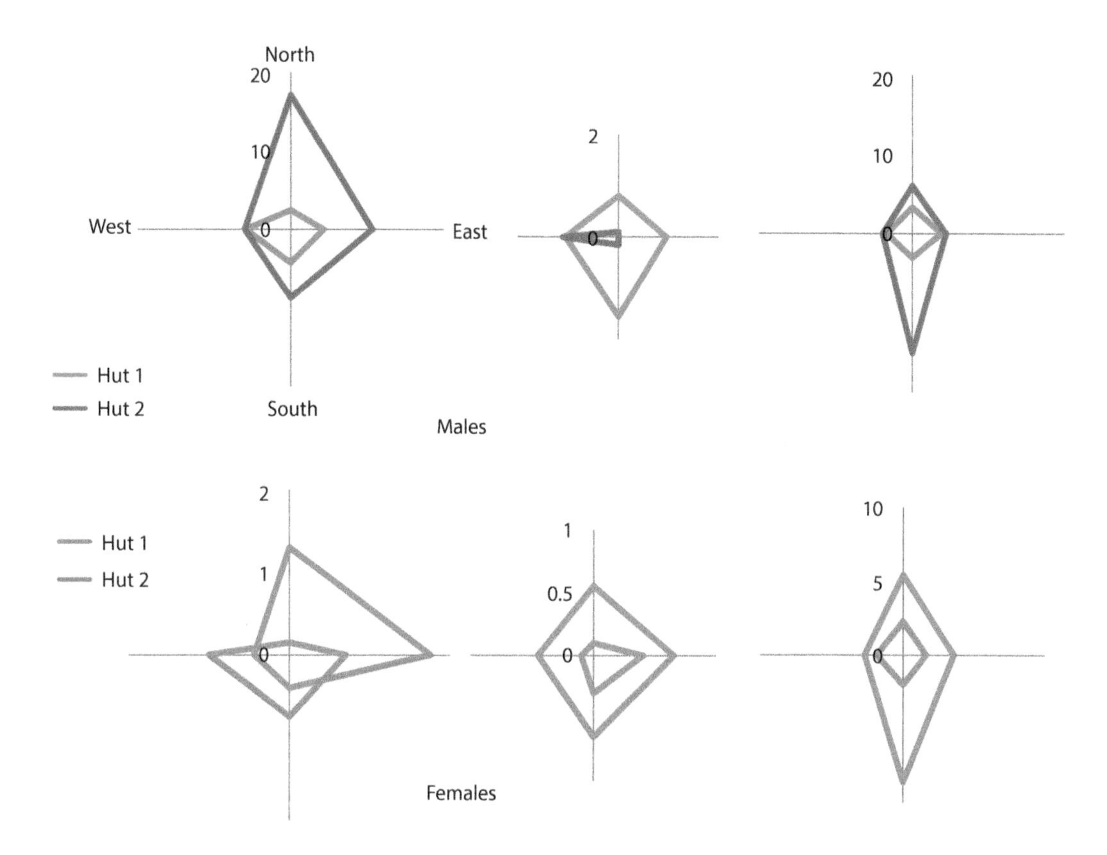

Figure 13.15 Exit direction of male and female *A. funestus* from two experimental huts. (From Charlwood, J.D. and Kampango, A., *Malar. World J.*, 3, 3, 2012.)

closed with a hinged thatched palm leaf (*makuti*) shutter. Above this there is a 2–4 cm gap through which mosquitoes can enter, even when the rest of the house is closed. In other circumstances this gap could be sealed with netting or some other material or could be isolated from the main part of the house by including a netting ceiling between the shutter and the roof. Thus, any or all of the sides can act as a door for the collectors and an exit route for insects. The floor of the hut was made of cement, which extends 50 cm beyond the walls, terminating in a 15 × 20 cm water-filled moat, designed to exclude ants.

Exiting mosquitoes were collected in mosquito nets placed horizontally against the different sides in which the louvres had been left open (Figure 13.15). The greatest exodus of gravid females occurred when relative humidity reached 85% and air temperature was 26.5°C. Most of the exit activity of unfed and fed females occurred during the period between 04:30 and 05:30 (i.e., before sunrise), whilst the greatest exodus of gravid females and males occurred during the first period of collection (17:30–18:30). The flexibility of the design allows for many other aspects of mosquito behaviour to be examined.

DRONES AND PHONES

Two tools that have yet to be fully exploited by entomologists working on disease vectors are 'drones and phones'. Despite their relatively high cost, many people now own 'smartphones', which can be connected via their local service provider (i.e., phone company) to the Internet.

Census data and mosquito collection data can be collected and sent across the web. This is being done but there are many other ways in which mobile phones can be used. This can be a two-way information channel because information on the location can be provided by the sender. A database will gradually evolve under these circumstances. The use of a camera on the phone to take photographs of houses and, more especially, breeding sites might enable a better description of habitats that are used by different vectors, leading to their easier location, and perhaps, elimination.

Drones too can play an important role in the location, identification and control of hard-to-reach breeding sites. They can provide important information of geography of an area that feeds into landscape ecology. What the particular features of any landscape are that give rise to mosquito populations remain to be fully explored. Drones and phones can help.

MOSQUITO MOUNTING AND PRESERVATION

In this day and age there is less attention given to the mounting and preservation of specimens for identification purposes than there was in the past. There is an attitude, perhaps, that a PCR from a piece of leg will provide more profound information on a mosquito than any morphological description. In the case of species complexes that have had their genome described, this is certainly true. The caveat is that they need to have their genome analysed, and this is not available for many species. Presently, it is possible to work in the field without needing to look particularly hard at the mosquito that you collect. The information on species identification, age (perhaps), and infection status can be determined without the need of a microscope. But, unless you get 'up close and personal' and really look at these mosquitoes, you will miss out on possible insights. The utility of a well-mounted, well-preserved, correctly identified reference set of mosquitoes available in all entomology laboratories cannot be overstressed.

Much of what follows is an amended version of the section on mosquito preservation by Belkin (1962) in his book *The Mosquitoes of the South Pacific* (Volume 1).

The quality of the material, its preservation and preparation, and the geographical and ecological data accompanying it determine to a large extent what a taxonomist can do with it and how reliable his data and inferences will be.

Material without accompanying data is worse than no material at all. It is, therefore, most important to have specimens properly labelled. The minimum is the location, date of collection, method of collection and name of collector. Other information, such as ecological notes, can be annotated by the use of a simple numbering system (such as year, month, day followed by an individual number). Thus, 25 specimens collected on 25 December 2015 will have the code 151225-1, 151225-2 until 151225-25. A collection from a given habitat is given a number to correspond with the data entered on a standardised record sheet, card or database file. The information available then includes the general data and a record of the material, its preservation and its rearings. The advantage of the lot system is that species collected together in the same habitat and at the same time are associated. The advantages of a uniform system are that all the material is associated throughout and that the data pertaining to the collection are recorded once.

Adult mosquitoes are so fragile, and in the Culicinae so easily denuded, that good material can be obtained only by picking up individual specimens with some type of suction tube or by inverting a vial on a resting mosquito. Material from light traps is often badly damaged and impossible to study, although a person familiar with the local fauna can usually readily identify even the majority of the females.

Mounting with Micropins

It has been customary to study adult mosquitoes preserved in the dry state and mounted either on a micropin (minuten) or on a heavy paper point attached to an ordinary insect pin (No. 3 or 2). Usually only freshly killed specimens make good micropin mounts. The specimen is best mounted by inserting the micropin between the mid-coxae with the specimen lying ventral side up; the micropin should go deep into the thorax but should not project through the mesonotum. Prior to pinning the specimen, the micropin should be mounted on a piece of pith or similar material attached to a No. 3 or 2 insect pin. The specimen is oriented vertically with the head up and the legs down and towards the large pin (Figure 13.16). The advantage of the micropin method is that all sides of the specimen are available for study; the disadvantages are that the micropin frequently corrodes and that the specimen tends to move on the pin after repeated handling and quite frequently becomes detached; furthermore, there is usually more shrivelling of the specimen than in the paper point method.

Mounting with Paper Points

Paper points should be punched or cut from thin but stiff card. Specimens should be freshly killed so that the legs and wings will not break when handled and can be moved or gently blown into the desired position. They can be attached in any desired position by bending the tip of the paper point if necessary before applying the adhesive. The safest orientation of the specimen is in a horizontal position with the legs directed toward the pin. The disadvantage of the paper point method is that one side of the thorax is not visible; however, this is usually not a serious problem if the specimen is mounted in such a manner that the tip of the paper point does not extend beyond the upper end of the sternopleuron. To create the paper point mount, the tip of the upper side of the paper point, already attached to the pin, is attached to a glass rod or matchstick dipped in glue and is then affixed to the properly oriented specimen under the low power of a stereoscopic microscope.

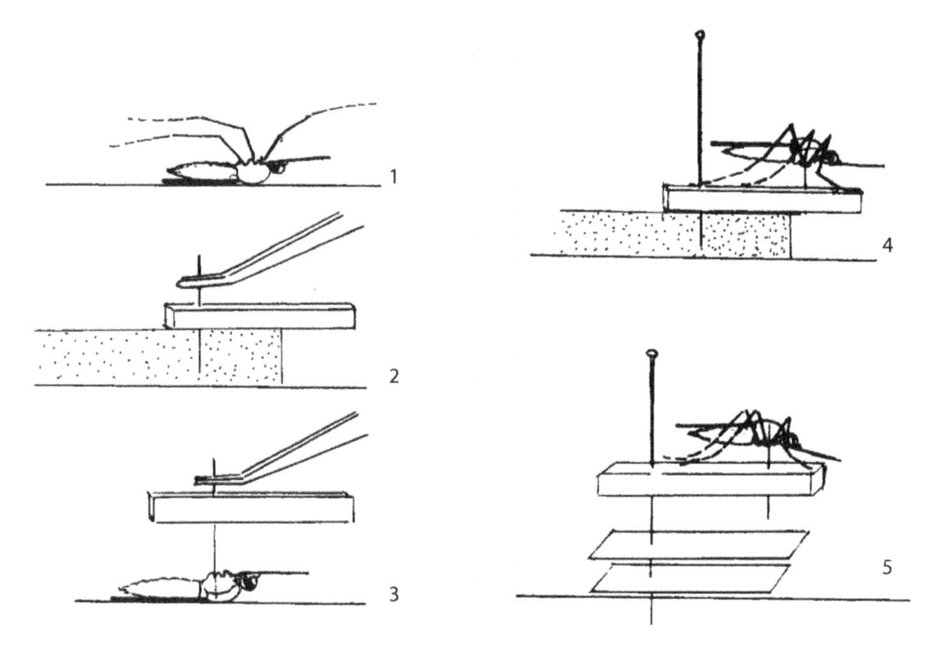

Figure 13.16 Mounting mosquitoes with micropins. Easier said than done! (From WHO, 1975.)

Mounted material is then kept in a tightly sealed specimen box. Care must be taken, particularly in the tropics, to dry the specimens; if this is not done, they will become mouldy and covered with fungus, making them worthless for study. It is advisable to keep a bag of silica gel crystals in the box so that the crystals are not loose and to protect the stored material from insect pests, particularly ants. For this, a strip of slow-release insecticide (such as a Vapona strip or a dog's flea collar) can be placed in the box.

COLLECTION OF IMMATURE STAGES

A much larger percentage of species can be collected as immature stages than as adults, because usually only the females that are attracted to bait or traps are encountered during surveys; these constitute only a small fraction of the mosquito fauna of almost every area in the world. Adults of many species are known only as a result of rearing of collections of immature stages, and association of the immature stages with the adults cannot be accomplished without rearing. Therefore, all or most of the immature stages collected should be brought alive into the laboratory in suitable containers with water from the habitat in which they were collected. If time is available it is a good practise to make mass rearings of part of every collection, to preserve some pupae and larvae of all instars and to make a few individual rearings from each lot, and all or some of the sublots as desired.

Frequently, only nearly mature fourth instar larvae and pupae are collected, while the younger instars and eggs are discarded. Following this time-saving custom may result in missing one or more species. Often, it is just not possible to rear all the larvae. Young instars can be usefully identified by PCR if the primers are available, as they are for the *A. gambiae* complex. The results can provide insights into the ecology of the two freshwater members of the complex, as they did in Ifakara, Tanzania (Table 13.2).

In this case, there was an excess of fourth instar larvae among the *A. gambiae* compared to *A. arabiensis*, leading to a significant difference overall. (The Chi-sq value is 10.1875; the *p*-value is 0.017.) There was no difference between the numbers collected for the first three instars. (The Chi-sq value is 0.11; the *p*-value is 0.95.) This would either mean that competition between the two species only occurred in the last larval instar or that a perturbation to the population had occurred in three instars (i.e., four or five days) prior to the collection. Five days prior to the collection a heavy rainstorm had eliminated a dense population of *A. arabiensis* from a nearby village. Assuming that the storm also affected the water under study, then the result would indicate that the *A. gambiae* were better able to resist the effects of the storm than the *A. arabiensis*.

It is quite easy to rear most species, except some temporary pool breeders, from the egg or from the first instar; in surveys an attempt should be made to rear at least a part of the younger instars. With fourth instar larvae, and, of course, pupae, no food need be added; with younger instars, small amounts of finely ground Tetramin© 'fish food' should be added to the water. It is easier to overfeed than starve larvae. If uneaten food remains after 24 hours, no more should be given and the condition of the larvae assessed.

Table 13.2 Number of *A. gambiae* Complex by Species and Instar Collected from Ifakara, May 1994

	Instar			
Species	**1st**	**2nd**	**3rd**	**4th**
An. gambiae	15	20	24	40
An. arabiensis	13	20	24	12

Source: Charlwood, J.D. and Edoh, D., *J. Med. Entomol.*, 33, 202–204, 1996.

After the adults emerge and before they are killed, they should be kept in a cage or in a covered jar of considerable size for at least 24 hours, preferably longer, to allow them to harden. A small wad of cotton saturated, but not dripping, with a 10% sugar solution or a moistened dried fruit (prune, raisin, apple) placed on the top of the jar often helps to produce sturdier and better-hardened specimens.

Individual rearings are essential to establish correlations between the immatures and the adults and between the two sexes of the adults. They are very simple to make but are tedious and time-consuming. Individual rearings are usually made from field-collected mature fourth instar larvae.

In mature larvae, feeding is not necessary if the original water from the breeding site is used; if the larvae are younger, a small amount of food should be added from time to time as in mass rearings, and it may be necessary to add distilled water to compensate for evaporation. Cast-off larval skins show the hairs used for identification purposes much more easily than do whole larvae. As soon as possible after pupation occurs, the cast larval skin is picked up with a lifter, briefly washed in a dish of tap water and transferred with its label to 75%–85% ethanol in an Eppendorf tube or preferably is mounted on a microscope slide for taxonomic evaluation. They can be mounted, in Berlese mounting medium, onto slides and covered with a cover slip that is then sealed with nail varnish to make in effect a permanent mount. In order to prevent air bubbles under the coverslip a drop of Berlese is added to the underside of the coverslip before it is placed on the specimen.

The pupa, with the remaining two labels, is then transferred with a pipette to a vial containing a small amount of tap water. Belkin (1962) continues, 'I normally use a loose plug of cotton to stopper the vial and insert the labels between the plug and inner wall of the vial. As soon as possible after emergence, the adult is picked up with a suction tube and transferred to a clean vial, which is provided with a strip of moist filter paper and a loose cotton plug; one of the remaining labels accompanies the adult. The pupal skin is picked up with a lifter, briefly washed in a dish of tap water, and transferred with the last label to 75%–85% ethanol in an Eppendorf tube or similar. The adult should not be killed for at least 24 hours to allow hardening of the cuticle to occur'.

CHAPTER 14

Laboratory Studies

WING MORPHOMETRICS

There is much to be learned from a mosquito's wings. In females, at least, wing length serves as a proxy for mosquito mass and mosquito mass may have important consequences for fitness and vectorial capacity. The distribution of sizes among emerging adults indicates whether or not competition between larvae has occurred, there being a skew towards larger insects if this was the case (Figure 14.1).

Competition appears to be greatest among tree-hole mosquitoes. Among anophelines, competition might be expected especially among larvae living in more temporary habitats, such as puddles. However, even with small pool breeders such as *A. gambiae*, the distribution among emerging adults tends towards normal. In other words, there is little evidence that competition between larvae is important.

In many animals, larger specimens are at a selective advantage compared to smaller ones. Although there is little evidence of competition among larvae, smaller emerging insects may have a lower survival rate as adults than larger females. They may also lay fewer eggs per oviposition cycle. Together, these reduce fitness and vectorial capacity. As a consequence, the effect of mosquito mass (i.e., wing length) on survival has been investigated in a number of studies. In São Tomé, the mean wing length of virgin females was significantly smaller than all other age groups combined (Figure 14.2). This implies that these small females were either dying or were leaving the study area.

In the majority of cases, however, little or no effect of size on survival has been demonstrated. For example, in Tanzania the cessation of breeding (and so the absence of newly emerged females) led to a month-long decline in the population. Insects alive at this time had survived for four weeks or more. If larger insects survive better than smaller ones one would expect that the mean wing lengths would rise. This was not the case; in fact, the wing length of these extreme survivors was no greater than that of insects collected at the start of the study (Figure 14.3).

Similarly, in Mozambique there was no evidence that the recently emerged *An. funestus* were smaller than gravid females, only 1.3% of the variation found in wing length being attributable to female state (d.f. = 1; F = 135.581; p-value < 0.01). In the warmest months, when both age groups were at their smallest, young females were, however, significantly smaller than gravid ones.

Wing size affected the mating pattern and pre-gravid feeding among *A. gambiae* from São Tomé (see Section 14.1). It has also been used to distinguish between members of different populations of the same species. For example, despite being genetically similar, populations of *Ny. darlingi* from two areas of Brazil had different wing-size distributions (Figure 14.4).

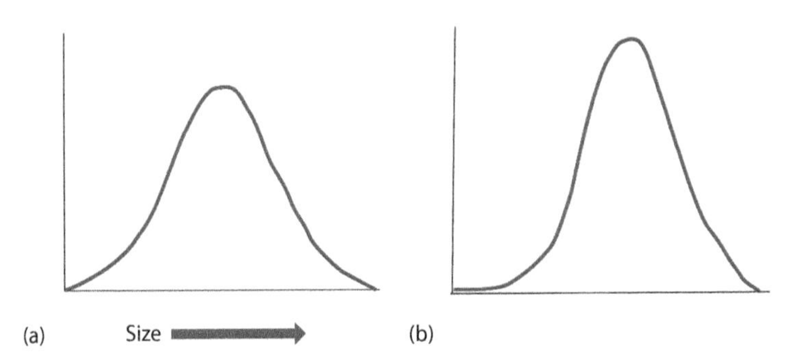

Figure 14.1 (a) The normal distribution and (b) a distribution with a negative skew. In the latter case the distribution arises because of a lack of small mosquitoes emerging.

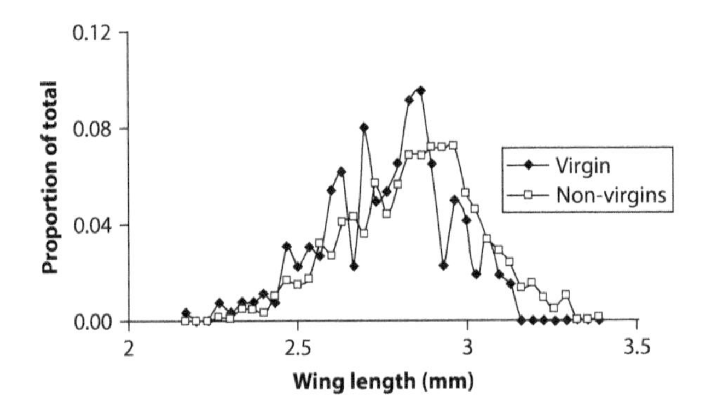

Figure 14.2 Mean wing length of virgin *A. coluzzii* and all other age groups combined collected from light traps in São Tomé ($n = 262$ for virgins and 1091 for other age groups). (From Charlwood, J.D. et al., *Malar. J.*, 2, 7, 2003c.)

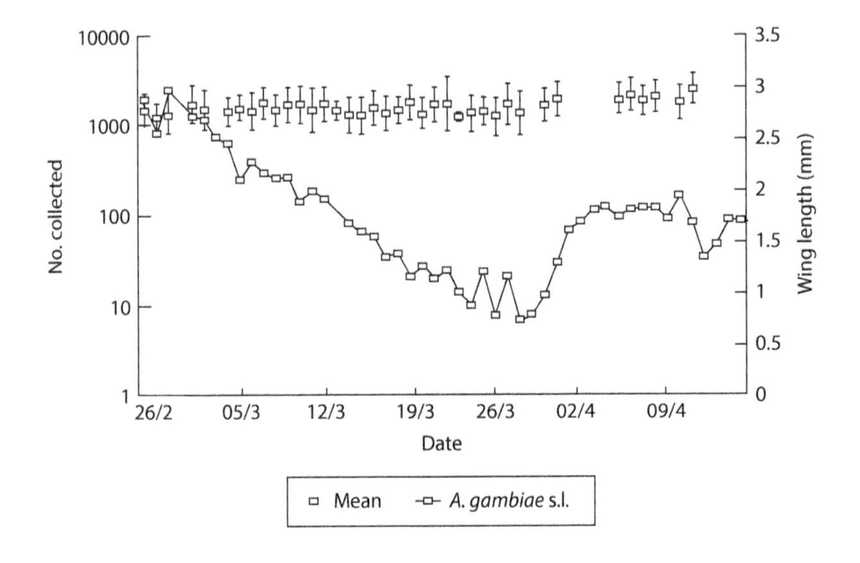

Figure 14.3 Wing length and number of *Anopheles arabiensis* collected from sentinel light traps between February and April 1991, Namawala village, Kilombero District, Tanzania. Error bars indicate standard deviation. (From Charlwood, J.D. et al., *Bull. Entomol. Res.*, 85, 37–44, 1995c.)

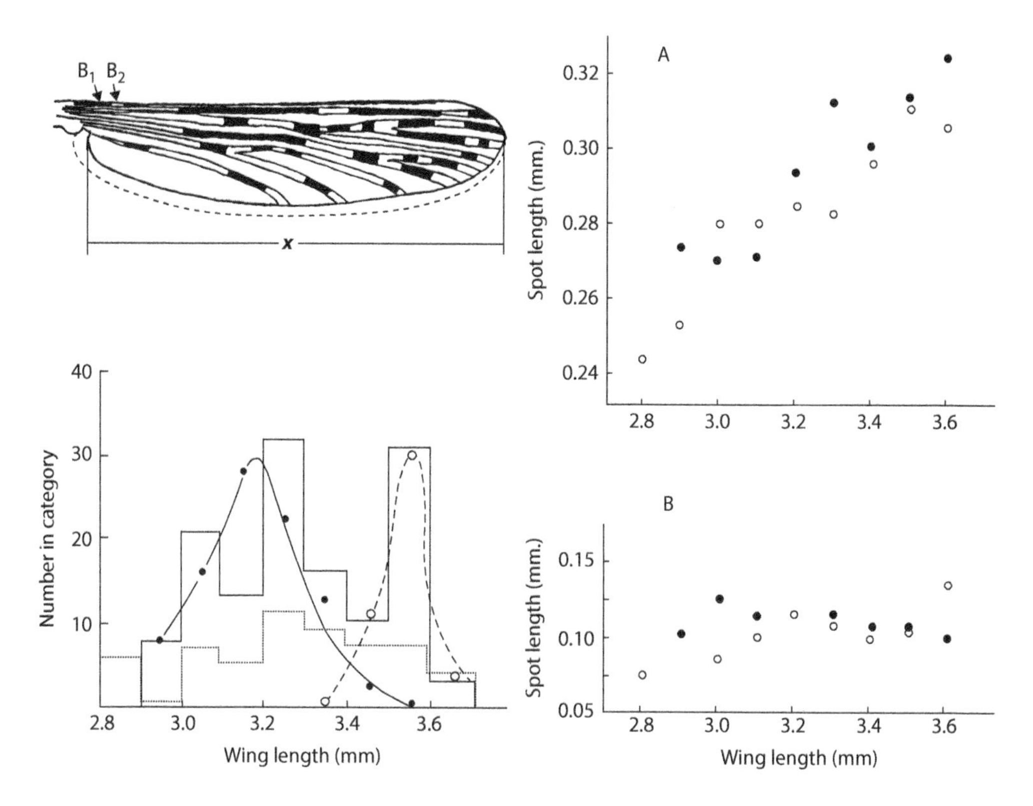

Figure 14.4 Wing lengths and spot ratios of *Ny. darlingi* collected from two sites in Brazil, Aripuana in the Matto Grosso (solid line) and 150 kms along the Manaus-Caracarai highway in Amazonas (dotted line). The two curves represent the normal distributions from the two sites using a graphical inflection method. Scatter plot A describes the relationship between spot length B_1, and scatter plot B describes the relationship between spot length B_2 and wing length (solid circles for Aripuana specimens and open circles for Amazonian specimens). (From Charlwood, J.D., *Mem. Inst. Oswaldo Cruz.*, 91, 391–398, 1996.)

Measuring lengths only these days is somewhat jejune. The study of wing morphometrics has gone beyond that. This does, however, require a suitable microscope with a digital camera. A technique for the measurement of different wing parameters in *Drosophila* (in particular trichome density) was developed by Dobens & Dobens (2013). This is what they have to say:

> Proper wing preparation and photography is critical for success in wing morphometry. We removed wings from 3- to 8-day-old adult flies (here we examine females only) using forceps and transferred five to six wings to a drop of Euparal mounting medium on a clean cover slip. The slip was overlaid by a clean microscope slide the Euparal allowed to spread, then the sandwich was flipped right side up and pressure from a probe used to spread and separate the wings, which removed large bubbles as detected under the dissecting microscope. Slides were baked for 24 hr at 65°C to slightly harden the Euparal and then weights were carefully laid on the cover slip to ensure flattening and further baked for another 24 hr. To visualize wing preparations, a Nikon TE-2000 with attached Colorview camera and Analysis image acquisition software microscope was used at either 200 or 40 with the condenser diaphragm fully closed to maximize contrast. After careful adjustment to focus on trichomes, photomicrographs (resolution 2080 · 1544) were collected.

Everything from then on is automated.

In Portugal, and with mosquitoes, it was done slightly differently. The following is from Vicente et al. (2011): 'the wings were kept in 5% KOH solution for 20 min to remove scales. The wings were then placed in 95% ethanol for <10 sec, after which they were transferred to cups containing

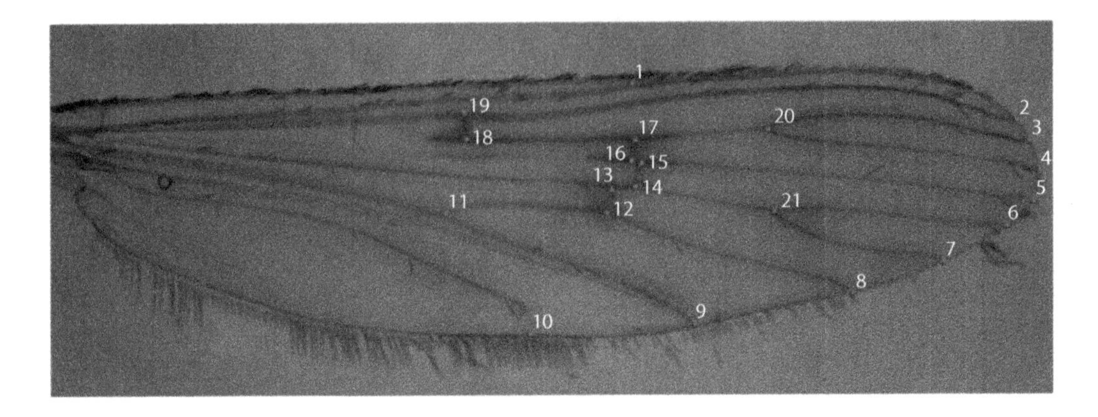

Figure 14.5 Wing of *Anopheles atroparvus* showing the 21 'landmarks' used in the morphometric analysis. (From Vicente, J.L. et al., *Malar. J.*, 10, 5, 2011.)

distilled water for washing. After staining, wings were mounted on labelled slides and coverslips with Entellan® A (Merck, Germany) medium. Slides were photographed using a Leica® MZ-7.5 stereoscopic zoom dissection microscope with a DC-300 digital camera system. In order to reduce the measurement error, specimens were digitalised twice and scored by the same person. The second session of measurement was conducted after specimens were removed and replaced under the stereomicroscope in order to take positioning error into account. Twenty one landmarks of the left wings of mosquitoes were used for the analysis, following the methods described in Rohlf and Slice' (1990) (Figure 14.5).

They were unable to find any clear pattern between the wing architecture and size and the origin of the samples from a 3000 km transect in Portugal.

The use of computer programmes (such as those described by Dobens & Dobens, 2013) should be applied to mosquito wings. Morphometrics may produce unexpected insights but until they are examined in greater detail we do not know.

NEAR INFRARED

One recently developed technique that shows promise is the use of near infrared absorption patterns displayed by different species and different aged mosquitoes (Mayagaya et al., 2009). The pattern changes due to changes in the internal and external composition of the insect. The constituents, which must be present at the parts per thousand level, may produce unique absorption spectra, which allows one to distinguish between closely related species (such as *A. gambiae* and *A. arabiensis*) and different aged mosquitoes. The technique is rapid, chemical free and nondestructive (so that the insect is preserved).

The validation of the technique is an ongoing process. It appears that the age of older mosquitoes (from known aged samples) may be underestimated (Figure 14.6). Nevertheless, although it may not have the absolute accuracy associated with other techniques for surveillance purposes, the fact that it is so much faster (meaning that large numbers of samples can be processed) and is easy to use means that it will probably become a widely used technique in the future.

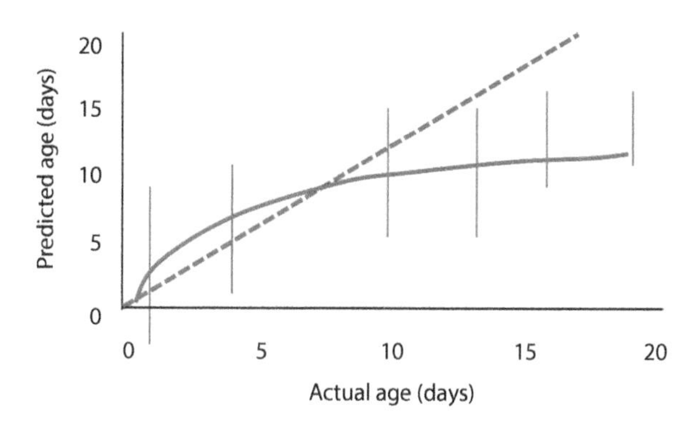

Figure 14.6 Actual versus predicted age of female *Anopheles gambiae* 1–19 days of age (n = 321) as determined from a cross-validation. The blue line shows the mean values whilst the red dotted line the correct fit. (Adapted from Mayagaya, V.S. et al., *Am. J. Trop. Med. Hyg.*, 81, 622–630, 2009.)

RESISTANCE MEASUREMENT

In order to investigate resistance in vector populations it is necessary to first obtain baseline susceptibility data for individual insecticides in a normal or 'susceptible' population of a given species.

A susceptible population is one that has not been subjected to insecticidal pressure and in which the presence of resistant individuals is either absent or rare. Baseline susceptibility is obtained by exposing nonresistant vectors to serial concentrations of a given insecticide or to serial time exposures at a single concentration, and plotting the percentage mortality against exposure on logarithmic-probability paper and enables a determination of the doses required to achieve various levels of mortality. This enables the dose that kills all individuals in a susceptible population to be determined. This dose is known as the 'diagnostic' or 'discriminating' concentration. Once this is known, it is sufficient to conduct resistance tests using this diagnostic concentration because any survivors after exposure to this concentration may be considered to be resistant. Diagnostic doses and diagnostic times for several insecticides have already been determined for some insect vectors from a number of geographic regions.

WHO Bioassay

The WHO bioassay test for the determination of resistance in mosquitoes was developed in the 1960s. It uses a kit of tubes available from the WHO. The tubes have a sliding plastic 'gate' that can be shut, opened to allow the entry of mosquitoes by an aspirator, or opened to connect to a second tube (Figure 14.7). Test papers with exact doses of insecticides, also available from the WHO, are used in the second tube.

Briefly, at least 20–25 healthy mosquitoes are aspirated into paper-lined individual holding tubes and kept for an hour. Any mosquitoes that have died in the interim are removed.

The exposure tubes, with a red dot, are lined with a sheet of insecticide-impregnated paper, while the green-dotted control exposure tubes are lined with oil-impregnated papers; papers are fastened into position with a copper spring-wire clip.

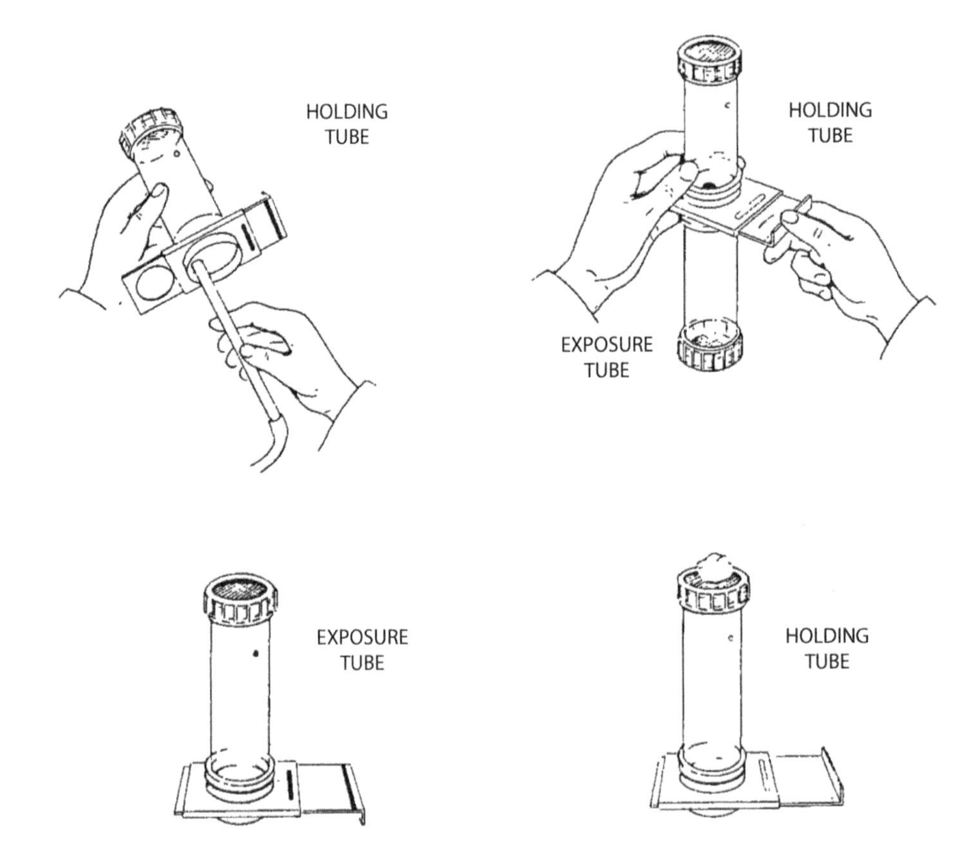

Figure 14.7 WHO test tubes for determination of resistance status in mosquitoes.

The empty exposure tubes are attached to the vacant position on the slides and with the slide unit open the mosquitoes are blown gently into the exposure tubes. Once all the mosquitoes are in the exposure tubes, the slide unit is closed and the holding tubes can be detached and set to one side. Mosquitoes are kept in the exposure tubes, which are set in a vertical position with the mesh-screen end uppermost, for a period of one hour. After an hour of exposure, the mosquitoes are transferred back to the holding tubes and the exposure tubes are detached from the slide units. A pad of a cotton wool soaked in sugar water is placed on the mesh screen end of the holding tubes, and the insects are kept for 24 hours at which time mortality in both exposed control mosquitoes is recorded.

Recent recommendations (WHO, 2013) include the monitoring of temperature and humidity during the test and holding period; a minimum of 100 mosquitoes should be tested (20–25 per test) with two sets of controls, making a total of 150 mosquitoes. Test tubes should be kept vertical for both test and holding period, and the test papers should only be used for a maximum of six times before being discarded.

The mortality of a sample in a test is calculated by dividing the number dead by the total sample size and multiplying by 100. The mortality among the control group is calculated in the same way. If the control mortality is above 20%, then the test is discarded. If it lies between 5% and 20%, Abott's formula (below) is used to obtain the corrected mortality among the test mosquitoes.

$$\text{Corrected mortality} = (\%\ \text{observed mortality} - \%\ \text{control mortality})/(100 - \%\ \text{control mortality})$$

Anything less than 98% kill of the test mosquitoes indicates the emergence of resistance and further investigation is required. This may include replicate tests, tests from other nearby localities (because resistance may be heterogeneous and differ between locations only a few kilometres apart) or biochemical tests.

If 90%–97% of the mosquitoes are killed, this indicates that resistant genes are present in the population.

Less than 90% mortality indicates that resistance is established in the population and action to ameliorate the situation should be undertaken with some urgency. This might mean changing the insecticide used for IRS or using an LLIN with an added synergist (such as PBO).

CDC Bottle Assay

The CDC bottle assay (Figure 14.8) measures the length of time it takes to kill a sample of mosquitoes exposed to a known concentration of insecticide. As such, it is not directly comparable to the WHO bioassay test (which measures mortality rates in mosquitoes exposed to a high concentration of insecticide for a fixed period of time), but both methods can detect resistance if it is present.

It is a relatively new method for monitoring insecticide susceptibility. It does not need the specially impregnated papers used in the standard WHO test; it is relatively simple and quick to carry out (because a 24-hour holding period is not required); it can be used with wild-caught females and can be used with various synergists for the testing of metabolic resistance mechanisms. It uses lower discriminating doses than the standard WHO test and so resistance may be detected earlier. The coating of the bottles with insecticide is, however, a difficulty. This is partly because the Acetone used as the solvent for the insecticide evaporates very quickly, making it difficult to coat the bottles in a uniform manner.

If the bottles are not to be used soon after coating them with insecticide, they should be kept dry with their caps off. When dry, they should be stored in a dark place (such as a drawer) with their caps off. Depending on the insecticide used, bottles can be stored from 12 hours to five days in this way.

Figure 14.8 250 mL Wheaton bottles as used in the CDC bottle assay.

The length of time that bottles can be stored depends on the insecticide. More than one batch of mosquitoes can be run in a single bottle in one day. The main factor limiting their reuse is moisture buildup with successive introductions of mosquitoes. If they are to be reused on the same day, bottles should be left to dry for 2–4 hours between tests. If the bottles are to be reused the following day, bottles with caps off can be left to dry overnight, protected from direct light. They should not be dried in ovens.

Resistance is assumed to be present if a portion of the test population survives the diagnostic dose at the diagnostic time.

Tunnel Test

Another test designed to determine the effect of resistance on the feeding behaviour of mosquitoes is the 'tunnel test'. This test uses animal bait, such as a guinea pig, placed at one end of a specially constructed tunnel. A fixed number of hungry mosquitoes are released at the other end of the tunnel and left overnight. They must pass through a holed treated or untreated piece of netting to reach the animal bait to feed. The following morning, both live and dead mosquitoes, blood fed and non-blood fed, are collected and counted from both sides of the holed net. As in the WHO bioassay, live mosquitoes are monitored for a further 24 hours to assess delayed mortality. Insects are classified according to whether they have passed through the net or not, whether they have fed and whether they have died in the meantime.

Cone Bioassays

It is important to know how long the insecticide is effective (either on nets or on walls following IRS) and for this cone bioassays are used. The duration of an insecticidal effect on walls depends on a number of factors, especially the material that the wall is made of (mud having a shorter effect than wood or cement). Cone bioassays provide information on the quality of the application of the insecticide (in the case of IRS). The duration of an insecticidal effect on nets may depend on the number of times they are washed and the manner in which this is done.

Following IRS, the first test should be undertaken within a few days after treatment. Where a previously sprayed area is due for another application of insecticide and no bioassay data are available, it is advisable to do the bioassay beforehand to ascertain the potency of the deposit present. The test should be carried out on an adequate scale and at regular intervals (about every three months). It is necessary to test and to evaluate separately the potency of the insecticide deposit on each main type of surface. At least 10 points, variously situated, should be chosen for testing on a given day. They should be distributed in several houses, with not more than three points in any one house. At least two controls should be used for every 10 tests. When the main objective is to determine the rate of loss of potency, there are advantages in using the same points throughout the series of tests. The points should therefore be marked carefully at the time of the first tests, and care should be taken to avoid rubbing or in any way impairing the deposit at these points during the performance of the tests.

Three- to five-day-old susceptible insects from a laboratory colony are used in these tests, which do not therefore measure the resistance status of the local population – this should be done before the application of any insecticide.

For many tests it is useful to have laboratory-reared insects derived from wild-caught females. The difficulty here is that many wild-caught mosquitoes are reluctant to lay eggs in insectary cages. A simple but effective remedy is to force them to lay in Eppendorf tubes as described by John Morgan and colleagues (2010). Briefly, collect blood-fed females and keep them in a cup or cage until they become fully gravid. Once gravid, the mosquitoes can be gently introduced individually into 1.5 mL Eppendorf tubes containing a 1 square cm filter paper inserted into the bottom of the tube.

The filter paper is moistened and excess water removed. The cap of the Eppendorf tube has three small holes (made with a syringe needle) to allow air into the tube.

Under these conditions, females – even species that rarely lay eggs in the laboratory – will often oviposit.

ENZYME LINKED IMMUNOSORBENT ASSAY (ELISA) FOR CIRCUMSPOROZOITE DETECTION

The protocol for circumsporozoite protein (CSP) ELISA devised by Robert A. Wirtz is published by the CDC and available from the MR4 website. It is the protocol described earlier and used in studies from Mozambique.

In the absence of a plate reader, three grades of positivity can be distinguished. Separate readings by two individuals improves accuracy. Disagreements can be resolved by reexamination of the plate by both observers together. It is better to run batches of mosquitoes sufficient to exhaust fresh solutions rather than testing small numbers with solutions that may be reaching the end of their shelf life. When the ELISA fails to work (i.e., the positive controls do not produce a colour change), it becomes difficult to know which of the solutions may not be working; therefore, several experimental runs may have to be performed in order to resolve the problem.

gSG6 ELISA

The prevalence of gSG6 protein, found in the saliva of *Anopheles*, can be used as a marker of exposure to biting females. It elicits an antibody response that is correlated with the level of exposure to *A. gambiae*. It is determined by ELISA. A protocol is as follows: Immulon 4 HX plates are coated overnight at 4°C with 50 μL of gSG6 (5 μg/mL). After washing with phosphate buffered saline (PBS)/Tween, wells are blocked (three hours, at room temperature, RT) with 150 μL of 1% w/v skimmed dry milk in PBS/T, washed again and incubated overnight at 4°C with 50 μL of serum (1:200) in a blocking buffer. In order to allow for standardisation of optical density (OD) values between day-to-day and interplate variation, sera should be analysed in duplicate with antigen and once without antigen (coating buffer only) and positive control sera (1:200 in PBST/Marvel). Following another overnight incubation at 4°C, plates are washed and incubated (three hours at RT) with 100 μL of polyclonal rabbit antihuman IgG/HRP antibody (Dako, 1:5000 in a blocking buffer). After washing, colourimetric development is carried out (15 min, RT in the dark) with 100 μL of OPD. The reaction is terminated by adding 25 μL of 2M H_2SO_4 and the OD_{492} determined using a micro-plate reader. IgG levels are expressed as final OD calculated as the mean OD value with antigen minus the OD value without antigen. Positive control sera is analysed to allow standardisation of OD values, and negative control sera – taken from individuals with no recent travel to malaria endemic countries – are used to calculate IgG seroprevalence. A cutoff for seropositivity of samples is determined as the mean OD of unexposed sera plus 3 standard deviations.

THE POLYMERASE CHAIN REACTION

A large amount of nucleotide variation has been shown to exist in genes when they are sampled from natural populations. Each novel nucleotide sequence is considered a novel allele. Variation among alleles consists primarily of synonymous substitutions that do not change the primary amino acid sequence. The majority of non-synonymous substitutions involve changes among amino acids of similar size, function or charge. This pattern of variation is consistent with the neutral theory of molecular evolution. This model assumes that most mutations are deleterious to the fitness of an individual and are removed through purifying selection.

In classical genetics, genes that fail to be transcribed or translated or to encode functional proteins segregate as recessive alleles relative to dominant alleles that encode a functional protein. Only individuals that inherit two copies of nonfunctional alleles will display the absence of gene function. Individuals with one (heterozygous) or two (homozygous) functional copies will display a normal phenotype.

PCR can quickly amplify a gene sequence represented once or only a few times in a large and complex mixture of genes. It directly amplifies a sequence in a few hours using oligonucleotide primers that anneal to conserved regions that flank a target gene sequence. The amplified gene is visualised on an agarose gel and can then be cloned or analysed with restriction enzyme digestion or even sequenced directly. It has a wide use and there is a considerable literature on its applicability to biological systems that are beyond the scope of this book.

CHAPTER 15

Global Heating – 'The Future Ain't What It Used to Be'

Global warming, or perhaps better global heating, in general, is something that we should all be worried about, and not just because of its possible effects on malaria transmission. In my time in the field I have often been literally at the 'cutting edge', so it is difficult for me to take a purely objective view – 'I have seen the enemy and we are it'.

As noted by Darlington in 1969, 'All man's progress has been made at the expense of damage to the environment which he cannot repair and could not forsee' (Darlington, 1969). Human activity has significantly altered the Earth's climate in less than a century. The rise in global CO_2 concentration since 2000 is about 20 ppm/decade, which is up to 10 times faster than any sustained rise in CO_2 during the past 800,000 years. Since 1970, the global average temperature has been rising at a rate of 1.7°C per century, compared to a long-term decline over the past 7000 years at a baseline rate of 0.01°C per century. Earth system dynamics can be described, studied and understood in terms of trajectories between alternate states separated by thresholds that are controlled by nonlinear processes, interactions and feedbacks. Because the effects of this catastrophe will be felt in a myriad of as yet unimagined ways, it is difficult to know how important the redistribution of the mosquitoes will be on human life. Other effects, such as a lack of food, are likely to be more important and devastating. The effects of the Anthropocene, – a new geological epoch based on the fact that human impacts on essential planetary processes, have become so profound that they have driven the Earth out of the Holocene epoch, in which agriculture, sedentary communities, and eventually, socially and technologically complex human societies developed. If a number of thresholds are crossed, the effects may have profound and irreversible impacts that prevent stabilisation and result in what has been called 'Hothouse Earth' (Steffen et al., 2018). Crossing thresholds would lead to a much higher global average temperature than any interglacial in the past 1.2 million years and to sea levels significantly higher than at any time in the Holocene. Agricultural systems and water supplies are especially vulnerable to changes in the hydro-climate, leading to hot/dry or cool/wet extremes. They are particularly vulnerable, because they are spatially organised around the relatively stable Holocene patterns of terrestrial primary productivity, which depend on a well-established and predictable spatial distribution of temperature and precipitation. This essentially means the end of life as we know it. This is not something new – popular science books from the 1970s, such as *Man and the Cosmos* by Ritchie Calder described the greenhouse effect and the possibility of positive feedback loops on the Earth's climate and its potential impact on human life.

Biogeophysical feedback processes within the Earth system coupled with direct human degradation of the biosphere may play a more important role than normally assumed, limiting the range of potential future trajectories and potentially eliminating the possibility of intermediate trajectories. There is a significant risk that these internal dynamics, especially strong nonlinearities in feedback processes, could become an important or perhaps even dominant factor in steering the trajectory of the Earth system over the coming centuries.

If the rate of climate change is too large or too fast, a tipping point can be crossed, and a rapid biome shift may occur via extensive disturbances such as wildfires, insect attacks or droughts. In some cases, such as widespread wildfires, there could be a pulse of carbon to the atmosphere, which, if large enough, could influence the trajectory of the Earth system. For some of the tipping elements, crossing the tipping point could trigger an abrupt, nonlinear response such as the conversion of large areas of the Amazon rainforest to a savannah or seasonally dry forest, while for others, crossing the tipping point would lead to a more gradual but self-perpetuating response, which may happen with large-scale loss of permafrost. There could also be considerable lags after the crossing of a threshold, particularly for those tipping elements that involve the melting of large masses of ice. Cascades could be formed when a rise in global temperature reaches the level of the lower temperature cluster, activating tipping elements, such as loss of the Greenland ice sheet or Arctic sea ice. These tipping elements, along with some of the non-tipping element feedbacks such as a gradual weakening of land and ocean physiological carbon sinks, could push the global average temperature even higher, inducing tipping in mid- and higher temperature clusters.

Recent, observation-based, estimates show rapid warming of Earth's oceans over the past few decades. This is approximately 40% faster than previously thought (Cheng et al., 2019). This warming has contributed to increases in rainfall intensity, rising sea levels, the destruction of coral reefs, declining ocean oxygen levels and declines in ice sheets, glaciers and ice caps in the polar regions.

According to a recent report by the IPCC, we have until 2030 before these catastrophes can be averted, and to do so temperatures should remain below the 2°C agreed upon by the Paris Agreement. The Earth system may already have passed one 'fork in the road' of potential pathways, ultimately affecting the habitability of the planet for humans and other life (as Paul Simon says, 'Sometimes even music cannot substitute for tears'). This is the background on which possible distributions of malaria vectors needs to be considered. Like everything else, global warming will affect the distribution of malaria vectors.

Tonnang et al. (2010) used the present distribution of *A. gambiae* and *A. arabiensis* to model the temperature and rainfall conditions favourable to both species and then developed predicted distributions under three different scenarios of climate change. In the first model, they assumed a rise in temperature of 2°C Africa-wide, and 10% increase of summer rainfall and 10% decrease in winter rainfall. In the second model, they assumed a 0.1°C rise in summer and winter maximum and minimum temperatures per degree of latitude, and a 10% increase in rainfall in summer and 10% decrease in winter. In the third model, they assumed a rise in temperature of 4°C Africa-wide, and 20% increase of summer rainfall and 20% decrease in winter rainfall. They also assumed that it is abiotic factors (i.e., climate) that determine the distribution of the mosquitoes rather than biotic ones (such as predators). They also produced maps based on the currently assessed distribution of the two species (Figures 15.1 and 15.2, which can be compared to Figure 1.12).

The results indicate that West Africa and some parts of central Africa might become less suitable for both *A. gambiae* and *A. arabiensis*. At the same time, the mosquitoes may move towards the eastern and southern regions of the continent.

An alternative model for the future distribution of *A. arabiensis* based on three slightly different future scenarios was provided by Drake and Beier (2014) (Figure 15.3). Scenario A1B implies a world of rapid economic growth, a population that peaks mid-century and a rapid introduction of energy-efficient technologies. Scenario A2A is a more heterogeneous world where growth and

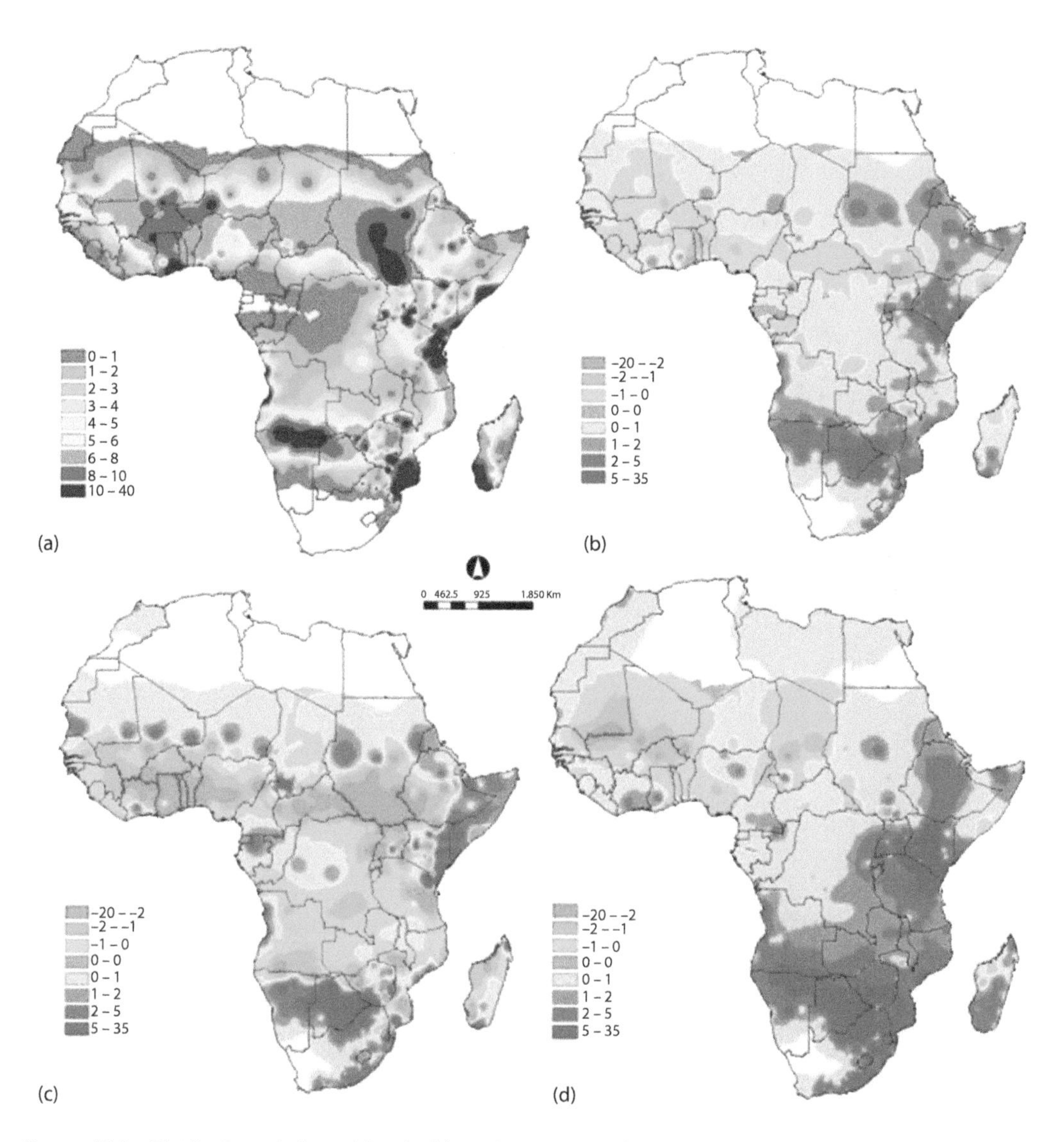

Figure 15.1 Distribution of *A. arabiensis* (a) under current climate; and illustrating species ranges shifts under (b) climate change model 1; (c) climate change model 2; and (d) climate change model 3. (From Tonnang, H.E.Z. et al., *Malar. J.*, 9,111, 2010.)

technology are unevenly distributed, whilst scenario B2A is a world with intermediate levels of development and more diverse technological change than in the A1 scenario. The results are, however, similar to the earlier analysis in that the distribution of *A. arabiensis* (a mosquito that becomes increasingly important as a vector due to its propensity to bite outdoors on animals – meaning that indoor control techniques may not affect it as much as the other African vectors) will be reduced as the world heats up.

Another phenomenon associated with global warming that will almost certainly affect malaria transmission is the increasing frequency and severity of El Niño/Southern Oscillation

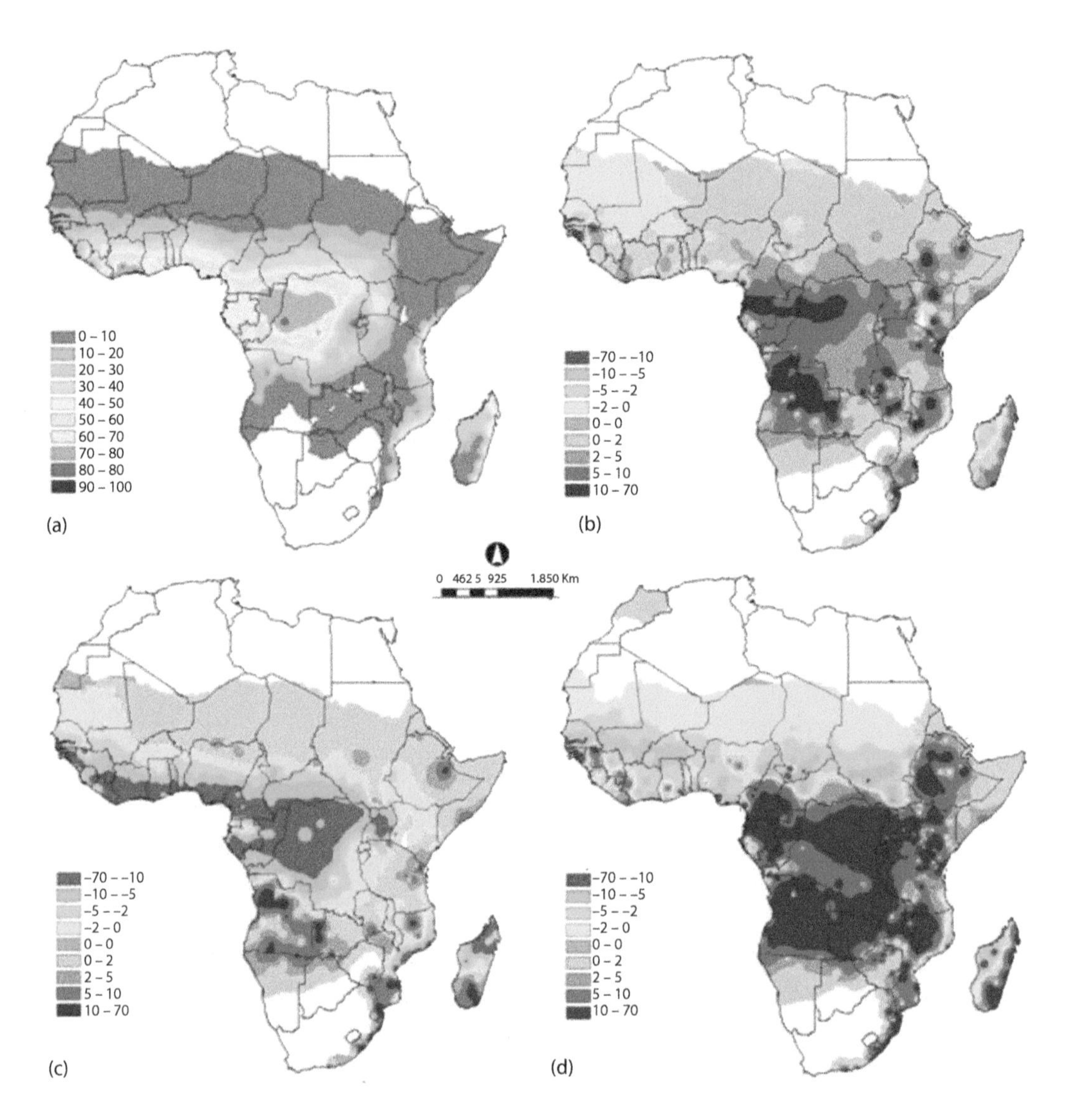

Figure 15.2 Distribution of *A. gambiae* (a) under current climate; and illustrating species ranges shifts under (b) climate change model 1; (c) climate change model 2; and (d) climate change model 3. (From Tonnang, H.E.Z. et al., *Malar. J.*, 9,111, 2010.)

(ENSO) events. El Niño (literally 'the Christ child' because these events occur in December) is the name given to a large-scale ocean-atmosphere climate phenomenon that is linked to periodic warming in sea surface temperature in the central equatorial Pacific. Sea surface temperatures in the Pacific greatly influence global atmospheric circulation, with a pronounced impact on global-scale tropical precipitation. Rainfall increases over the Peruvian coast, equatorial East Africa and the Pacific Islands whilst severe drought conditions occur in the Western Pacific, northeast Brazil and southern Africa. These conditions can have a significant effect on vector-borne diseases including malaria. More severe ENSO events will exacerbate these effects when they occur. It is

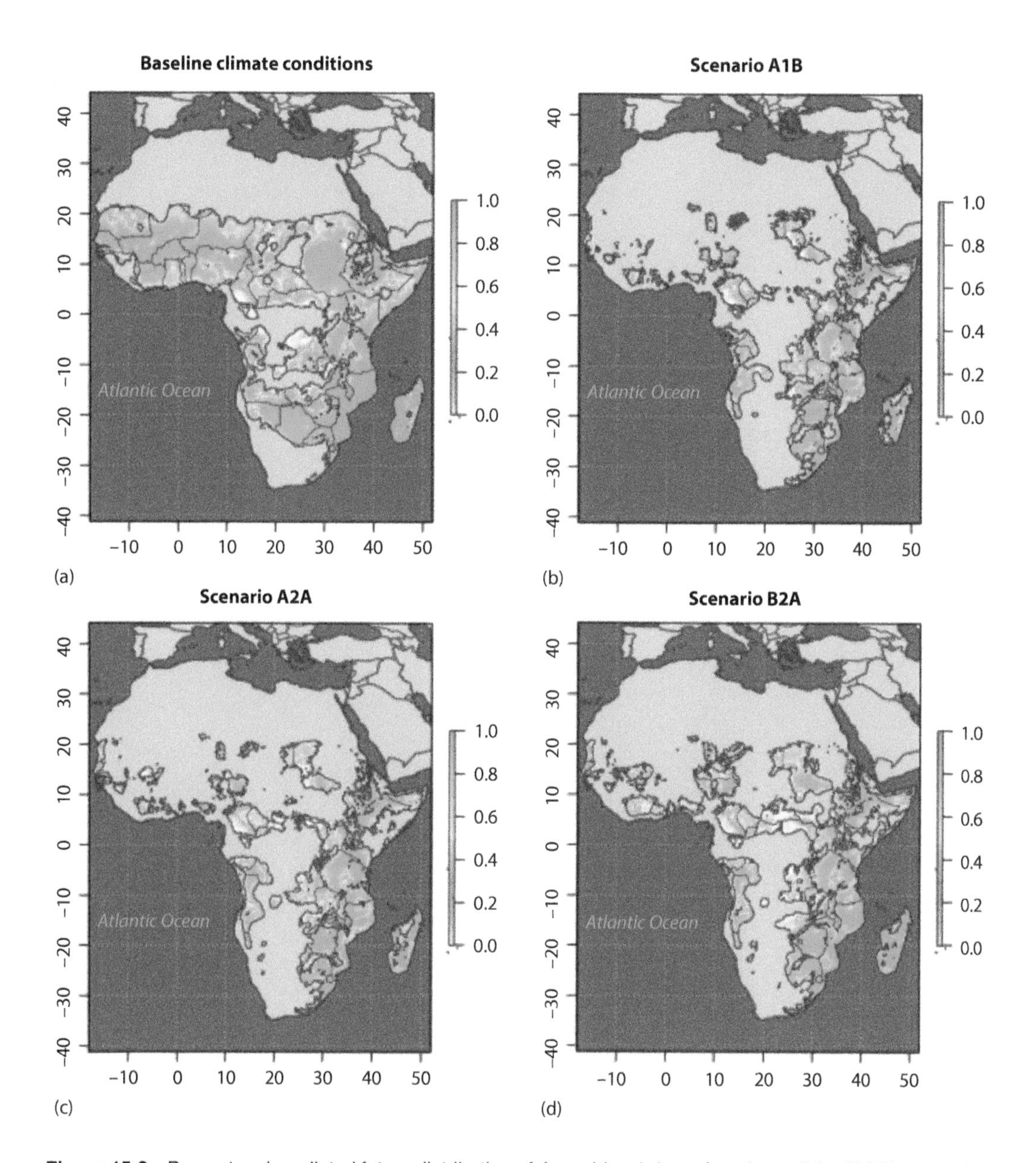

Figure 15.3 Present and predicted future distribution of *A. arabiensis* based on three of the IPCC's scenarios for the future. (From Drake, J.M. and Beier, J.C., *Malar. J.*, 13, 213, 2014.)

possible that some vectors will be more affected than others. For example, *A. funestus* may have suffered a population crash and gone through a genetic bottleneck (as evinced in its genetics) in the past due to drought conditions.

As pointed out by the COP24 report for the WHO (2019) 'The severity of the impact of climate change on health is increasingly clear. Climate change is the greatest challenge of the twenty-first century, threatening all aspects of the society in which we live, and the continuing delay in addressing the scale of the challenge increases the risks to human lives and health'.

CHAPTER 16

Some Case Histories

MOZAMBIQUE

Malaria remains a serious problem in Mozambique. According to UNICEF, it is the leading killer of children, contributing to around 33% of all child deaths and overall more deaths have been attributed to it (28.8%) than to any other single cause, including HIV/AIDS. Studies on the ecology of the vectors and the epidemiology of the disease are limited. Here, we describe studies from two villages in the province of Inhambane that have quite different epidemiologies. The two villages of Furvela and Linga Linga lie opposite each other by the Indian Ocean (Figure 15.1). They have decidedly different malaria transmission patterns despite their proximity and despite the main vector in both villages being *Anopheles funestus*.

Furvela

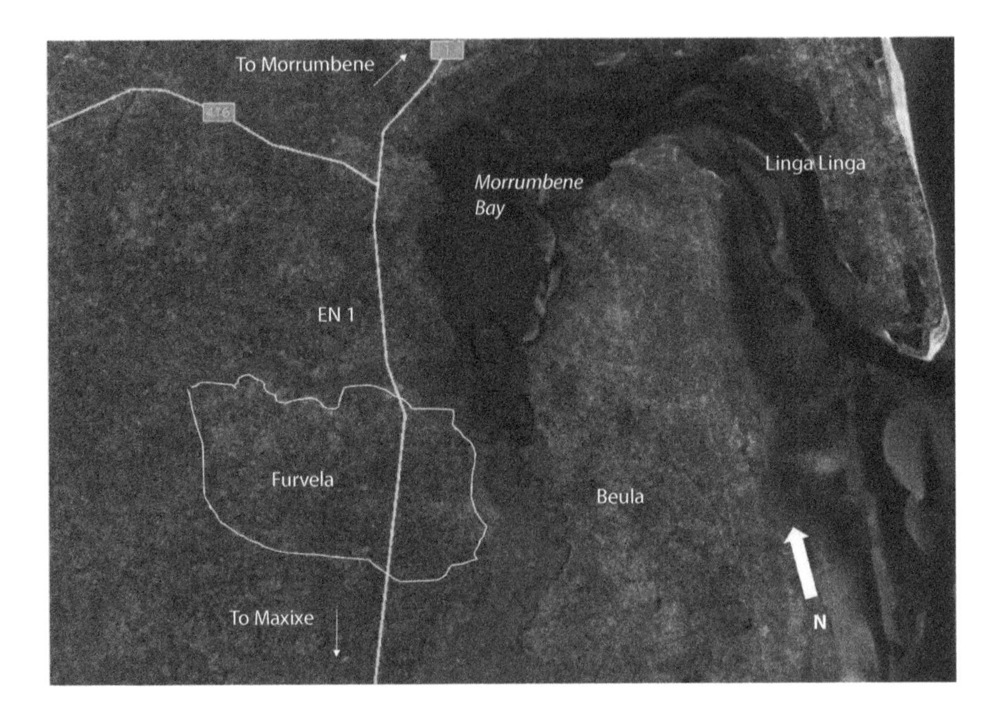

Figure 16.1 Map of Furvela and Linga Linga. EN1 is the main highway from Maputo to the north of the country. (© Google Earth; Image 2016 ©CNES/Astrium.)

Furvela (24°43′S, 35°18′E), (Figure 16.1) some 475 km north of the capital Maputo, straddles EN1, the road that goes from the south to the north of the country. The approximately 5 × 4 km village is bordered on two sides by the alluvial plain of two river systems. Both provide ample sites for the larvae of *A. funestus*. The Furvela River valley to the north of the village in particular has a considerable amount of local irrigation that provides a large and relatively stable number of small canals, typical *A. funestus* breeding sites. The Inhnanombe river, to the east of the village, consists largely of beds of reed (*caniço*), used in housing, and sugar cane, used in the production of local alcohol. The Inhnanombe does not flow as fast as the Furvela and is not used for washing or bathing.

Houses in the village are generally made with *caniço* walls and palm thatch roofs. Although most houses don't have windows, the majority have a gap between the roof and walls at either end of the house. Doors and door frames are also generally badly fitting; hence, mosquitoes can easily gain access to the inside of the house. Other styles of house include those with zinc sheets for the roof and those made of concrete blocks (which do have windows). Houses are built either in family compounds of three to six houses, or as relatively evenly spaced individual homes. Houses, which are often just a single room, have separate kitchens. The life span of an individual house, from just 2 years to more than 10, depends very much on the ability and care of the constructor, and on the relative effects of termites and bad weather.

Most people practise subsistence agriculture. Maize, manioc, groundnuts, cashew and coconut are the main crops. Sugar cane is extensively grown in the valley. In the winter (June to September) the plots in the valley are used for market gardening (cabbage, lettuce, carrots, etc.) and in the wet season may be used for the cultivation of rice. They may be used at other times to grow maize. There are only a sufficient number of cattle in the village for ploughing purposes. Where they exist, they are kept singly or in twos or threes.

Drinking water in the village comes from a limited number of wells or pumps close to the valley edge. Clothes, and often bodies, tend to be washed in the Furvela River. A single rainy season occurs from October to March during which approximately 1200 mm of rain falls, mostly in February and March. Daily mean temperatures vary between 18°C in July and 30°C in December. *Anopheles funestus* is the most abundant mosquito in the village, although *A. gambiae* complex and other culicines are also present. *Anopheles funestus* s.s. is the only member of the funestus group that has been identified by PCR from the village. Both *A. gambiae* (86%) and *A. arabiensis* (14%) as well as a small number of *A. merus* have been identified from the village, again by PCR.

Malaria is holoendemic in Furvela. Highest prevalence rate, of circa 80% determined by microscopy, occur in children 2–4 years of age (Figure 16.2). This, more or less, means that all children in that age group are likely to harbour malaria parasites. By the time children are 10 years old, prevalence rates are around half those in infants. In a clinic established in the village most attendees were infants younger than 2 years of age. Thus, it is likely that many of the babies attending the clinic were experiencing their first attack of malaria. This was a pattern seen throughout the duration of a project established in the village from 2001 to 2010 (Figure 16.3).

People attending the clinic had their temperature taken and were asked about the duration of symptoms as well as mosquito net usage. Most children were bought to the clinic on the second day of being ill (Figure 16.4a), but children with fever were most likely to attend as soon as they were ill

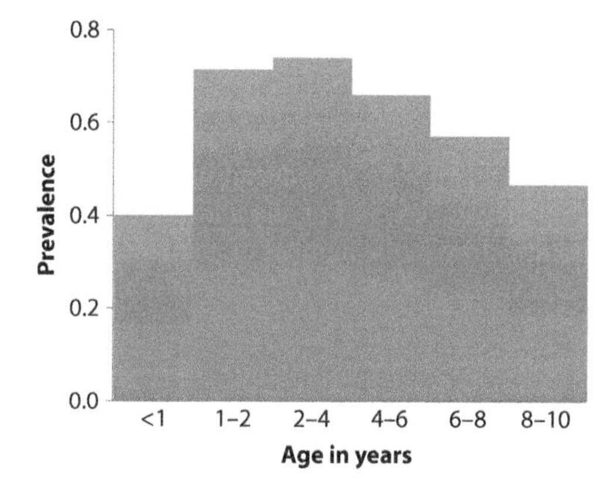

Figure 16.2 Prevalence of *Plasmodium falciparum* in a random sample of 435 children from Furvela 2007.

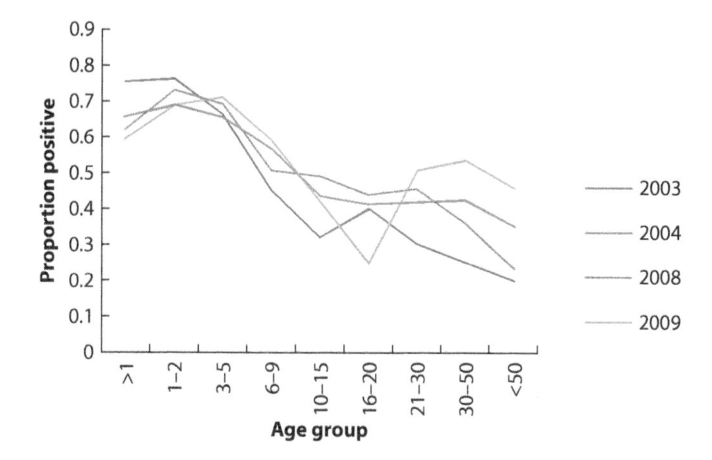

Figure 16.3 Incidence of malaria cases, determined by microscopy, from the Furvela clinic by year of sample.

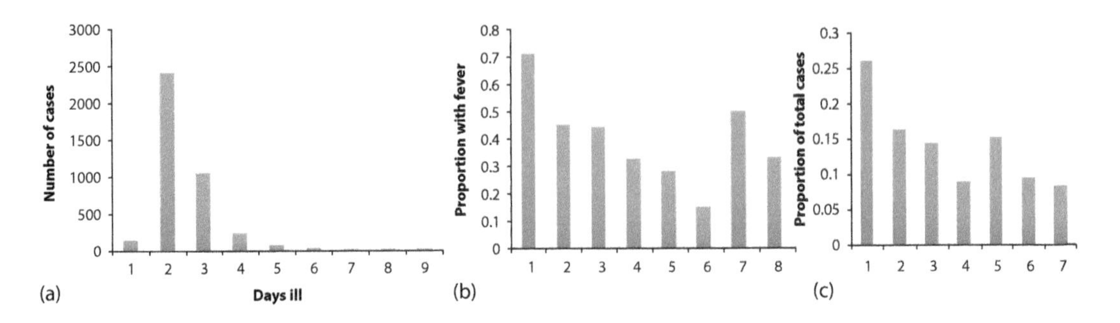

Figure 16.4 (a) Number of days ill among attendees to the clinic. (b) Proportion of cases attending the clinic with fever (c) proportion of cases with an f4 infection.

(Figure 16.4b). These children were also most likely to have a high-density infection (Figure 16.4c). Thus, although babies with fever only constituted a small proportion of cases, mothers were able to identify the danger and brought them to the clinic without delay. Otherwise, they probably waited to see if the illness would disappear by itself before attending, but they rarely delayed beyond the second day of symptoms.

The number of cases attending the clinic varied by month and by year. In 2003, cases were exceptionally high – this was associated with a higher than average sporozoite rate among the mosquitoes. Cases were lower in subsequent years with a reduction in the number of attendees during the cooler months of the year. In 2007, a *cordon sanitaire* of mosquito nets (LLINs) impregnated with a pyrethroid (alpha-cypermethrin) were introduced into all houses up to 300 m from the valley on both sides of the Furvela River (i.e., in Furvela and Jogo, the village on the opposite bank). There was a rapid, but temporary, drop in the number of attendees following the introduction of the nets. In subsequent months and years, the number of cases of malaria increased, however, and even exceeded pre-intervention levels (Figure 16.5).

This reduction or lack of effect was probably not a result of the nets losing their effect against susceptible mosquitoes because bioassays, using a colony of *A. arabiensis* from Maputo, resulted in 100% mortality even two or three years after use in the field (Figure 16.6).

Although the nets worked against susceptible mosquitoes, resistance had reared its ugly head and, despite there only being a limited exposure to the insecticide (many houses did not have nets and insecticide was not used for agricultural purposes), the local *A. funestus* was hardly affected by

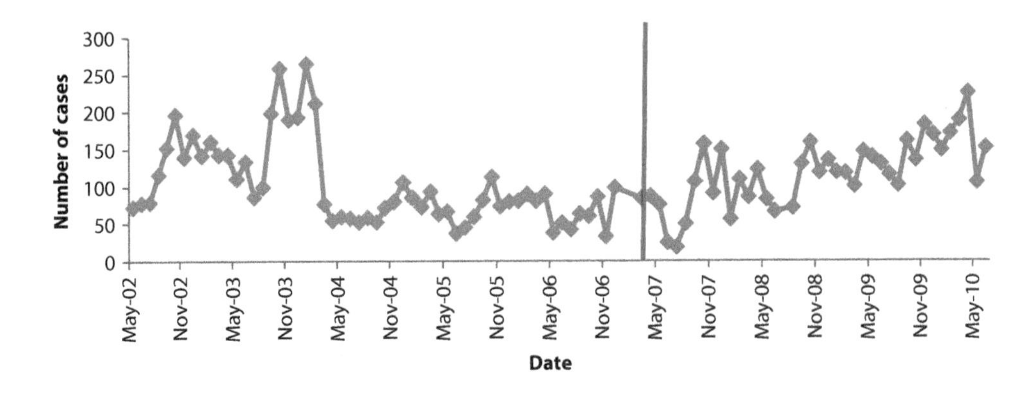

Figure 16.5 Number of attendees at the Furvela clinic by month. The vertical line represents the time that a cordon sanitaire of nets was introduced into the village.

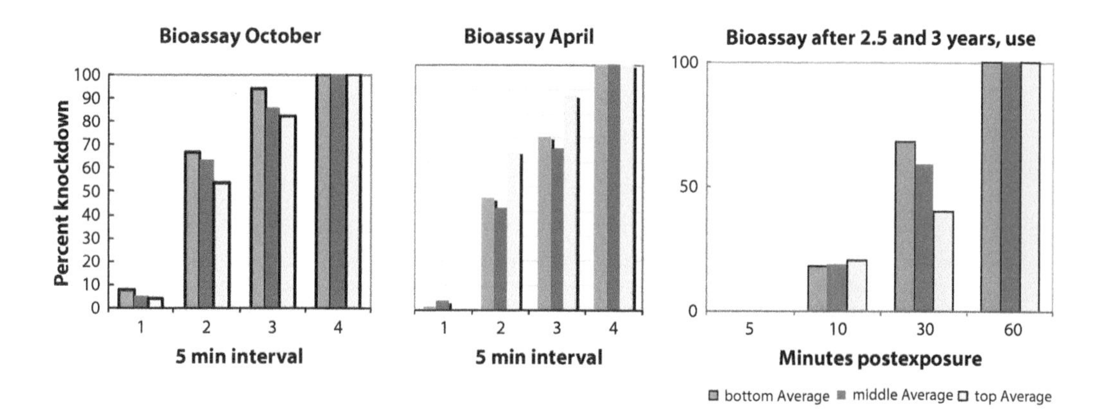

Figure 16.6 Bioassay results from three different areas of a random sample of nets tested against a susceptible colony of *A. arabiensis*.

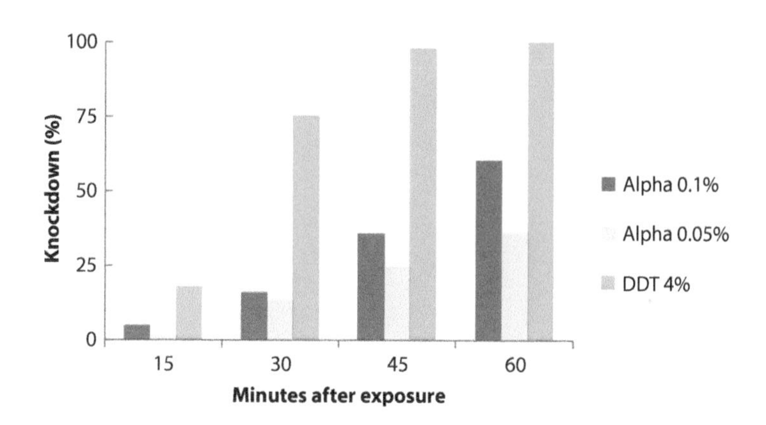

Figure 16.7 Knockdown rates of *A. funestus* from Furvela exposed to alpha-cypermethrin and DDT.

Table 16.1 24-Hour Mortality Rates Following Exposure to DDT and Alpha-Cypermethrin

Insecticide	% Mortality (s.d.) After 24 hours
Alpha-cypermethrin 0.1%	58.6 (16)
Alpha-cypermethrin 0.05%	51.3 (8.5)
D.D.T.	100 (0)

pyrethroids. It was, however, still susceptible to DDT, being both knocked down (Figure 16.7) and killed by DDT but not by alpha-cypermethrin at either 0.05% or 0.1% concentrations (Table 16.1).

In Furvela, 301,705 female *An. funestus* were collected in 6043 light-trap collections, 161,466 in 7397 exit collections and 16,995 in 1315 resting collections. The equivalent numbers for *A. gambiae* s.l. are 72,475 in light traps, 33,868 in exit collections and 5333 from indoor resting collections. Outdoor resting collections failed to produce any mosquitoes. Other anopheline species collected in light traps included 5776 *Anopheles tenebrosus*, 725 *Anopheles letabensis*, 22 *Anopheles rufipes*, 5 *Anopheles squamosus* and 1 *Anopheles pharoensis*. A further 219 *A. tenebrosus* and 5 *A. rufipes* were collected exiting houses.

A total of 41,792 *A. funestus* and 9431 *A. gambiae* s.l. collected in light traps and 22,323 *A. funestus* and 6860 *A. gambiae* s.l. from exit collections were analysed for sporozoites using an ELISA.

Both numbers of mosquito, in particular *A. gambiae* s.l., and sporozoite rates increased with increasing temperatures (Figure 16.8a and b); at the same time the number of positive malaria cases at the clinic increased as the sporozoite rate increased (Figure 16.8c).

A linear relationship between sporozoite rate and number of cases at the clinic is obtained when a month delay is added to the number of cases to account for the prepatent period of the parasite in the human host (Figure 16.9). This may be because the rates, although high, were nevertheless lower than those observed in Idete in the Kilombero Valley, Tanzania. In the latter village, there was not a linear relationship between EIR and incidence in children younger than one year of age but there was a plateau of sorts, indicating that saturation had occurred (although it was never 100%) and that some immunity had already developed in such children (see 'Malaria notes').

Numbers of *A. gambiae* s.l. increased tenfold over the range from 20°C to 30°C whilst numbers of *A. funestus* merely doubled (Figure 16.10). The effect of temperature on the size of *A. funestus* has been described elsewhere in this tome. Male mosquitoes can also easily be collected resting inside and leaving houses at dusk.

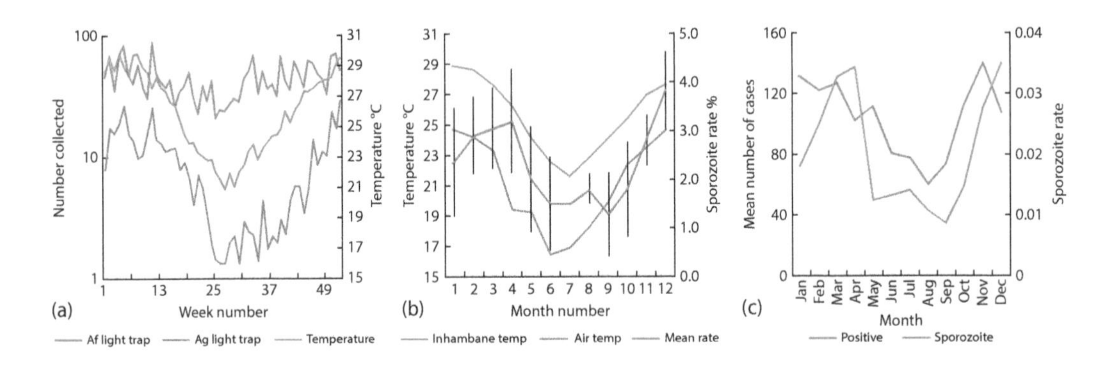

Figure 16.8 (a) Mean number of *A. funestus* and *A. gambiae* s.l. by ISO week collected in light traps and mean air temperatures measured in the village (note the log scale in mosquito numbers); (b) mean sporozoite rates in *A. funestus* and temperature measured in the town of Inhambane and in the weather station in Furvela by month. Bars indicate 95% confidence intervals; (c) mean number of positive cases of malaria attending the clinic by month and sporozoite rates (as a proportion of those tested).

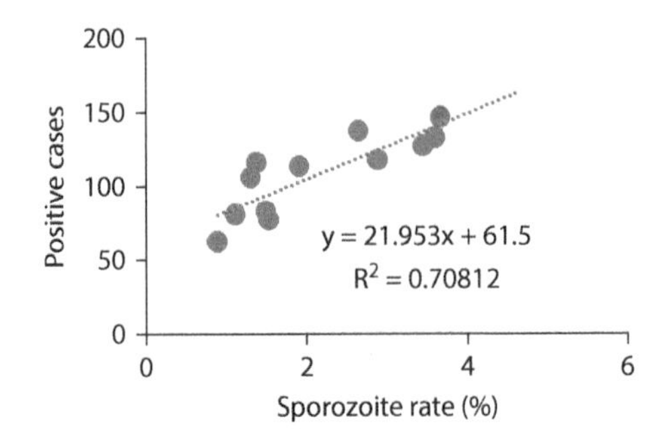

Figure 16.9 Relationship between sporozoite rate and malaria cases at the clinic (+1 month lag to account for the prepatent period).

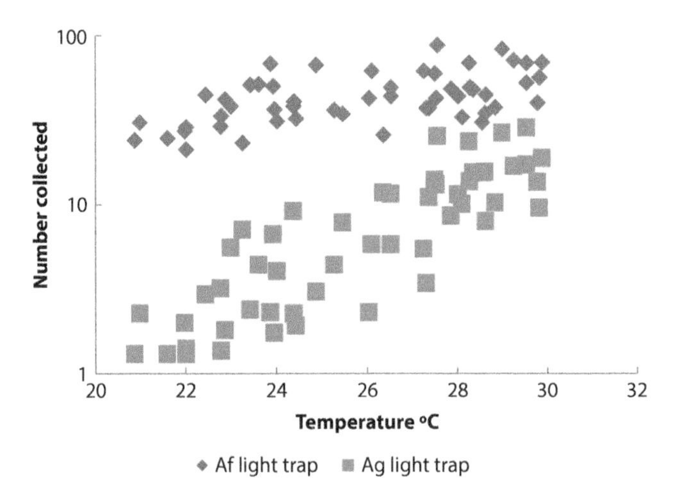

Figure 16.10 Numbers of *A. funestus* and *A. gambiae* s.l. females collected in light traps by mean air temperature recorded in the village.

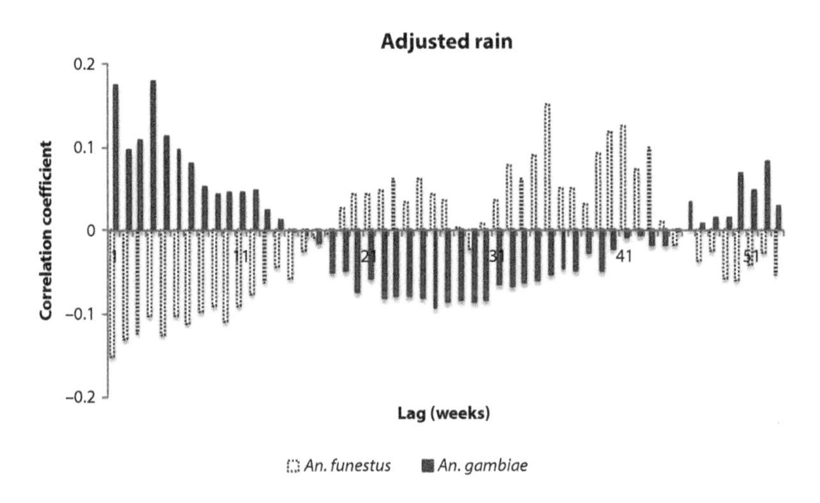

Figure 16.11 Pearson correlation coefficient between weekly mean number of male mosquitoes attempting to leave the open door of houses from the village of Furvela and adjusted rainfall measured at Maxixe, 21 kms to the south, by weekly lag. (From Charlwood, J.D., *J. Vector Ecol.*, 36, 382–394, 2011.)

Numbers of male *A. gambiae* s.l. were positively correlated with rainfall for lags up to 14 weeks, but these relationships were weaker and less regular than those of temperature (Figure 16.11). Results were similar for females. Thus, in Furvela it is temperature that drives the mosquito population dynamics and it is the mosquito that drives the malaria.

Historical fluctuations in effective population size (*Ne*) can be inferred from the genomes of extant individuals. From such data, a decline in the *A. gambiae* from Kenya was inferred to have occurred before the wide-scale use of nets for control. In Tanzania, a population decline was observed in the absence of any intervention and in Furvela a similar decline was seen (Figure 16.12).

What was responsible for this decline remains unknown! It goes to show that we (or at least the present author) still have a lot to learn about these mosquitoes.

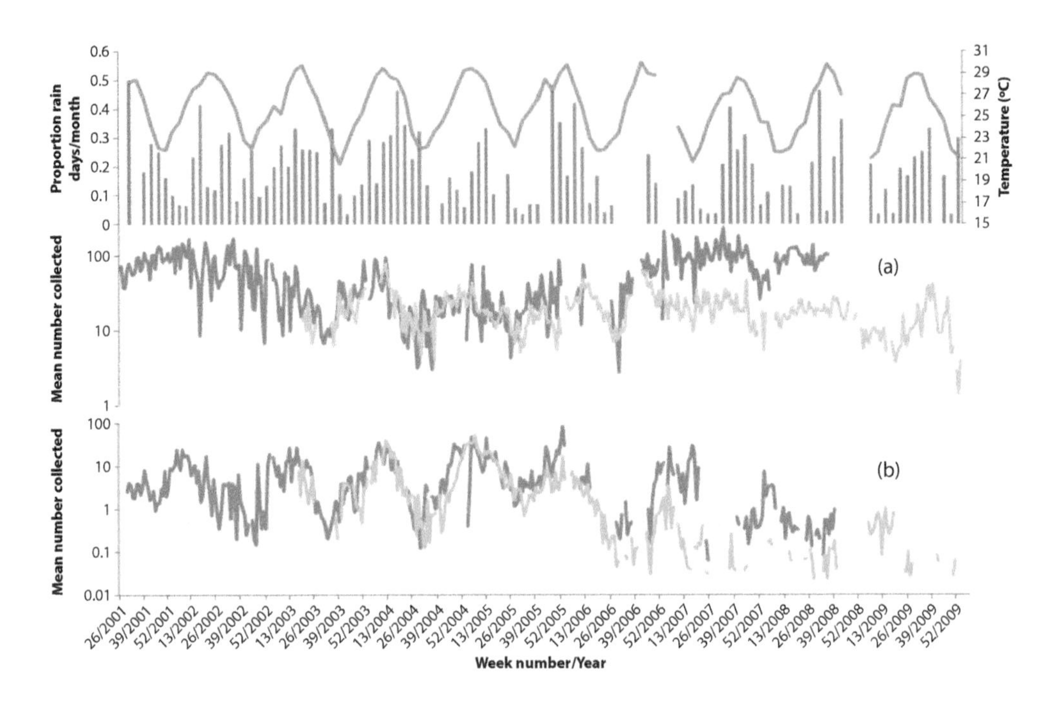

Figure 16.12 Rainfall (histogram), mean air temperature (orange line) and numbers of (a) *A. funestus* (dark blue – light trap and green exit collection) and (b) numbers of *A. gambiae* s.l. (dark blue light trap, light blue exit collection) from Furvela, 2001 to 2010. Note the log scale of the mosquito numbers and the decline in the *A. gambiae* s.l. (more than 80% of which were *A. gambiae*).

Linga Linga – Land of 'Born Not Made'

Bordered on one side by the Indian Ocean and on the other by the Morrumbene Bay, and 8 km from Furvela, the approximately 2 × 7 km peninsula of Linga Linga (it means 'try try' in the local language) is a sandy finger of land that has a very different epidemiology of malaria from Furvela. There is an area of uninhabited bush circa 1.5 km long at the neck of the peninsula making it a virtual 'ecological island'. Fruit trees, notably cashew and marula, are grown in addition to large numbers of coconut palms. Some manioc and beans are cultivated in a limited area of the peninsula. Apart from a seasonal pond close to the middle of the peninsula (Figure 16.13) and a permanent lake at its northern end, naturally occurring standing water is almost nonexistent. The water table is relatively high, however, and in the dry season people dig shallow wells to obtain water for their crops.

Most of the houses in Linga Linga consist of just a single room (400 of 467 – or 86% -enumerated in a census); only 4% (19/467) had more than two rooms. Most (87%) were built of palm or reed (*caniço*) (Figure 16.14) and, as in Furvela, most had a gap between the roof and the walls, thus providing access to mosquitoes. Three-quarters of the houses had only one or two inhabitants.

Densities and species composition of mosquitoes varied according to location on the peninsula. At its southern tip, *Culex quinquefasciatus* predominated with very few *A. funestus* being collected whilst in the middle of the peninsula the situation was reversed (Table 16.2).

A negative binomial regression indicated that numbers of *A. funestus* females decreased with distance from the pond but increased with the number of inhabitants (distance from pond regression coefficient (r) $= -0.018$, $p = 0.01$, number of inhabitants' $r = 0.567$, $p = 0.01$). The density of *A. funestus* males also decreased significantly with distance but decreased with the number of inhabitants (distance $r = -0.009$, $p = 0.05$; inhabitant's $r = -0.511$, $p = 0.001$).

Figure 16.13 The semi-permanent seasonal pond in Linga Linga that acts as a major source of larvae of *A. funestus* in the wet season.

Figure 16.14 The author's house (a JDC design) on Linga Linga is made from locally available material, reed (*caniço*) and palm thatch (*makuti*) but, despite only having two rooms, is larger than most of the houses on the peninsular.

Table 16.2 Comparison of Collection Results Between the Southern End and the Middle of the Peninsula

	Exit		Light Traps	
Zone	A Tip	B Middle	A Tip	B Middle
N	56	47	12	5
A. funestus females	0.07	5.40	0.08	0.00
A. funestus males	0.02	6.57	0.08	0.00
Cx. spp females	5.09	1.87	18.17	29.60
Cx. spp males	1.04	3.43	0.42	0.00

Note: Numbers indicate density per collection. The number of collections is indicated by n.

Unlike the situation in Furvela, densities of *A. funestus* in Linga Linga followed the rain. Numbers in both light traps and exit collections were at their peak shortly after peak rainfall (Figure 16.15).

At low densities a high proportion of the females were gravid (Figure 16.16), implying that they were gonotrophically discordant and possibly aestivating.

Malaria incidence at a clinic established by the project also followed the rain and mosquito density but, as might be expected, with the highest correlation between rainfall and cases occurring with a lag of nine weeks (Spearman correlation coefficient between incidence and weekly rainfall = 0.34, $p = 0.0002$) (Figure 16.17).

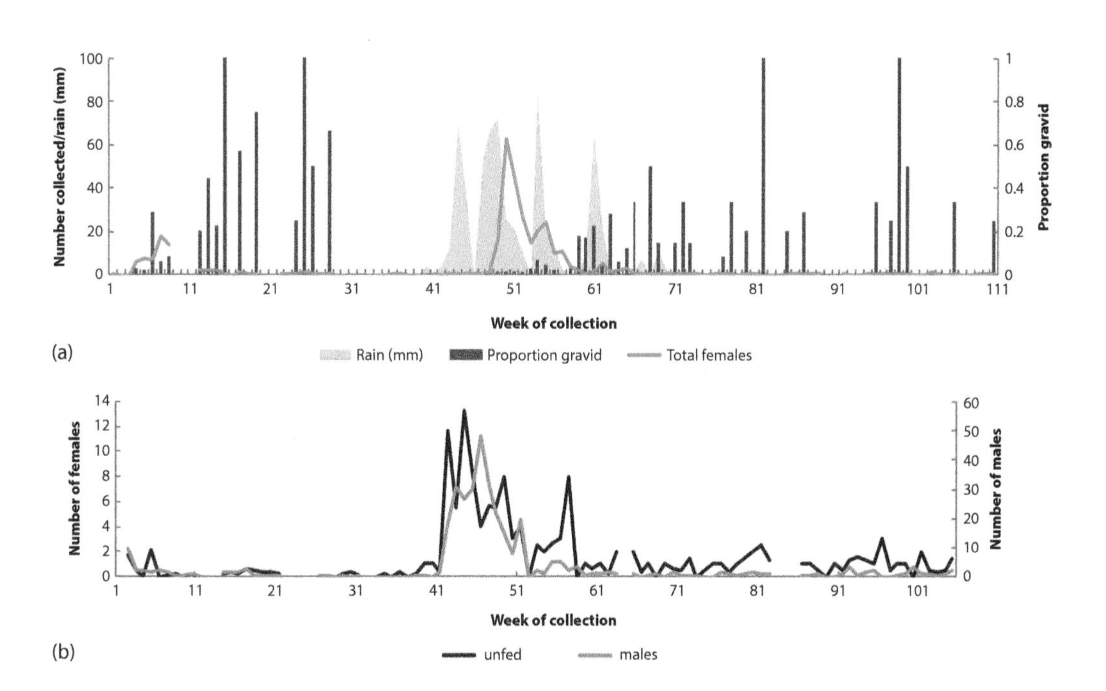

Figure 16.15 (a) Numbers of *A. funestus* collected in light traps, rainfall and the proportion of females that were gravid from Linga Linga. (b) Numbers of unfed females and males in exit collections from Linga Linga. (From Charlwood, J.D. et al., *Malar. J.*, 12, 208, 2013a.)

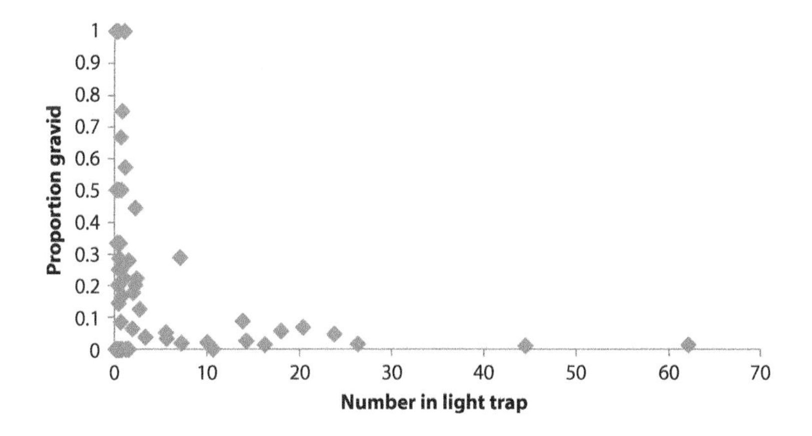

Figure 16.16 Proportion of female *A. funestus*, collected in light traps, that were gravid by number collected, Linga Linga.

Annual prevalence surveys were undertaken in Linga Linga from 2007 to 2011. People were invited to attend at a series of locations on the peninsula where thick and thin blood films were taken as well as information on sleeping habits obtained (Figure 16.18).

Prevalence varied from year to year and may have been influenced by rainfall. In 2007 the year with the lowest prevalence also had the lowest amount of rain, whilst the highest prevalence occurred in 2009, the year with most rain (Figure 16.19).

The incidence of fever (temperature ≥37.5°C) was greatest in babies and infants (Figure 16.20a) whilst parasite prevalence was greatest in the 5- to 9-year-olds (Figure 16.20b)

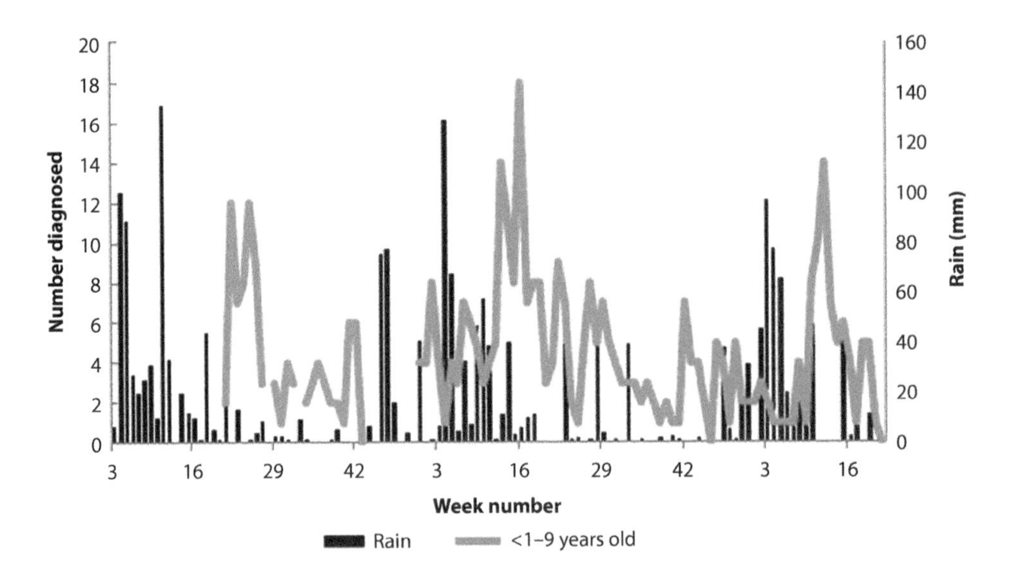

Figure 16.17 Incidence of malaria among 2- to 9-year-olds and weekly rainfall, Linga Linga. (From Charlwood, J.D. et al., *Peer J.*, 1370, 2015a.)

Figure 16.18 Prevalence survey from Linga Linga, 2008.

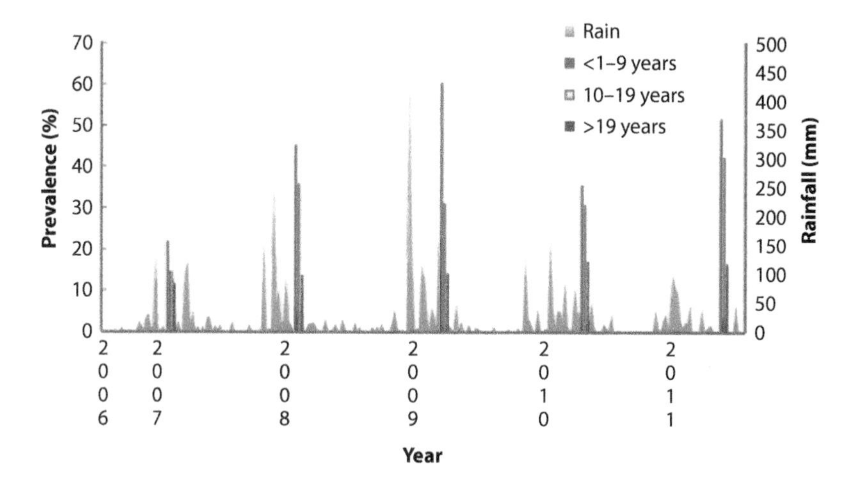

Figure 16.19 Prevalence of *Plasmodium falciparum* by age group and rainfall (measured in Maxixe) for the years 2007–2011, Linga Linga. (From Charlwood, J.D. et al., *Peer J.*, 1370, 2015a.)

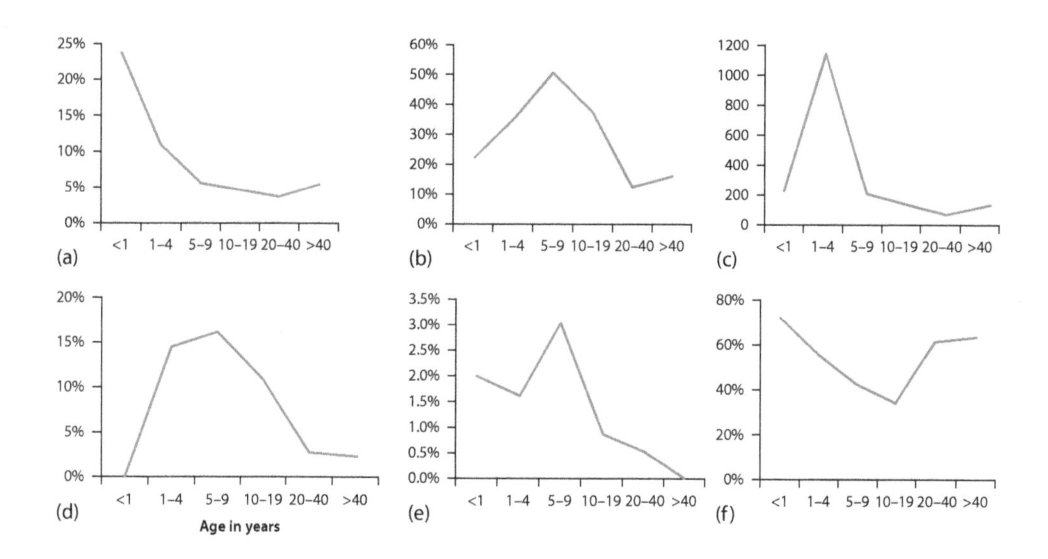

Figure 16.20 Age dependence and malariometric indices, Linga Linga. Prevalence surveys (a) fever, (b) prevalence of *P. falciparum,* (c) median *P. falciparum* density, (d) prevalence of *P. falciparum* gametocytes, (e) *P. malariae* (f) used net. (From Charlwood, J.D. et al., *Peer J.*, 1370, 2015a.)

(compare that to the prevalence in Furvela). Median densities were, however, highest in the 1- to 4-year-old age group (Figure 16.20c) whilst gametocyte prevalence peaked in 10- to 19-year-olds (Figure 16.20d) and *P. malariae* (Figure 16.20e), although always low, was highest in the 5- to 9-year-old age group. Ten- to 19-year-olds, in addition to having the highest prevalence of gametocytes, were also the least likely to use a bed net (Figure 16.20f). This combination would have enhanced transmission to the mosquito.

During the course of the project a variety of interventions were introduced whenever they became available. At the start of the study only 183 (19%) people from 58 (12%) households used a bed net.

A small number of nets, kindly donated by Vestergaard-Frandsen, were given to households with two children in 2007 and to all households in 2008 when more nets were available (Figure 16.21).

During distribution, the names of heads of households were called out and they came and could choose from three colours of net that were on display. Pink nets were the favoured colour but there was little difference between blue and green. These days you can have any colour of net as long as it is blue! (Figure 16.22).

Figure 16.21 Bed net distribution in 2008, Linga Linga.

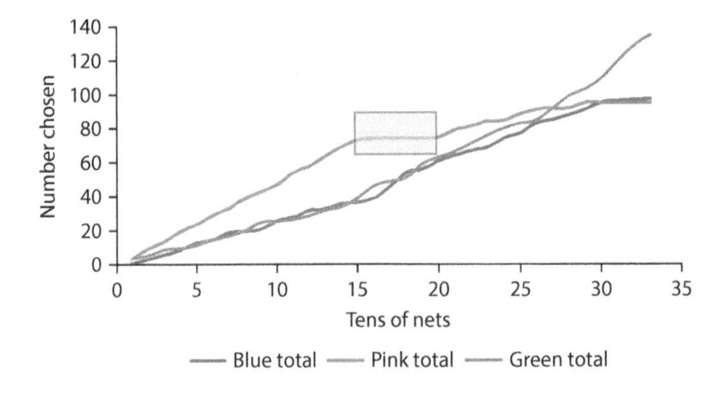

Figure 16.22 Uptake of nets by colour. The grey rectangle represents a period when pink nets were not available (hidden in the bottom of the sack!).

In 2009, a clinic dispensing ACTs for treatment was inaugurated at a central point on the peninsula.

The prevalence of gametocytes dropped from 39.5% (135 out of 342) in *P. falciparum* positive slides before the opening of the clinic to 14.7% (33 out of 224) once it had opened ($\Sigma^2 = 22.6$, $p = <0.05$). Treatment with ACT significantly reduces infectiousness of individual patients with uncomplicated *falciparum* malaria compared to previous first-line treatments. Rapid treatment of cases before gametocytaemia is well developed may enhance the impact of ACT on transmission. Reducing risk factors may also reduce transmission. We were able to identify a variety of risk factors, some of which can perhaps be reduced.

Fever and malariological indices among residents attending the clinic varied with age (Figure 16.23). The risk of fever was at a maximum in 1- to 4-year-old children. As in the prevalence surveys, it declined with age but in this case more slowly (Figure 16.23a). Overall peak diagnosis and peak positivity occurred in the 5- to 9-year-old age group (Figure 16.23b and c). Thus, the accuracy of the diagnosis was greatest in this age group. As in the prevalence, reported bed net use among residents attending the clinic was lowest among 10- to 19-year-olds (Figure 16.23d). People using a net the night before reporting ill were, however, as likely to have malaria as those who did not – of the 820 people who reported using a net who were diagnosed and tested for malaria, 506 (62%) were positive, whilst of the 194 tested who did not use a net, 120 (62%) were positive for malaria.

Many of the houses in Linga Linga had thatched ('green') roofs (Figure 16.24a). People living in such houses were at greater risk from malaria than those living in houses with tin roofs. There was

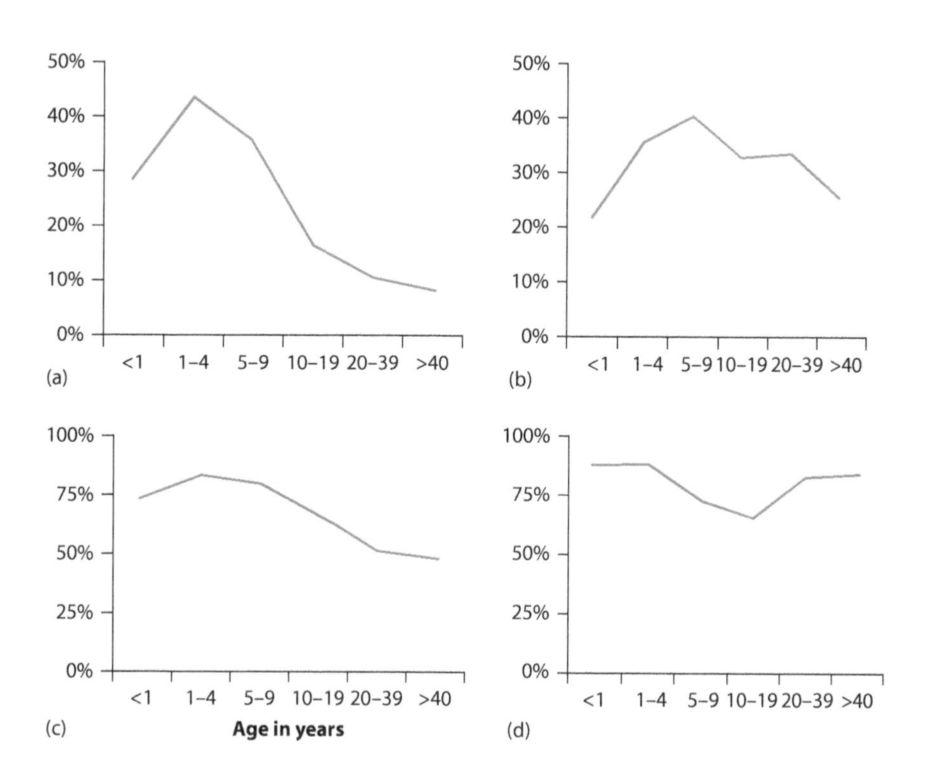

Figure 16.23 Malaria incidence among residents in Linga Linga: (a) fever, (b) diagnosed *P. falciparum*, (c) confirmed *P. falciparum* (d) proportion used net. (From Charlwood, J.D. et al., *Peer J.*, 1370, 2015a.)

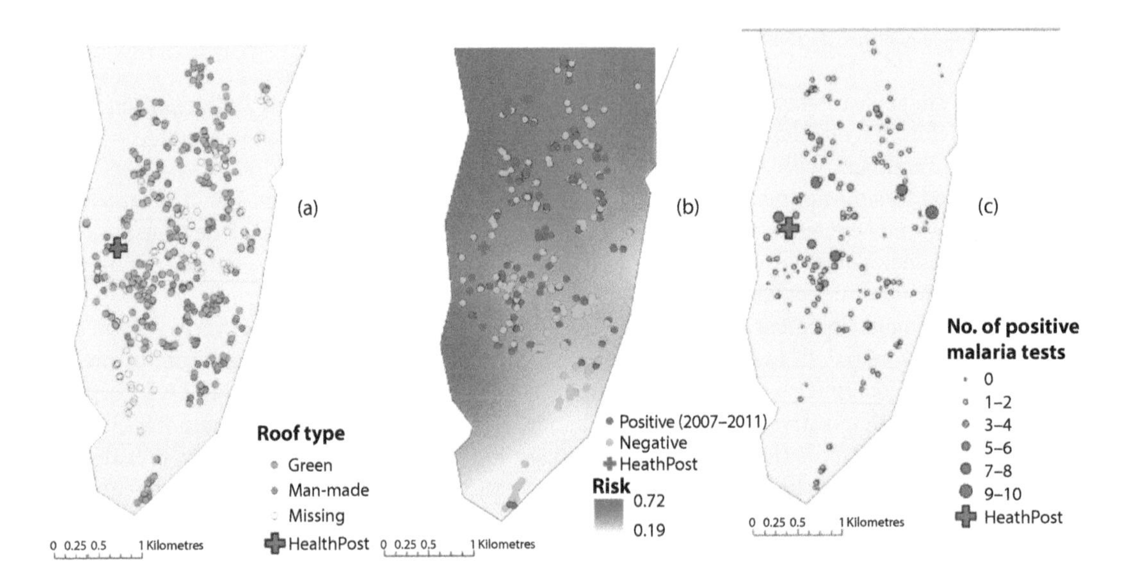

Figure 16.24 (a) Map of Linga Linga showing the distribution of houses recorded in the census according to roof type; (b) spatial pattern in malaria prevalence, after accounting for observed risk factors, fitted to the individual data; (c) distribution of people attending the Health Post by location according to the relative number of positive malaria tests performed. (From Charlwood, J.D. et al., *Peer J.*, 1370, 2015a.)

also an area of lower risk in the southeast of the study region and an area of higher risk in the north and west of the study area (Figure 16.24b).

There was no significant clustering of cases attending the clinic, although by mapping the number of visits per household and weighting these values by the number of people in the household (obtained from the census data), there was evidence that those living away from the clinic were less likely to attend (Spearman correlation coefficient between distance to clinic and number of visits per person per household = −0.1492, $p = 0.0031$) (Figure 16.24c).

In general, we have no reason to suppose that isolation causes the clinical epidemiology of malaria in Linga Linga to differ from that on continental Africa. Malaria was the most common diagnosis for children younger than 10 years of age attending the clinic. Fever peaked in the 1- to 4-year-olds, but the proportion of attendees diagnosed with malaria was greatest in the 5- to 9-year-olds. Diagnosis was also more accurate in children younger than 10 years of age than in older age groups, most of whom were women. It is likely that these were mothers or caregivers of sick children who also asked to be tested for malaria when they brought their sick children to the clinic.

The proportion of resident attendees younger than 9 years of age who were diagnosed with malaria decreased significantly from 48% in 2009, to 35% in 2010 and 25% in 2011 (for residents under 1 year of age $\chi^2 = 10.5$, $p = 0.005$; for 1- to 4-year-olds, $\chi^2 = 24.4$, $p \geq 0.000$, and for 5- to 9-year-olds, $\chi^2 = 5.92$, $p = 0.52$). At the same time, there was a shift in the peak age of cases from 1- to 4-year-olds to 5- to 9-year-olds (Figure 16.25).

Thus, together the interventions (LLINs and ACTs) appeared to have a major impact on incidence and morbidity. Maintaining the interventions has, unfortunately, been a problem (the clinic closed down due to lack of funds in 2012) – but in the future it is hoped that they will be reinitiated and that malaria will no longer be a major health hazard on the peninsula.

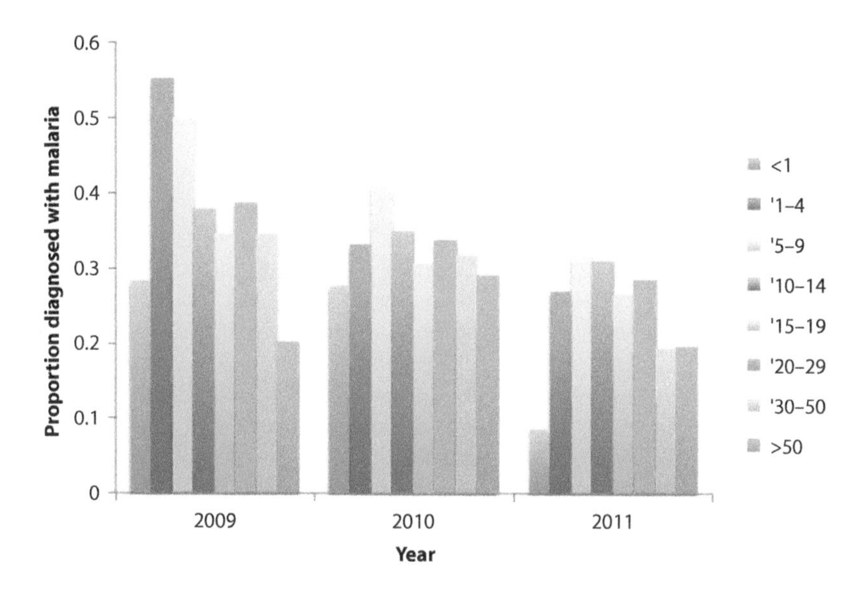

Figure 16.25 Proportion of patients attending the Health Post diagnosed with malaria 2009–2011, Linga Linga, Mozambique. (From Charlwood, J.D. et al., *Peer J.*, 1370, 2015a.)

SÃO TOMÉ AND PRÍNCIPE

If malaria can be eliminated from anywhere, it is islands – largely because they have much less immigration of mosquitoes or malaria than equivalent continental landmasses and the transmission picture is likely to be relatively simple. For the same reasons, islands may also act as natural 'laboratories' where novel control techniques might be employed. One such technique is the use of genetically modified mosquitoes' refractory to malaria transmission. Should this ever be attempted then background information on the genetic structure of the vectors at the release site should be determined. This was attempted on the islands of São Tomé (836 km^2) (Figure 16.26) and the smaller Príncipe (Figure 16.27) in the Gulf of Guinea, 240 kms off the coast of Gabon. The archipelago consists of two inhabited islands and a number of uninhabited islets. They form part of a chain of volcanic islands that mark a major fault line in the Earth's crust (with Mount Cameroon, and the islands of Bioko and Annobon being the other volcanoes in the chain). The islands have a mountainous topography, with many peaks in São Tomé rising above 1500 m, particularly in the west, central and southern parts of the island. The southern part of the island is heavily forested, while the northern part is flatter and humid, savannah, landscapes prevail.

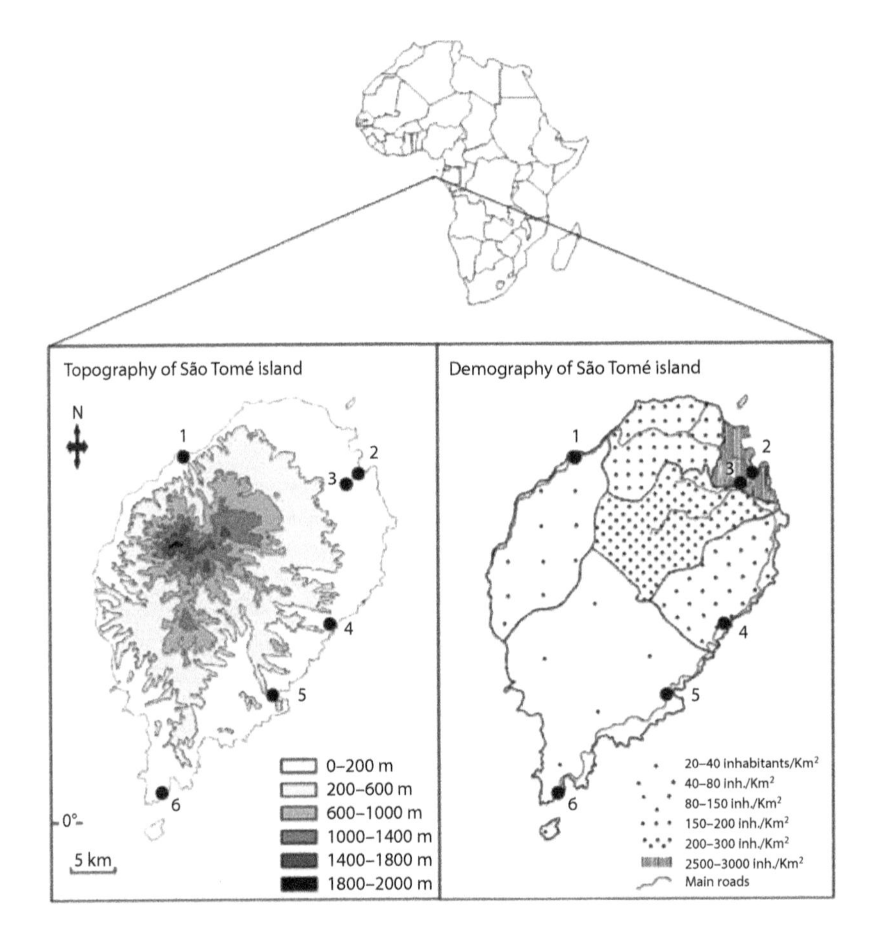

Figure 16.26 The island of São Tomé. Geographic location, topography, demography and collection sites. (1) Neves: capital of Lembá district, where the port for fuel importation is located. (2) Riboque: one of the main suburbs of São Tomé city, the capital of the country. (3) Bobo Forro: peripheral village of São Tomé city, in the most populous district Agua Grande. (4) Ribeira Afonso: main town of Cantangalo district. (5) Angolares: capital of Caué, the largest but least populated district of the island. (6) Porto Alegre: the most extreme southern settlement, a former colonial farm (*roça*). With the exception of Porto Alegre, where most houses are ground-level and brick-built, wood houses built on stilts up to 2 m prevail in all other collection sites. (From Pinto, J. et al., *Heredity*, 91, 407–414, 2003.)

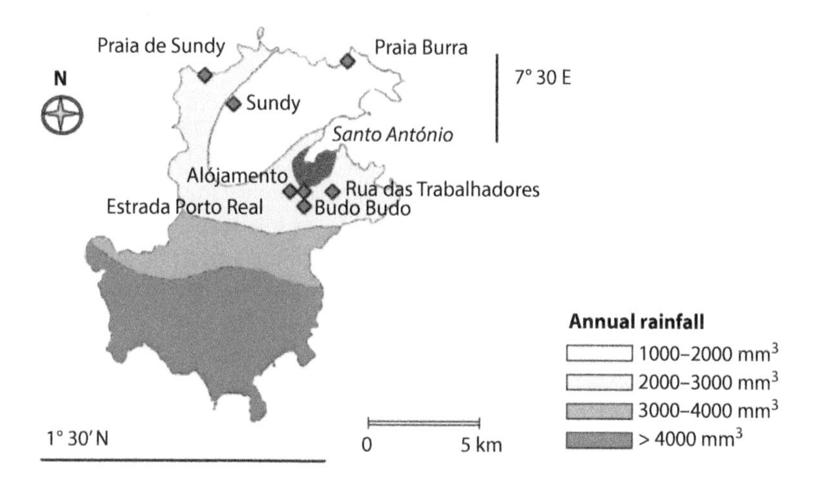

Figure 16.27 The island of Príncipe showing collection sites and rainfall distribution. (From Hagmann, R. et al., *Malar. J.*, 2, 9, 2003.)

The climate is equatorial with an average annual temperature of 25°C and an average humidity of 80%. Average rainfall varies between 500 and 2000 mm³ in the north and 3000 and 7000 mm³ in the south. There are two dry seasons, *gravana* (June–August) and *gravanito* (January), when rainfall is reduced but rarely absent.

When the islands were discovered in 1471 they were uninhabited. *Anopheles coluzzii* probably arrived in the late fifteenth century when the islands were colonised, because malaria was a health problem from the early days of colonisation. Indeed, becoming the governor of the islands was considered to be a death sentence. Two major waves of colonisation have occurred on the islands. The first took place between 1500 and 1521 when 2500–5400 slaves were imported every year into the islands to provide labour for the Portuguese sugar cane producers. Sleeping sickness was a major problem on the island of Príncipe and annual mortality rates were astounding. The tsetse vector was eventually controlled by an enterprising plantation owner who had his workers attach a sticky patch on the backs of their shirts. Flies attacking the workers were stuck on them and the population declined to virtual extinction. This was possible because of the low reproductive rate of the fly compared to mosquitoes.

By the end of the sixteenth century, sugar production had decreased and the islands were largely used as a *fuelling station for ships travelling* from Africa to Brazil. Little human migration occurred for 200 years. This lull was broken when coffee and cacao production was started on the islands; these crops needed cheap labour to be economically viable. The Portuguese introduced indentured labour, mainly from the Cabo Verde Islands and Angola, into the islands; by the 1860s more than 1000 labourers were being transported to the islands annually, resulting in a second wave of migration. In the early twentieth century the island was used during a solar eclipse as one of the sites for observations designed to test Einstein's theory of relativity.

The islands are exceptional in that four of the five human malarias are present. Five anopheline species, including the other major malaria vector *A. funestus*, were recorded in 1946 but by the 1990s, following an intensive control campaign using DDT for indoor residual spraying, only *A. gambiae* s.l. and *A. coustani* were recorded (the latter species in very low numbers from a restricted number of sites). The control campaign reduced the incidence of malaria to zero for more than a year and reduced indoor densities of mosquitoes to virtually zero. In a recent survey on the archipelago, 21 species across 7 genera were collected (Loiseau et al., 2018). The only malaria vector collected was *A. coluzzii*. Interestingly, half of the *A. coluzzii* analysed had apparently hybridised with *A. gambiae*, with introgression of the insecticide resistance gene, L1014F. The invasive species *A. albopictus* was also recorded in these surveys (previously the only *Stegomyia* mosquito recorded had been *A. aegypti*).

The control programme in the 1980s was interrupted after three years and, following the introduction of a chloroquine-resistant parasite from Angola, the islands suffered a major epidemic of malaria with many people dying as a result. Indeed, the epidemic was so deadly that in the Millennium Edition of the famous *Guinness Book of World Records* São Tomé was considered to be the most dangerous place on Earth!

Presently, the human population on São Tomé is unevenly distributed throughout the island with more than 60% of the island's 140,000 inhabitants living within 10 km of the island's capital city, São Tomé. Most of the remaining population lives in small villages along the main roads close to the coast. In the mountainous and forested south and southeast, human settlements are scarce or absent and the southwestern part of the island is almost completely uninhabited.

The picture of human settlement is similar on the smaller island of Príncipe (Figure 16.27). Approximately half of the island's 6000 inhabitants live in the island's only town, San Antonio, whilst the remaining people live in fishing villages, with 100–300 inhabitants located wherever there is a suitable beach and supply of fresh water, especially in the north of the island. The nearest road can be several kilometres distant.

The majority of people of Cabo Verdean descent live in the presently dilapidated former coffee/cacao estates or *roças* where they practise subsistence agriculture and gain some cash from a declining cacao industry. Although *roças* can be reached by road transport, there are few motor vehicles. This also makes *roças* several hours distant from the hospital in San Antonio. Otherwise, as in São Tomé, people tend to live in relative isolation alongside the one paved road on the island.

People living in San Antonio have a wider variety of occupations among them, selling salted fish to São Tomé. When they have sufficient fish, the traders take these to São Tomé by one of two boats that usually make a 36-hour return journey between the islands. In addition to the boats, there is a thrice-weekly plane service capable of transporting 20 people at a time between the islands. Hospital emergencies can be transported to São Tomé by the Portuguese air force, which remains stationed on the archipelago.

In the late 1990s a comprehensive study was undertaken to elucidate factors involved in malaria transmission on the islands. Studies included the analysis of the genetic structure and behaviour of the vector to transmission dynamics of the parasite.

Early on in these studies it was found that the FOREST form of *A. coluzzii* (formerly known as the M form of *A. gambiae*) was the only member of the *A. gambiae* complex on the islands. It was largely restricted to areas 200 m below sea level. This is probably due to a combination of a lack of hosts and to the steep topography of the islands, which mean that there is considerable runoff from even moderate rainfall. Despite relatively high biting rates the sporozoite rates on the two islands were of the order of 0.5% and 0.2%. The human blood index (HBI) was considerably lower on the islands than on continental Africa. This was probably due to a combination of host preference and difficulties associated with house entry. Many houses on the islands are built on stilts in order to reduce the effect of floods and to avoid mosquitoes. They are mostly built of wooden planks with a narrow piece of wood (a '*ripa*') over any gaps. The use of stilts also makes house entry more difficult for the *A. coluzzii* but has a lesser effect on the *Cx. quinquefasciatus* (Table 16.3). This meant that exposure to vectors could be up to 18 times lower between houses situated just 5 m from one another.

The mosquitoes on the island of Príncipe had a lower daily survival rate compared to the mosquitoes on São Tomé (0.801–0.818 compared to 0.834–0.849 in São Tomé). This resulted in prevalence rates that were lower in Príncipe compared to São Tomé. Fifty-three percent of people were positive from São Tomé and 35% were positive from Príncipe in 1997. In Príncipe in a larger sample from 1999, 32% were positive for *P. falciparum* by PCR, but only 3.2% were positive for *P. malariae* and 1/185 specimens was positive for *P. vivax*. In São Tomé malaria prevalence peaked in the 5- to 9-year-old-age group (Figure 16.28).

Plasmodium falciparum was the most common parasite. During initial surveys in 1997, significant differences were found in parasite formulae determined by OM and PCR (Table 16.4). Mixed infections, determined by PCR, especially with *P. malariae*, were more common than expected

Table 16.3 Mean Numbers of *A. coluzzii* and *Cx. quinquefasciatus* Collected Indoors and Outdoors from Houses Built on the Ground or Stilts in Riboque, São Tomé

Mean Number Collected per Man-hour (95% CI)	Outdoor Ground Level	Veranda	Indoor Ground Level	Indoor Upper Level
A. gambiae	16.51 (12.8–21.2)	8.10 (5.5–11.7)	3.58 (2.9–4.4)	2.38 (1.7–3.3)
Cx. quinquefasciatus	5.11 (3.9–6.6)	5.14 (3.5–7.4)	6.18 (5.1–7.4)	4.23 (3.2–5.5)

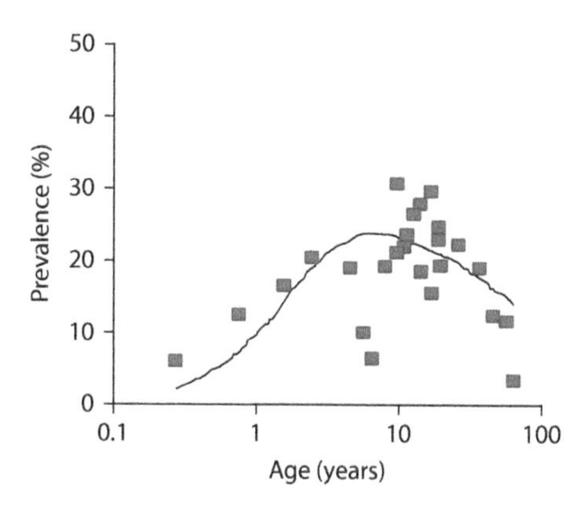

Figure 16.28 Parasite prevalence by age on Príncipe. (From Pinto, J. et al., *Acta Trop.*, 76, 185–193, 2000a.)

Table 16.4 Parasite Formulae from São Tomé Determined by Optical Microscopy and PCR

	P. falciparum	*P. malariae*	*P. vivax*	*P. ovale*
OM	92.4	6.2	0.7	0.7
PCR	82.2	10.2	4.8	2.8

by chance but this may have been due to the long clearance time (44 years!) for the latter species. For an unknown reason, many European travellers to São Tomé returned with *P. ovale* despite this being the rarest of the parasites found in cross-sectional surveys.

Malaria prevalence was positively correlated with *A. coluzzii* density (Spearman's rho = 0.6, $p = 0.02$, 2-tailed). Because mosquito density did not follow a normal distribution, a logarithmic regression ($R^2 = 0.58$, $p = 0.001$) was used to describe the relationship between parasite prevalence and mosquito density (Figure 16.29).

There is evidence from Papua New Guinea and Tanzania that having an asymptomatic infection may provide a measure of protection against developing a new infection and that the greater the number of clones that are present the greater the protection. This was examined in São Tomé in a three-month study in 1999. Blood slides were taken from 491 residents of an area of the capital, Riboque, and the incidence of disease among them followed up for three months. One hundred ninety-four (40%) of the samples were microscopy positive for *P. falciparum*, three for *P. vivax* and one each for *P. ovale* and *P. malariae*. Three hundred twenty-one (65%) of the samples were PCR-positive for *P. falciparum*. PCR-positivity increased from a minimum of 31% in children younger than 2 years of age to 82% in children between 10 and 14 years (Figure 16.30). Prevalence decreased somewhat with age after the age of 15. There was a clear tendency for the sensitivity of microscopy to decrease with age. Overall, 759 distinct *P. falciparum* infections were detected by RFLP-PCR.

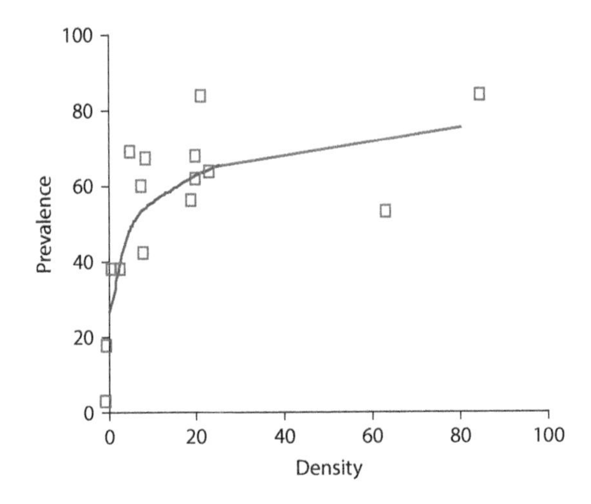

Figure 16.29 The relationship between malaria prevalence and mosquito density from São Tomé. Prevalence (P) percentage of individuals positive for malaria in each locality; density (D) number of *A. coluzzii* females caught per human hour in each locality. (From Pinto, J. et al., *Acta Trop.*, 76, 185–193, 2000a.)

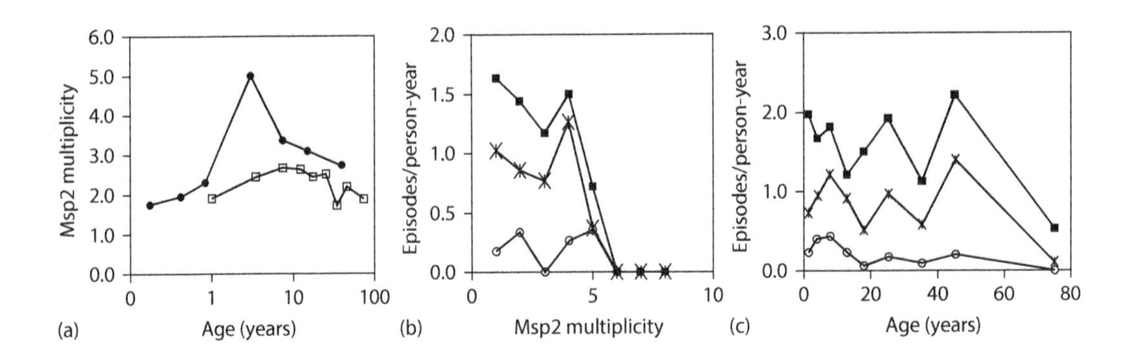

Figure 16.30 (a) Prevalence of *P. falciparum* by age; thin line microscopy; thick line, PCR. (b) Mean multiplicity of infection by age; upper curve Kilombero (Tanzania), lower curve, Riboque. Multiplicity is the mean number of msp-2 genotypes detected in each PCR sample. (c) Incidence rate of self-reported fever by multiplicity in baseline survey; solid squares all episode; * *P.* falciparum positive episodes; open circles *P. falciparum* positive episodes (density >5000 parasites/μL) (From Müller, D.A. et al., *Acta Trop.*, 78, 155–162, 2001.)

Out of 43 different *msp-2* alleles, 12 (28%) belonged to the FC27 allelic family and 31 (72%) of alleles belonged to the 3D7 allelic family (Table 16.5).

Islands like São Tomé are possible sites for the testing of the effect of releasing genetically modified insects' refractory to malaria. For this, some information on its effective population size (N_e) and the degree of isolation of the island population from the mainland is necessary. It is also interesting to know where the original population came from. In order to estimate the origin of the population, haplotypes (i.e., a group of genes that behave as single units and are inherited together) of different populations from Africa were investigated and compared to the haplotype found on São Tomé. Results are shown in Figure 16.31.

Table 16.5 Frequency Distribution of Multiplicity of Infections

Multiplicity	Frequency
1	94
2	96
3	56
4	47
5	12
6 or more	10
Total	**315**

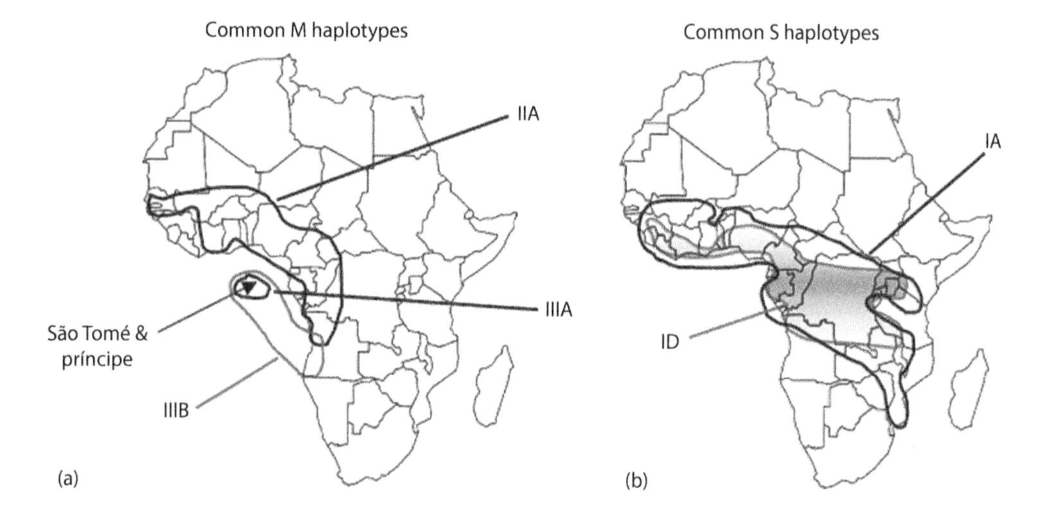

Figure 16.31 Haplotypes of M and S form (i.e., *A. coluzzii* and *A. gambiae*) from continental Africa compared to the haplotype IIA found on São Tomé. (From Marshall, J.C. et al., *Evol. Appl.*, 1, 631–644, 2008.)

The only haplotype that matched that found in São Tomé comes from an Angolan population. It is therefore probable that this was the source of the local population. Slaves were also taken to STP from Angola so it is likely that the mosquito arrived with them. Thus, although in many regions *A. coluzzii* is associated with drier environments, it also thrives in the humid environment of São Tomé. A founder effect, in which the established population is derived from a small number of genotypes that do not cover all of the available genotypes in the species, is evident in the population from the islands.

Major demographic events leave genetic signatures that could be used to gain important information about the impact of vector control interventions. Using microsatellite data, there was significant differentiation in the genetic structure of the two islands, but there was no evidence of any genetic bottleneck occurring as a result of the eradication campaign undertaken in the early 1980s. In a subsequent study, significant population differentiation revealed by the Fixation index (F_{ST}) and the analogous estimator for microsatellite data R_{ST} was found between the southernmost collection site, Porto Alegre, and northern localities. The Porto Alegre population was an 'island within an island'. The observed patterns of population substructure are probably the result of restrictions to gene flow in the less inhabited, more densely forested and mountainous south. In all localities, the

A. coluzzii appeared to be experiencing a demographic expansion consistent with a relatively recent (ca. 500 years) founder effect.

Entomological studies on the behaviour and population dynamics of the mosquito on the two islands indicated that survival rates in the larvae were low and consistent with other areas of Africa (Figure 16.32).

Mating takes place for a limited period of time in swarms that form shortly after sunset (Figure 16.33). The time that swarming started was related to light intensity rather than any other factor (Figure 16.34).

Although smaller females appeared to feed and mate successfully, there was a deficit of them in older age groups, implying either that they had left the study area or had died in the meantime (Figure 16.35a). Smaller females took a blood meal before mating (Figure 16.35b).

Small size did not, however, appear to influence the males' chances of mating because males collected in copula as they (and the female) fell out of a swarm was the same as that of the swarming males or males collected from resting collections (Figure 16.35c). Mated status of the female did not affect biting times (Figure 16.36).

A total of 5697 and 1567 host-seeking *A. coluzzii* were dissected from São Tomé and Príncipe, respectively. Also, 1649 (29%) of the insects dissected from São Tomé and 605 (39%) from Príncipe were taking their first blood meal. Overall, approximately half of these were virgins (802, 49% from São Tomé; 294, 49% from Príncipe).

The proportion of mosquitoes feeding as virgins in landing collections remained relatively constant throughout the year. Only 1 of 1158 recently mated females had two mating plugs. All nulliparous insects with ovaries at Stage II and all parous insects dissected had mated but none had

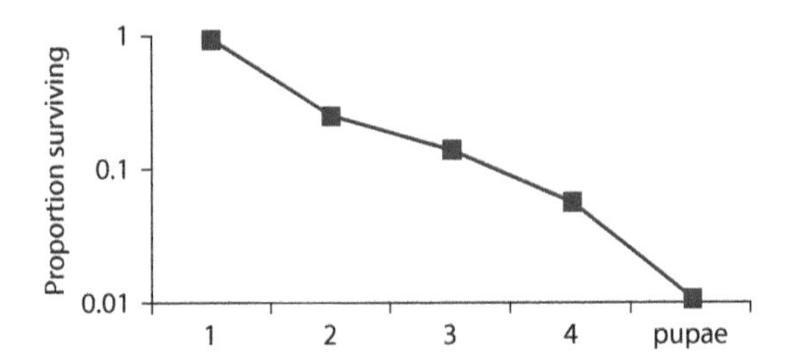

Figure 16.32 Larval survival of A. *coluzzii* derived from removal sampling. From Riboque, São Tomé. (From Charlwood, J.D., May the force be with you: Measuring mosquito fitness in the field. Chapter 5, in *Proceedings of a Workshop on Transgenic Mosquitoes*, Scott, T. and Takken, W. (Eds.), Wageningen, the Netherlands, 2003b.)

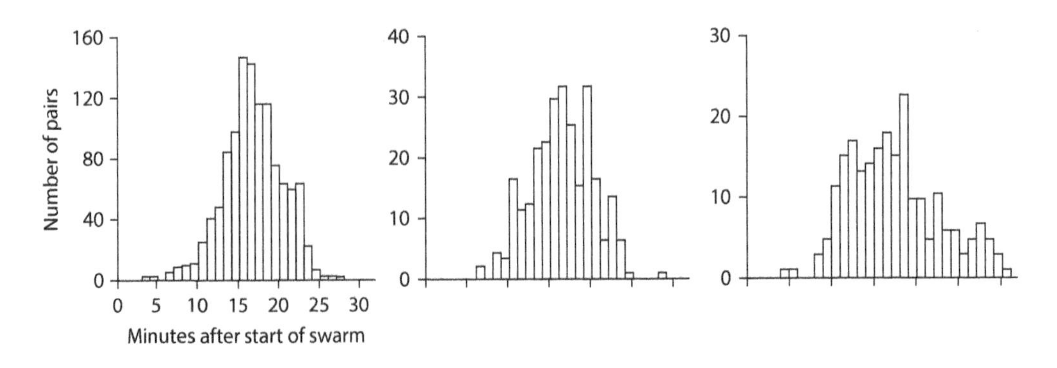

Figure 16.33 Mating times in *A. coluzzii* from Riboque, São Tomé, relative to the start of the swarm.

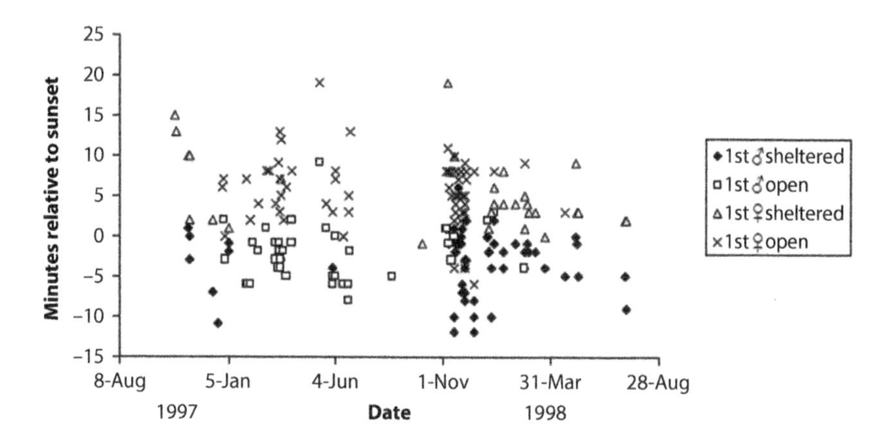

Figure 16.34 The initiation of swarming of *A. coluzzii* relative to the time of sunset, Riboque, São Tomé.

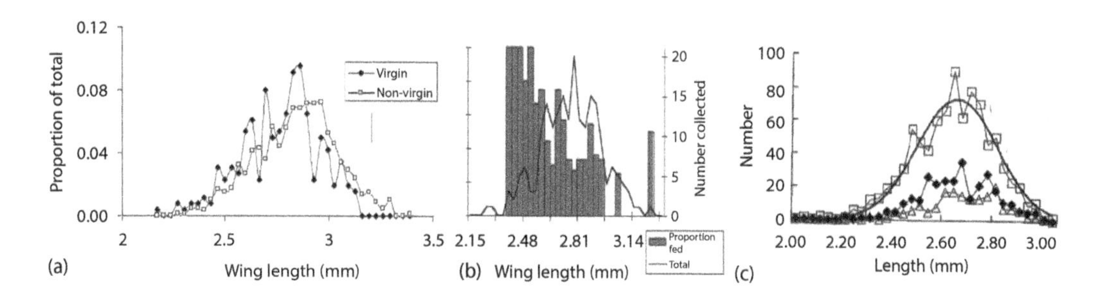

Figure 16.35 (a) Wing length of virgin and mated *A. coluzzii* females collected from landing collections; (b) proportion of females collected in copula that had taken a previous blood meal as a function of wing length; (c) wing length of males collected from swarms, caught resting or leaving swarms in copula. (From Charlwood, J.D. et al., *Malar. J.*, 2, 7, 2003c.)

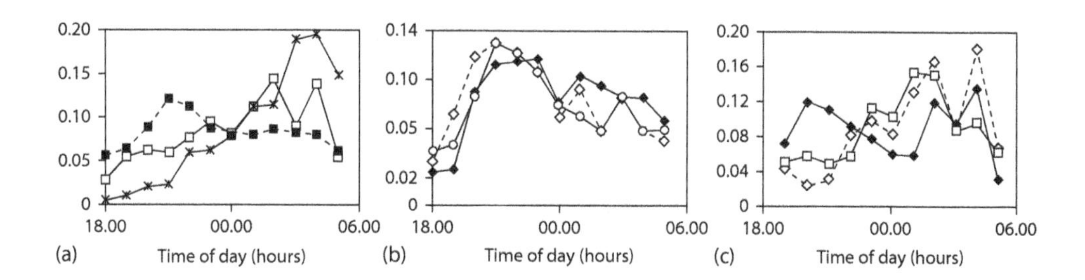

Figure 16.36 (a) Biting times of *A. coluzzii* from São Tome (filled squares) and Príncipe (open squares) and a composite figure from mainland Africa (*) and (b) the cycles for virgin (open diamonds), 'with plug' (filled diamonds) and mature (open circles) from São Tome (c) from Príncipe. (From Charlwood, J.D. et al., *Ann. Trop. Med. Parasitol.*, 97, 751–756, 2003b.)

a mating plug. Estimated survival rates from São Tomé were approximately 84% per day but rates from Príncipe, at 80%–82% per day, were significantly lower ($p = <0.001$), resulting in the expected life span of the mosquito from São Tomé being approximately twice that from Príncipe. This difference is sufficient to explain the lower parasite prevalence in Príncipe compared to São Tomé.

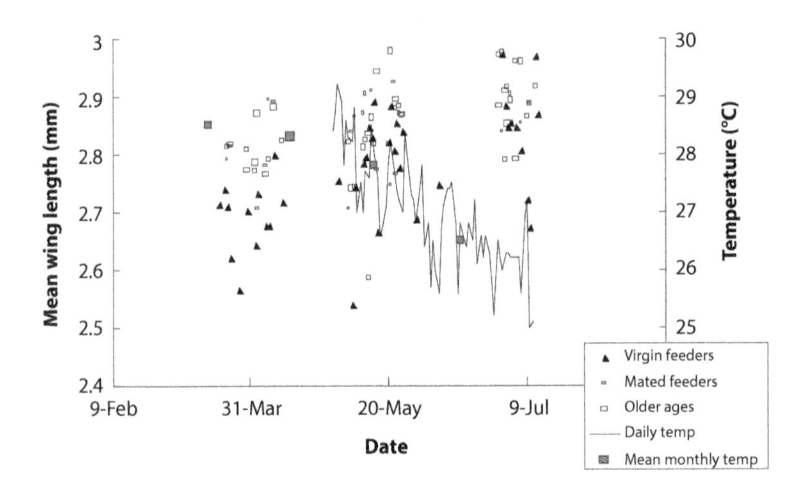

Figure 16.37 Wing length of female *A. coluzzii* according to physiological age and temperature (measured at the airport 5 kms from the collection site). (From Charlwood, J.D. et al., *Malar. J.*, 2, 7, 2003c.)

As temperatures decreased during June and July the size of the mosquitoes increased (Figure 16.37), a phenomenon observed elsewhere. Nevertheless, despite being larger than the mosquitoes in the hot season which mated before feeding, the smaller mosquitoes still took a blood meal before mating, indicating that there is more involved than size per se in this behaviour. Numbers of *A. coluzzii* decreased during this time of the year, due to the absence of rain and consequently of potential larval habitats. At the same time, the numbers of *Cx. quinquefasciatus* increased (Figure 16.38).

Another factor influencing vectorial capacity is host choice. In previous work the human blood index of the local *A. coluzzii* (from an admittedly small sample) was considered to be 0.96. The HBI of mosquitoes collected in 1998 was, however, only 0.27, the other hosts being dogs and pigs (Table 16.6). This was probably because many of the mosquitoes took blood meals outside, in part perhaps, because of an innate tendency to do so and in part because house entry was difficult.

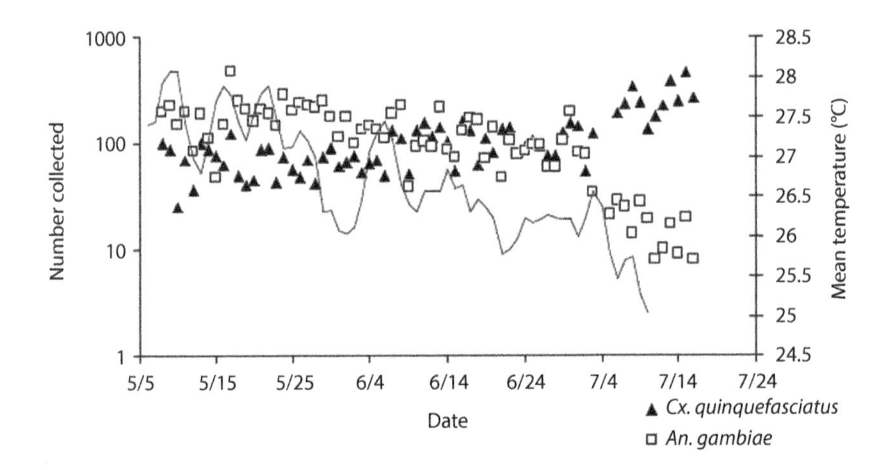

Figure 16.38 Numbers of *A. coluzzii* and *Cx. quinquefasciatus* collected in a sentinel light trap according to daily temperature from 5 May to 24 July 1999. (From Charlwood, J.D. et al., *Malar. J.*, 2, 7, 2003c.)

Table 16.6 ***Anopheles coluzzii*** **Blood Meal Source in Riboque São Tomé According to Collection Method and Location of Sample**

Host	Light-Trap Collections		Resting Collections Indoors		Resting Collections Outdoors	
	n	%	n	%	n	%
Human	379	87.3	161	83.4	87	20.6
Dog	11	2.5	1	0.5	185	43.8
Pig	0	0.0	10	5.2	84	19.9
Chicken	1	0.2	0	0.0	0	0.0
Goat	0	0.0	0	0.0	0	0.0
Rat	0	0.0	0	0.0	0	0.0
Total	391	90.0	172	89.1	356	84.3
Mixed feeds						
	22	5.1	21	10.9	45	10.7
Negative/other						
	21	4.8	0	0.0	21	5.0
Total	434	100.0	193	100.0	422	100.0
HBI	0.92		0.94		0.27	

Source: Sousa, C.A. et al., *J. Med. Entomol.*, 38, 122–125, 2001.

Table 16.7 ***Anopheles coluzzii*** **Captured Resting Indoors in 52 Rooms and in 33 Outdoor Resting Sites, Riboque, São Tomé**

Sites	Unfed	Engorged	Half-Gravid	Gravid	Total	Males Total
Indoors (PSC)	9	113	4	0	126	6
Outdoors (OR)	269	172	91	169	701	575

Source: Sousa, C.A. et al., *J. Med. Entomol.*, 38, 122–125, 2001.

Thus, only a few gravid insects were collected in resting collections (by pyrethrum spray catch, PSC) indoors compared to outdoor collections (Table 16.7).

The insertion of an HBI of 0.27 instead of 0.96 in the formula for vectorial capacity (C),

$$C = \frac{ma^2 p^n}{-\log_e p}$$

where m = density relative to humans, a = man-biting habit, p = probability of the mosquito surviving through 1 day, n = length in days of the extrinsic cycle of the parasite, while retaining the other figures, derived previously has the effect of reducing the estimate from 44.83 to 3.60. The revised estimate of vectorial capacity thus explains the low sporozoite rate observed and tallies better with the observed meso-endemicity of malaria on the island.

Despite the low rates of entry into houses by the mosquito, the relative proportion of positive blood slides among people presenting with fever at a community health post in Riboque in the town of São Tomé decreased when impregnated bed nets were increasingly used on the island (Figure 16.39).

In 2003, a malaria eradication program was initiated on the archipelago with emphasis on Príncipe. IRS with the pyrethroid alpha-cypermethrin was undertaken, intermittent preventive treatment for pregnant women and treatment with ACTs initiated. Prevalence rates dropped (Figure 16.40) and the island was heading towards eradication by 2009. Of the 5609 residents interviewed in 2009, only 273 were unprotected.

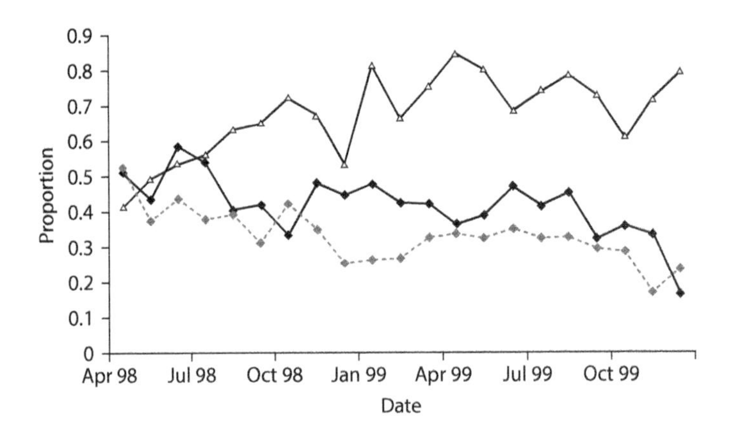

Figure 16.39 Proportion of people reporting with fever who had a positive blood slide. Prevalence in non-net users – solid squares; prevalence in net users – grey squares and open triangles – the proportion reporting using a bed net the previous evening. (From Charlwood, J.D. et al., *Trans. R. Soc. Trop. Med. Hyg.*, 99, 901–904, 2005.)

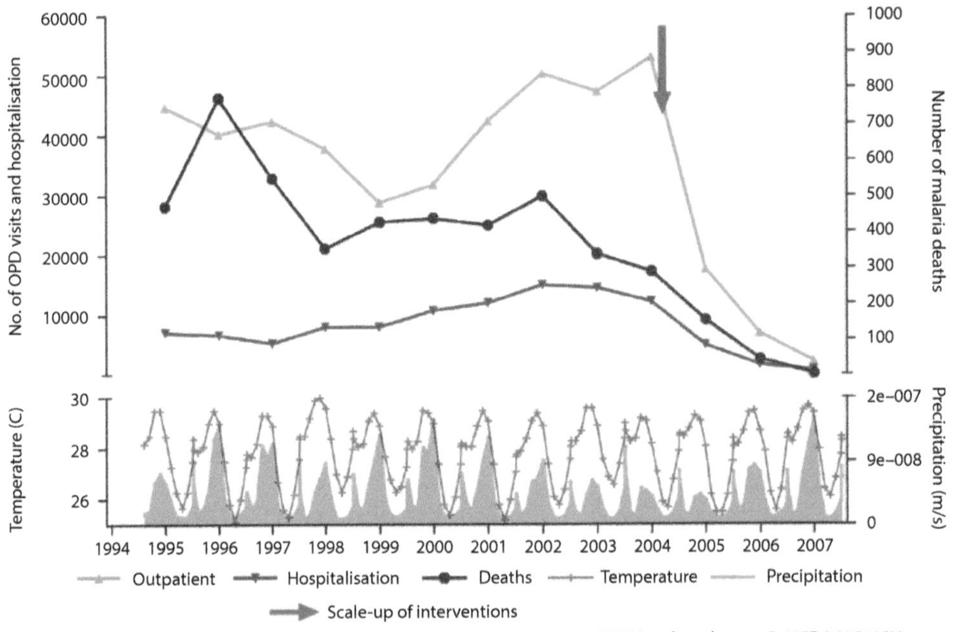

Figure 16.40 Malaria cases, deaths inpatient data, temperature and rainfall from São Tomé and Príncipe, 1995–2007. The arrow depicts the time of scale-up of the interventions. (From Teklehaimanot, H.D. et al., *Am. J. Trop. Med. Hyg.*, 80, 133–140, 2009.)

Estimated costs were between $7 and $10 per person per year (Lee et al., 2010). This is considerable, but in the long run if the disease can be eliminated from the island and the island can be kept malaria free, it may be worthwhile.

One way of keeping the islands malaria free is to ensure that arrivals to the island are not infected. One potentially interesting way of screening arrivals might be to use 'sniffer' dogs trained to detect people with parasites, as is being developed by Prof. Steve Lindsay and colleagues (Kasstan et al., 2019).

PAPUA NEW GUINEA

The island of New Guinea, consisting of Irian Jaya (a province of Indonesia) and the independent nation of Papua New Guinea, is the second-largest island in the world. It is located in an area of the globe where tectonic plates meet, the so-called Pacific Rim. Thus, seismic events are relatively common. In addition to the main island there are numerous other islands, many of which are also sizeable.

There are more than 700 separate languages spoken in New Guinea. In many situations, individual villages will speak a distinct language. This is an indication of the relative isolation between the different tribal groups. Many people live in traditional villages.

This isolation also affects the local mosquito fauna.

Malaria is endemic on the islands. Prior to the introduction of wide-scale interventions (notably the recent introduction of LLINs), prevalence rates were as high as anywhere in continental Africa. There are six major vectors of malaria and filariasis in PNG. These include: *Anopheles farauti* (formerly *A. farauti #1*), *A. hinesorum* (formerly *A. farauti #2*), *A. farauti* #4, *A. farauti* #6 (the major vector in the highland river valleys >1500 m above sea level), *A. koliensis* and *A. punctulatus* (Cooper et al., 2009).

The species differ in their favoured larval habitat. *Anopheles punctulatus* is most commonly found in open sunlit pools (similar to those colonised by *A. gambiae* and *A. arabiensis*). *A. farauti* can be collected from brackish pools close to the coast, whilst *A. koliensis* occurs in semipermanent habitats at the forest margin and in habitats such as water-filled bomb craters left by the last war. *Anopheles punctulatus* is the most tolerant of high temperatures. In general, *A. koliensis* could be collected in villages with either of the other two species but *A. punctulatus* and *A. farauti* were rarely encountered together (Figure 16.41). Following a blood meal, *A. koliensis* apparently disperses more widely than *A. farauti* because in villages where there were high biting densities of both species only *A. farauti* was collected in resting collections.

The Papua New Guinea Institute of Medical Research established a research program to study malaria and its vectors in the coastal region of Madang in the 1980s. The study area consisted of a narrow coastal shelf covered with secondary forest, village gardens or coconut plantations. One to six kilometres inland there was a range of steep-sided forested hills rising to approximately 400 m. Numerous small streams and rivers cross the whole area, and the larger Gogol River bisected the study area near its southern boundary. People from the Amelé tribe lived in the study area. They were fishermen and subsistence farmers, mainly growing a variety of root crops such as manioc. Pigs were commonly kept and in one village at least (Maraga, Figure 16.42) outnumbered people. On the coast, houses are made from coconut leaf stems for the walls and the fronds for the roof. Most houses are built on stilts circa 1 m off the ground. In many places, the houses at the edge of the village are used by adolescent boys (boy houses). These houses serve a double purpose. They reduce the stress associated with growing up

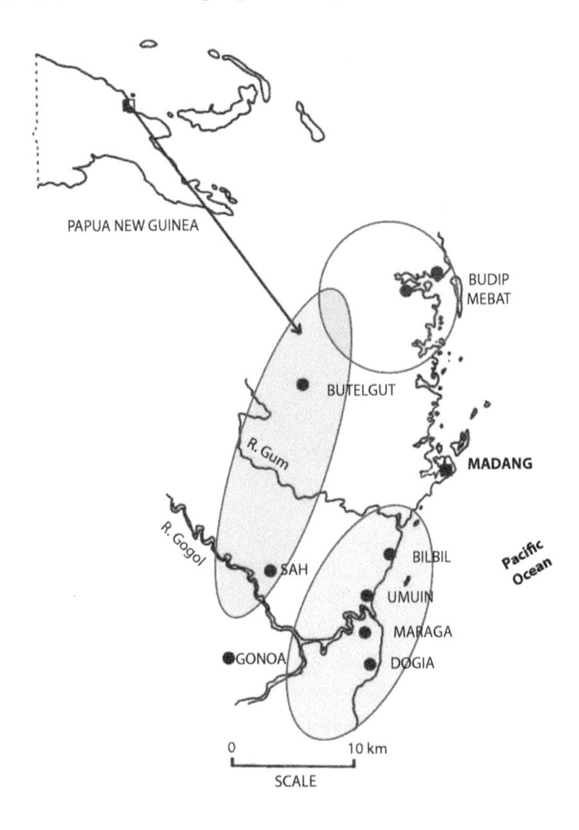

Figure 16.41 Map of the PNG IMR study area showing the approximate distribution of the predominant members of the *A. punctulatus* complex (*A. punctulatus* in blue; *A. koliensis* in green; and *A. farauti* in pink).

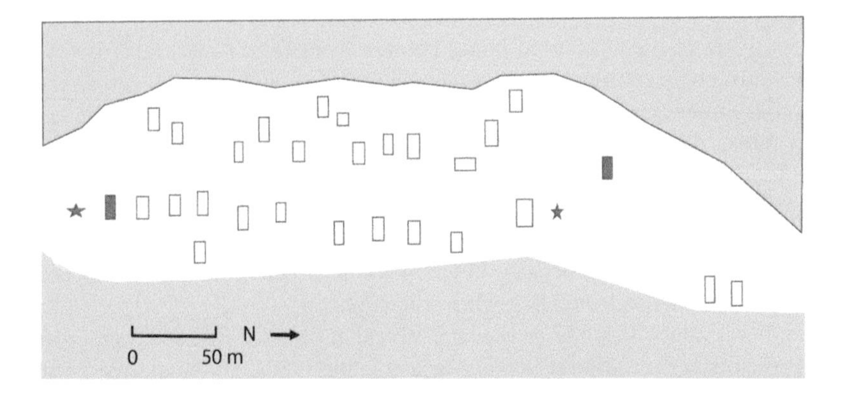

Figure 16.42 The village of Maraga (population 125) The stars denote release sites of mosquitoes. The rectangles are houses and filled rectangles the 'boy' houses of the village.

and should the village be attacked by another tribe, the first house the attackers would reach would be the one inhabited by the young 'warriors', the people best able to defend the village.

The annual rainfall is more than 300 m there, being more or less marked wet and dry seasons.

At the time of the studies reported here, there had been no vector control undertaken for a number of years. Previously, houses had been sprayed with DDT (at 2 grams/m^2). This reduced the numbers of *A. punctulatus* and *A. koliensis* but had little effect on the more exophilic *A. farauti.* Spraying had stopped, in part due to the complaints of the villagers who found that the roofs of their houses were being eaten by a pest (a mite) that had previously been controlled by a spider. Both were killed by the DDT. As with any other predator-prey system the predator took a lot longer to recover than the pest, which then feasted on the roofs, causing them to leak.

Different villages had different mosquito populations and therefore different epidemiologies. How isolated these mosquito populations might be were investigated in a series of experiments. Capture-recapture studies designed to determine survival rates of the different species and the duration of the oviposition cycle were conducted in the different villages. In order to do this, hungry mosquitoes (from landing collections) were blood-fed, marked with fluorescent powder and released. They were recaptured when they were once again attempting to feed (i.e., one or more complete oviposition cycles later). The human blood index (HBI) of *A. farauti* collected resting outdoors was also examined for mosquitoes from different villages. As expected, villages with many pigs, especially Maraga, had a lower HBI (Table 16.8).

When given the choice, 80 (70%) of 115 *A. farauti* fed on a dog rather than a human (Figure 16.43).

Table 16.8 Number of *A. farauti* Tested and Number Positive for Human Blood, Madang

Village	Number	%
Budip	37/44	84
Bilbil	1356/2709	50
Umuin	62/106	58
Maraga	237/2777	09
Dogia	86/242	36

Figure 16.43 An attractive dog and an unattractive human.

Because the pigs in Maraga acted as an alternative host for the local *A. farauti*, despite a very high biting rate, prevalence rates were lower than expected (Figure 16.44).

The survival rate of the mosquitoes from Maraga, determined by dissection, was also lower than that from neighbouring villages (Figure 16.45). This was probably in part due to the distance that the mosquitoes had to fly between the breeding site and the village. The site in Maraga was separated from the village by 1 km of forest whilst in most of the other villages it was adjacent to it.

The effect of the distance to the oviposition site was also reflected in the duration of the oviposition cycle. In Maraga it was significantly longer than in the nearby village of Umuin (Table 16.9). What was also apparent was that, where they occurred together, *A. punctulatus* had a longer cycle than *A. koliensis* whilst *A. koliensis* had a longer cycle than *A. farauti*.

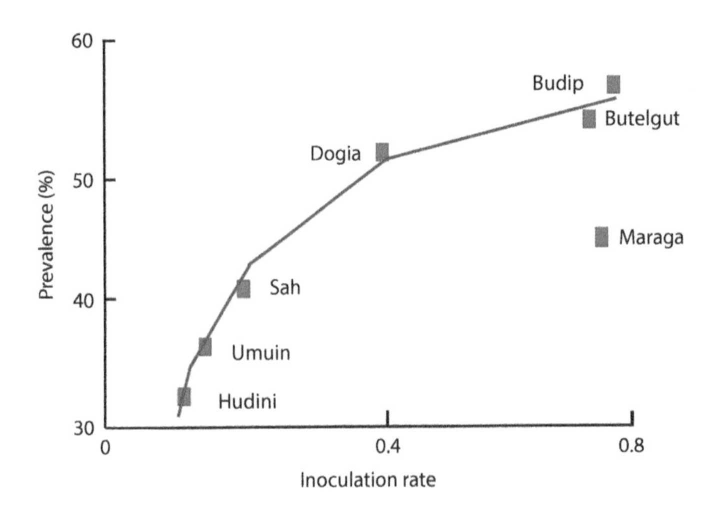

Figure 16.44 Prevalence rates in different villages in Madang by estimated inoculation rate of *P. falciparum*. (From Birley, M.H. and Charlwood, J.D., *Parasitol. Today*, 3, 231–232, 1987.)

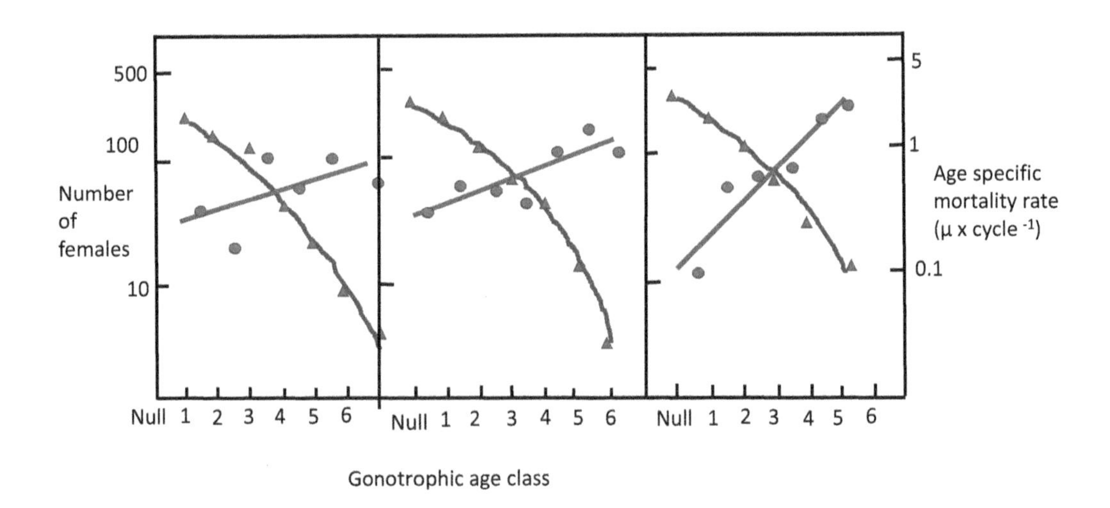

Figure 16.45 Mortality rates in *A. farauti* obtained from three villages in Madang. (From Charlwood, J.D., *J. Med. Entomol.*, 23, 361–365, 1986b.)

Table 16.9 Oviposition Cycle Duration (in Days) by Species and Village, Madang, PNG

	A. punctulatus	*A. koliensis*	*A. farauti*
Maraga	–	–	3.0 (0.67)
Umuin	–	2.9 (1.00)	2.1 (0.35)
Mebat	2.7 (0.66)	2.4 (0.50)	–
Butelgut	3.7 (1.00)	3.2 (0.60)	–

Source: Charlwood, J.D. et al., *Bull. Entomol. Res.*, 76, 211–227, 1986a.

By marking insects collected biting indoors separately from those collected outdoors, it was also possible to show that distinct subpopulations did not exist in these species – something worth knowing for control purposes.

In another experiment, mosquitoes were taken from Agan village to Maraga and released while at the same time other mosquitoes made the reverse journey. Recapture rates were compared to those of mosquitoes released in their home village. Surprisingly, transported mosquitoes were subsequently recaptured in their original village, but none of those released in their home village were recaptured in the other village. The oviposition cycle of transported mosquitoes was also longer (in the first cycle after release). This led to the idea that the mosquitoes had some spatial memory of their original emergence/oviposition site.

The use of mosquito nets to avoid the 'slings and arrows' of mosquito bites has long been advocated – indeed, Ronald Ross describes how on his trips to Mauritius he slept under a net whilst the macho soldiers who travelled with him didn't. They of course got malaria while he remained malaria free. Studies in Madang showed that, as might be expected, using nets reduced the number of *Anopheles* in a room. Numbers of *Cx. quinquefasciatus* were less affected by the introduction of nets into a room two, previously unprotected, adolescent boys slept (Charlwood, 1986) (Table 16.10). This reflects a difference in the hunting strategies of the two mosquitoes. *Culex quinquefasciatus* are 'sit and wait' predators, so once in the vicinity of a host are less likely to leave. Anophelines, on the other hand, if they fail to feed within a short time of locating a host will leave in search for a more accessible meal.

The idea of treating fabric with repellent was championed by Carl Schreck. Some of his DEET impregnated jackets, as a means to avoid leishmaniasis, were tested during a jungle warfare training programme in Manaus in the late 1970s (unfortunately, the trees ripped them to shreds). The development of synthetic pyrethroids in the early 1980s meant that nets could be treated with insecticide, which would then make them a killing agent against mosquitoes (Schreck & Self, 1985; Curtis & Lines, 1985). Early tests with experimental huts were encouraging (Darriet et al., 1984) and they also were effective against bedbugs (Cimicidae) (Charlwood & Dagoro, 1987), but they needed to be tested both for their effect on malaria and for their effect on malaria vectors on a

Table 16.10 Numbers of *Anopheles* and *Culex* from Two Rooms Before and After Nets Were Used by the Occupants

	Room 1		Room 2	
	Anopheles	*Culex*	*Anopheles*	*Culex*
Mean/night (% blood fed)	No net		With net	
	133 (75)	53 (72)	26 (74)	18 (48)
	With net		With net	
	23 (58)	65 (43)	27 (86)	41 (43)

Source: Charlwood, J.D., *Trans. R. Soc. Trop. Med.*, 80, 958–961, 1986a.

village scale. A study was therefore undertaken in one of the study villages in Madang where *A. farauti* and *A. koliensis* were relatively common.

The village of Agan consisted of 12 houses and had 64 inhabitants. During the experiment, one house fell down but three more were built. One person died but a baby was born. All-night landing collections were undertaken for 25 consecutive nights before and 21 nights after the introduction of the nets to cover all sleeping places in the village. The nets in this case were cotton and were impregnated on site. Capture-recapture experiments and resting collections were also performed before and after the introduction of the nets. A sample of the collected mosquitoes was dissected to determine parity and hence survival rate. In the 46 nights of collection, 27,372 *A. farauti* and 928 *A. koliensis* were caught and 2602 *A. farauti* and 656 *A. koliensis* were dissected.

Because collections were made on a daily basis, an attempt was made to determine survival rates by time series analysis. A suitable cross-correlation was evident for the *A. farauti* before the introduction of the nets. This was lost, however, after the nets were introduced (Figure 16.46). In other words, the population changed from a relatively synchronous one to an asynchronous one once the nets were in place. The nets produced a number of other changes on the ecology of the mosquitoes. The time of biting of both the *A. farauti* and *A. koliensis* changed from a peak after midnight to a peak before midnight (Figure 16.47), and the oviposition cycle (determined from the capture-recapture experiments) increased from 2.3 (+/− 0.4) days before the nets to 2.5 (+/− 0.5) days after they had been introduced.

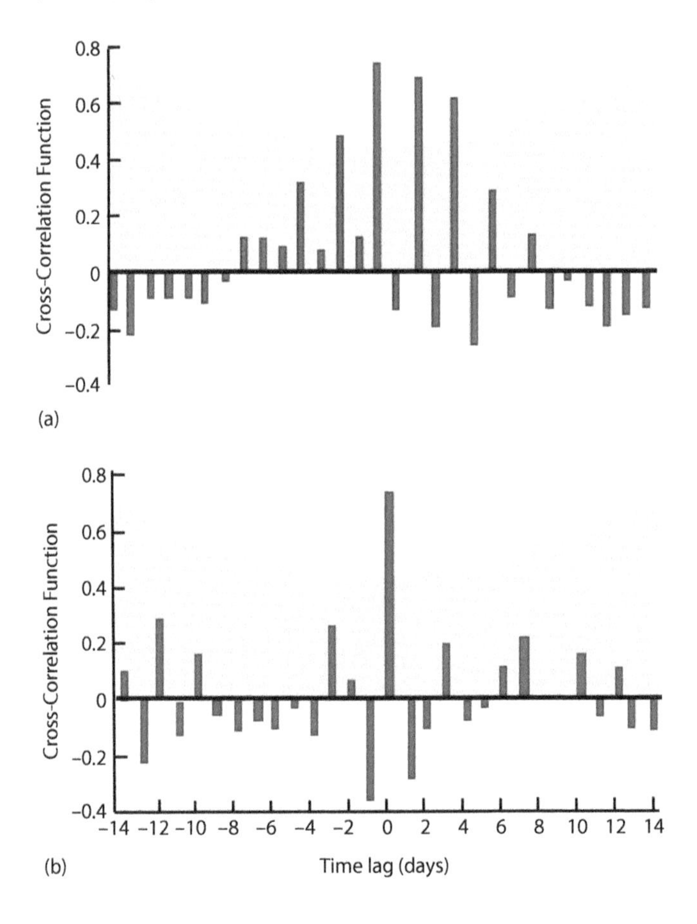

Figure 16.46 Filtered cross-correlations of the total and parous *A. farauti* collected in landing catches before and after the introduction of the impregnated bed nets into Agan village. The shaded section indicates the 9S% significance levels. (From Charlwood, J.D. and Graves, P.M., *Med. Vet. Entomol.*, 1, 319–327, 1987.)

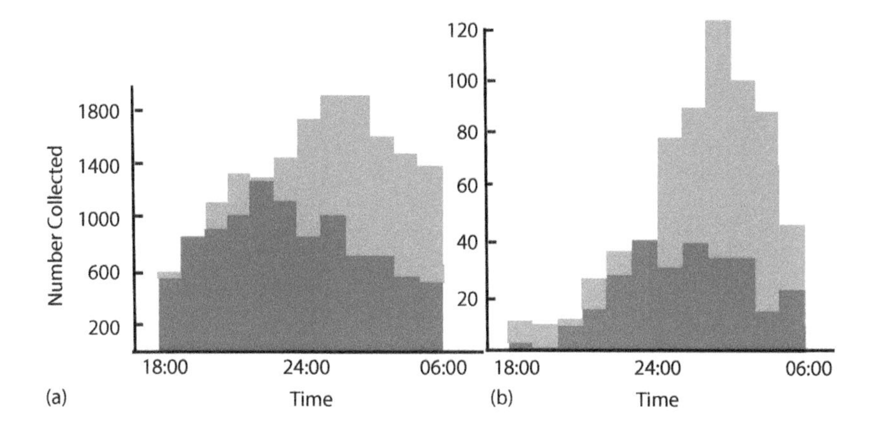

Figure 16.47 Biting cycles of *A. farauti* and *A. koliensis* from Agan village, Papua New Guinea before (orange histogram) and after (blue histograms) the introduction of pyrethroid impregnated nets into the village. (a) *A. farauti* (b) *A. koliensis*. (From Charlwood, J.D. and Graves, P.M., *Med. Vet. Entomol.*, 1, 319–327, 1987.)

The mean number of mosquitoes caught in resting collections also decreased, whilst the proportion of mosquitoes that fed on humans also decreased (Table 16.11).

The estimated population size from the capture-recapture experiments used two models: a simplified Lincoln index that ignored the nulliparous fraction of the population and a more realistic model developed by Alan Saul (1987). Saul's model allows an assessment of recapture rate, A, independent of survival rate.

For the Lincoln index:

$$\text{Estimated total} = \text{Total marked } \Sigma \text{ parous}/\Sigma \text{ recaptured}$$

The estimates of the total population size differed considerably (Table 16.12), especially after the nets had been introduced. This is because the collectors caught a larger proportion of the total population when nets were used. The estimated daily survival rate (determined from the recapture experiments also decreased, but not dramatically (Table 16.12).

Thus, the nets, overall, had a significant impact on the vector; in a series of experiments in six other villages the prevalence and incidence of malaria decreased significantly when nets were introduced (Graves et al., 1986).

Malaria is not the only vector-borne disease of significance in Papua New Guinea. In the East Sepik Province close to the Irian Jaya border, filariasis (*Wucheria bancrofti*) was also a major problem. In the villages in which we worked, 68% of people had microfilaraemia, with 64% of adults between the ages of 31 and 40 having obstructive disease being manifest as hydrocoele or

Table 16.11 Number of *A. farauti* and *A. koliensis* in Resting Collections Before and After the Introduction of Impregnated Mosquito Nets into Agan Village, Madang, PNG

	A. farauti		*A. koliensis*	
	Before	**After**	**Before**	**After**
Total	807	127	1.40	0.02
Mean	6.45	1.41	–	–
Blood fed	5.62	0.64	–	–
Human/animal ratio	2.12	0.57		

Table 16.12 Estimated Population Size and Daily Survival Rates of *A. farauti* Before and After the Introduction of Impregnated Mosquito Nets in Agan Village, Madang, PNG

	Population Size		Daily Survival Rate	
	Before	After	Before	After
Lincoln Index	22,451	13,330	0.70 (0.2)	0.65 (0.1)
Saul model	18,288	1864		

elephantiasis. There is the possibility that filarial developing in the flight muscle of the mosquito will affect dispersion and survival. This was investigated in a series of capture-recapture experiments with the local vector, *A. punctulatus*. Of 313 unmarked females dissected, 225 were uninfected, 58 (18.5%) had a single infection, 26 (8.3%) had double infections (i.e., had taken two infective blood meals) and 4 (1.3%) had a triple infection. These numbers fit a Poisson distribution, which therefore indicates that prior infection did not affect the chance of a mosquito picking up a new one.

Among recaptured mosquitoes, there was an overall infection rate of 57% and an infective rate of 7% (which was higher than the rates seen in unmarked insects, but this was not surprising because the marked population was an ageing one). One marked mosquito was collected five days after release, 1.8 km from the release site. It contained one L2 and 43 sausage-stage larvae. Thus, in that experiment at least, infection with *Wuchereria bancrofti* did not appear to affect survival.

Only very few people in the study villages used bed nets. Given that impregnated nets worked with *A. farauti* in Madang, a small experiment was also undertaken to determine how effective such nets might be against *A. punctulatus*. Impregnated and untreated nets were introduced into a series of houses and rotated on a daily basis, and the numbers of mosquitoes collected resting from the walls of rooms were compared to the situation where there were no nets in use. Overall, there was a 30-fold reduction in the number of mosquitoes collected when impregnated nets were in use compared to the control situation and a 14-fold reduction when untreated nets were used. The nets had been treated with 0.5 gms permethrin/m^2 and it appeared that at least part of the effect was because the net acted as a repellent rather than just a killing agent.

GHANA – ALL GENERALISATIONS ARE FALSE

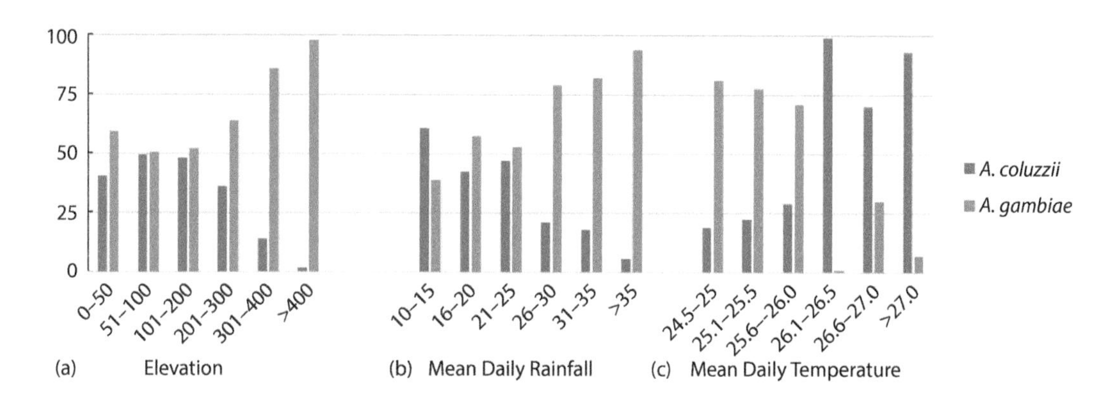

Figure 16.48 Distribution of *A. gambiae* (in red) and *A. coluzzii* (in blue) from southern Ghana in 2011. Numbers refer to the number of samples tested. *Anopheles coluzzii* predominates in rice growing areas close to the coast. (From Clarkson, C.S. et al., *Nat. Commun.*, 5, 4248, 2014.)

The whole of Ghana, with an estimated population of 29 million, is at risk of malaria. It accounts for 4% of the global burden and 7% of the malaria burden in West Africa. Despite malaria-attributable mortality declining significantly, from 19% in 2010 to 4.2% in 2016, in 2015 it was responsible for 19% of all recorded deaths in Ghana. The principal vectors are *Anopheles gambiae* and *A. coluzzii*. They occur sympatrically in the Gold Coast of Ghana (Figure 16.48).

In recent years the proportion of *A. coluzzii* with genes that confer resistance to pyrethroids, originally derived by introgression from *A. gambiae*, has increased even though the proportion of hybrids has remained low (Clarkson et al., 2014). As occurs in the Gambia, the distribution of *A. gambiae* and *A. coluzzii* in Ghana is largely based on ecological differences; *A. gambiae* occupies higher, wetter and cooler areas than *A. coluzzii* (Figure 16.49).

A short series of collections, designed to elucidate the ecology of the mosquitoes, were undertaken in the village of Okyereko (5° 24.87′N, 0° 36.25 W), some 70 km to the west of Accra. The village, 5 kms from the coast, consisted of 110 relatively rundown cement houses and was bordered on two sides by extensive rice fields (Figures 16.50 and 16.51).

Larval populations in rice fields (generally the primary habitat for *A. coluzzii*) consisted of both *A. coluzzii* and *A. gambiae* whilst puddles in the village (the archetypal habitat for *A. gambiae*) were occupied exclusively by *A. coluzzii*. This is not so surprising because of 235 mosquitoes identified to

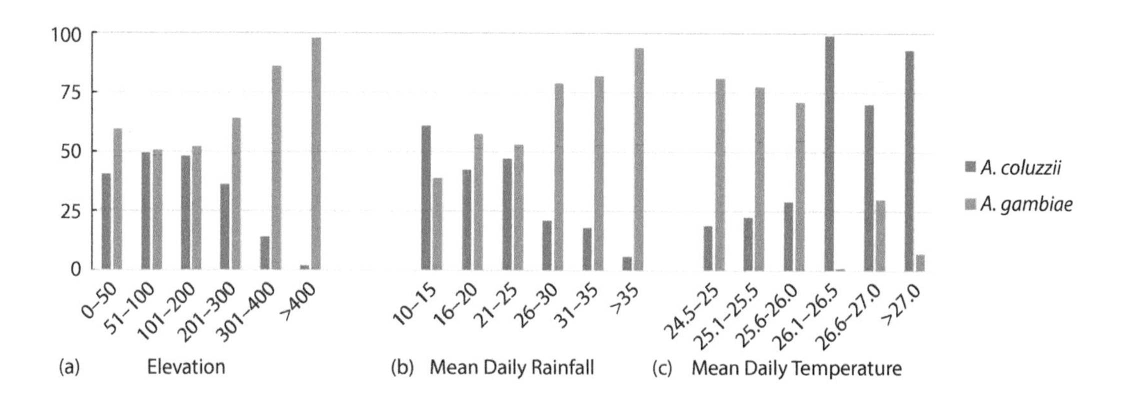

Figure 16.49 Relative distribution of *A. coluzzii* and *A. gambiae* in Ghana according to (a) elevation, (b) mean daily rainfall and (c) mean daily temperature. (From de Souza, D. et al., *PLoS One*, 10, 1371, 2010.)

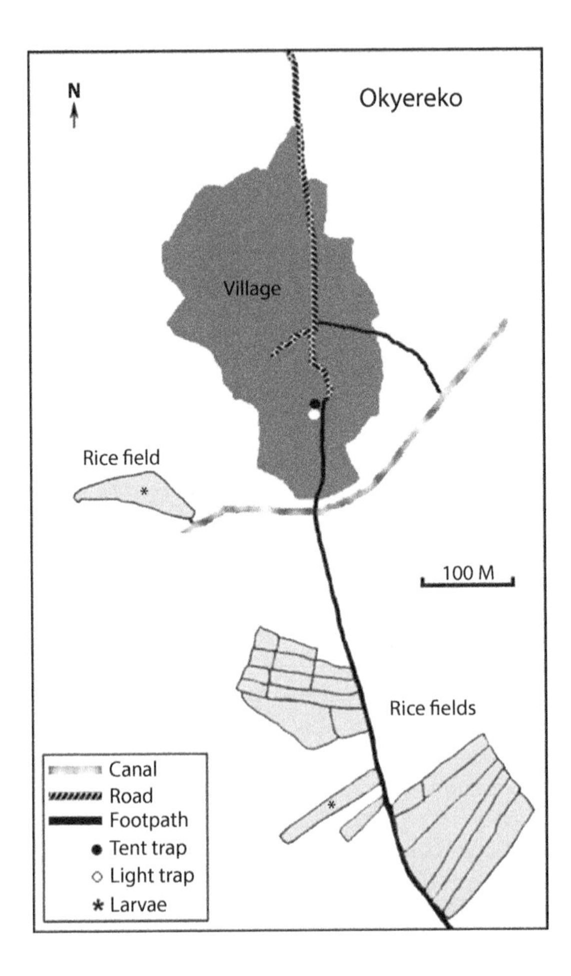

Figure 16.50 Map of Okyereko showing a sample of the rice fields (the other fields were not mapped). The fields with large populations of mosquito larvae are indicated by an asterisk.

species in 2009, 230 (98%) were *A. coluzzii*, 3 were *A. gambiae*, 1 was a hybrid and 1 was *A. melas*. All four *A. gambiae* specimens, including the hybrid, were collected indoors. The picture remained much the same in 2010. Of 107 *A. gambiae* complex females collected in a light trap and identified to species, 17 were *A. gambiae* (16%) and the remainder were *A. coluzzii*. Similarly, 10 of 116 (9%) mosquitoes collected in a tent trap were *A. gambiae*. No statistical difference in the samples collected was observed. A sample of *A. gambiae* larvae collected from puddles in a fallow rice field (Figure 16.49) showed a similar composition, with 8% of the sample comprising *A. gambiae* (*A. gambiae*, $n = 6$; *A. coluzzii*, $n = 74$), whereas a sample collected from a puddle in the village was composed entirely of *A. coluzzii* ($n = 48$). Moreover, all of the individuals collected in the middle of the village by sweep net from swarms or as mating pairs and identified to molecular form were *A. coluzzii* (110 males, 8 females). Thus, a limited amount of *A. coluzzii* mating occurred within the village itself, but swarming sites for *A. gambiae* remained undiscovered.

For *A. coluzzii*, markers can have swarms of limited size over them (i.e., there must be a 'packing' limit). When populations increased, so did the number of places where the insects were seen to swarm. This may result in some insects swarming in non-optimal places or in places otherwise used by *A. gambiae* which, because they do not appear to use markers but perhaps use an 'arena', may not have such a limit (Manoukis et al., 2009). Because there is little evidence for close-range

Figure 16.51 A 'new' *niche?* Rice fields in Ghana, now largely occupied by *A. coluzzii* where it occurs, are a human-made niche for mosquito larvae.

discrimination between the species, the end result might be an increase in the likelihood of cross-mating. Under these circumstances, the number of hybrids (and type of cross) that occur would depend on the dynamics of the most common species and the period during which optimal markers are super-populated and superfluous males swarm over non-optimal markers. It also implies that the proportion of hybrids in an area varies seasonally.

One of the main objectives of the study was to examine the behaviour of newly emerged females and to determine factors that might affect the population dynamics of the mosquitoes. Given the great preponderance of *A. coluzzii* in the village, it was considered that this was the only member of the *A. gambiae* complex we were dealing with.

In both 2009 and 2010 the population of *A. coluzzii* increased during the study from fewer than 200 insects per night in a single tent trap to more than 1000 (Figure 16.52).

On rainy nights mosquitoes were caught in greater numbers inside houses compared to the tent trap, although the total collected was as expected. This implies that on rainy nights people are more exposed to mosquitoes indoors than they otherwise would be (Figure 16.53a).

Short-term environmental effects were not limited to rainstorms. Lower than expected numbers of recently mated females were collected on nights when wind had disrupted swarming. The wind did not apparently affect survival, because on the succeeding night a greater proportion of recently mated mosquitoes were collected (Figure 16.53b).

In 2009, a rainstorm of 148 mm on 11 June, which heralded the rainy season, had a small negative effect on the number of *A. coluzzii* collected. By 19 June, following two further nights when 20 mm of rain fell, the ground was saturated. Numbers of mosquitoes were reduced following a storm of 127 mm on 19 June. There was subsequently greater variability in numbers collected, perhaps due to flushing of larvae as a result of this storm.

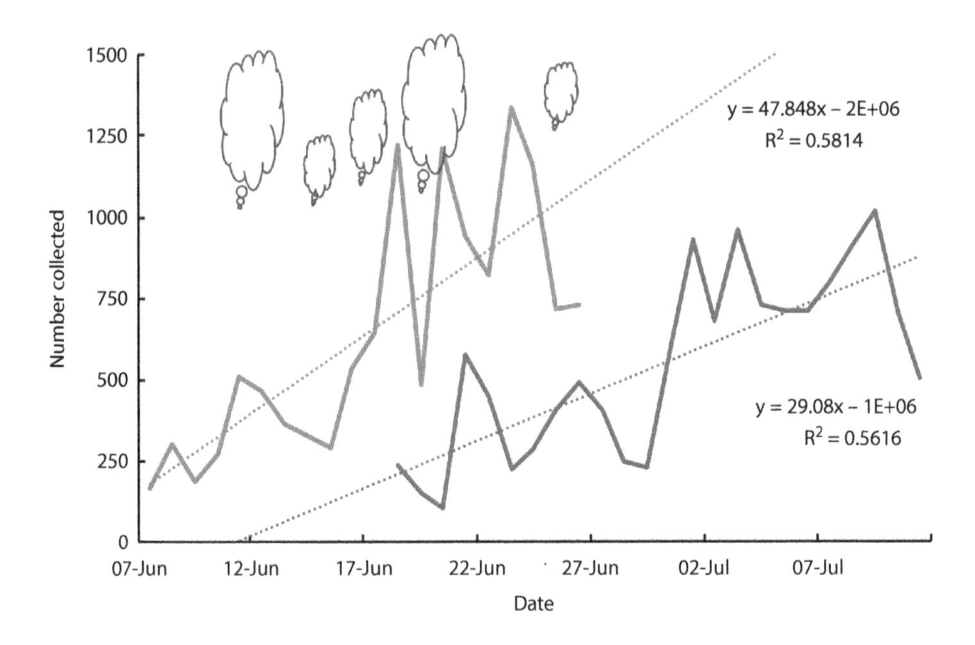

Figure 16.52 The number of *A. gambiae* s.l. (largely *A. coluzzii*) collected from a single tent trap in Okyereko village in 2009 (orange line) and 2010 (blue line). Clouds indicate nights of rainfall in 2009.

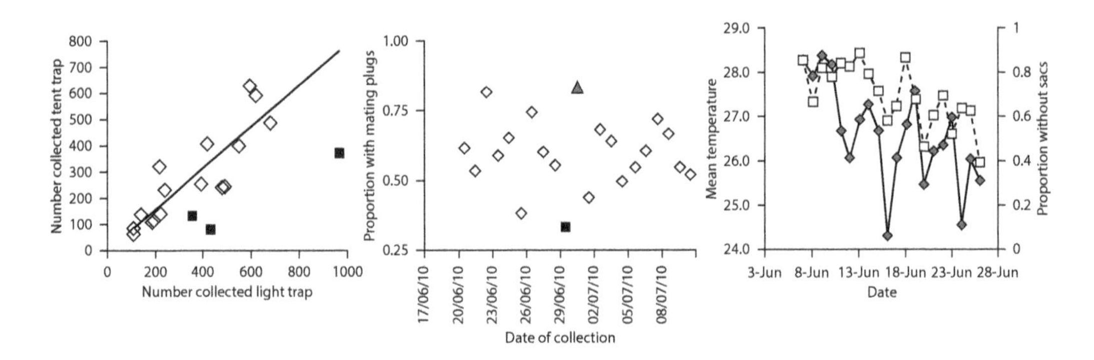

Figure 16.53 (a) Relationship between the number of *A. coluzzii* collected in tent and light trap on nights with (solid square) and without rain (open diamond, June 2009; (b) proportion of newly emerged *A. coluzzii* with mating, June/July 2010. Solid square collection during the night after wind had dispersed swarms, solid triangle, collection the following night; (c) proportion of parous *A. coluzzii* returning to feed without ovariolar sacs and mean daily temperature (solid line with filled diamonds). (From Charlwood, J.D. et al., *Bull. Entomol. Res.*, 101, 533–539, 2011.)

During the study, temperatures also dropped (Figure 16.53c) from a mean of 28.4°C on 9 June to 24.4°C on 24 June. At the same time, the proportion of parous mosquitoes without sacs also decreased – giving rise to a lowered estimation of the duration of the oviposition cycle.

In 2010 the most severe flooding of the previous 19 years occurred in the days following a day of heavy rain on 21 June. This inundated much of the surrounding area and many of the rice fields close to the village (Figure 16.54). The water in the flooded fields took 10 days to disperse and the affected fields were not brought into production during the study; nevertheless, despite an immediate drop in numbers, the population of *A. coluzzii* increased 4.5 times (from a mean of 163 per night in the first three nights of collection to a mean of 740 in the last three nights). The population of *A. pharoensis* increased tenfold at this time (from 1.3 per night to 13.3 per night).

Figure 16.54 Flooded houses on the outskirts of Okyereko village following heavy rain on 21 June 2010.

The increase in recently emerged insects was more regular than it was in the older age groups (Figure 16.55), implying that other environmental factors affected the adults once they had emerged but that larvae were developing independently of environmental peturbations.

In both 2009 and 2010, recently mated females with evidence of a previous blood meal ('plug-blood' insects) were rare or absent. In 2009, 192 of 311 (62%) recently emerged insects dissected from the light trap and 354 of 574 (62%) from the tent trap had mating plugs, indicating that the two traps were sampling the same population.

The overall parous rate in 2009 was 0.33 and in 2010 was 0.36. Cross-correlations between virgin and recently mated females were significant at a lag of one day. Significant cross-correlations occurred at four days between first- (virgin and plug) and second-feeding insects (Nulliparous females with ovarioles at Stage II).

The population of *A. coluzzii* in Okyereko differs from that in São Tomé in a number of respects. The ecological 'templet' (Southwood, 1977) for the two populations differed because, in Okyereko, feeding sites were distant from emergence sites, whereas in São Tomé they were within a few metres of one another. Pairs in copula were rare (in relation to the size of swarms) in Okyereko but were commonly seen leaving swarms in São Tomé; insects that fed as virgins and returned to feed with a mating plug (*plug-blood*) were nonexistent in Okyereko, but, although relatively under-sampled, were found in São Tomé. Further, mated females with ovaries at Stage I but without a mating plug (N-I) were common in Okyereko but were not recorded in São Tomé. Thus, the environment would appear to be a major influence on the ecological characteristics of this important vector of malaria.

The additional distances involved, and the extra energy requirements that this entails, may be responsible for the differing ecologies, but how this could be used against the mosquito is uncertain. In Okyereko, controlling the larvae in the rice fields at a distance from the village, perhaps by intermittent irrigation, may be useful.

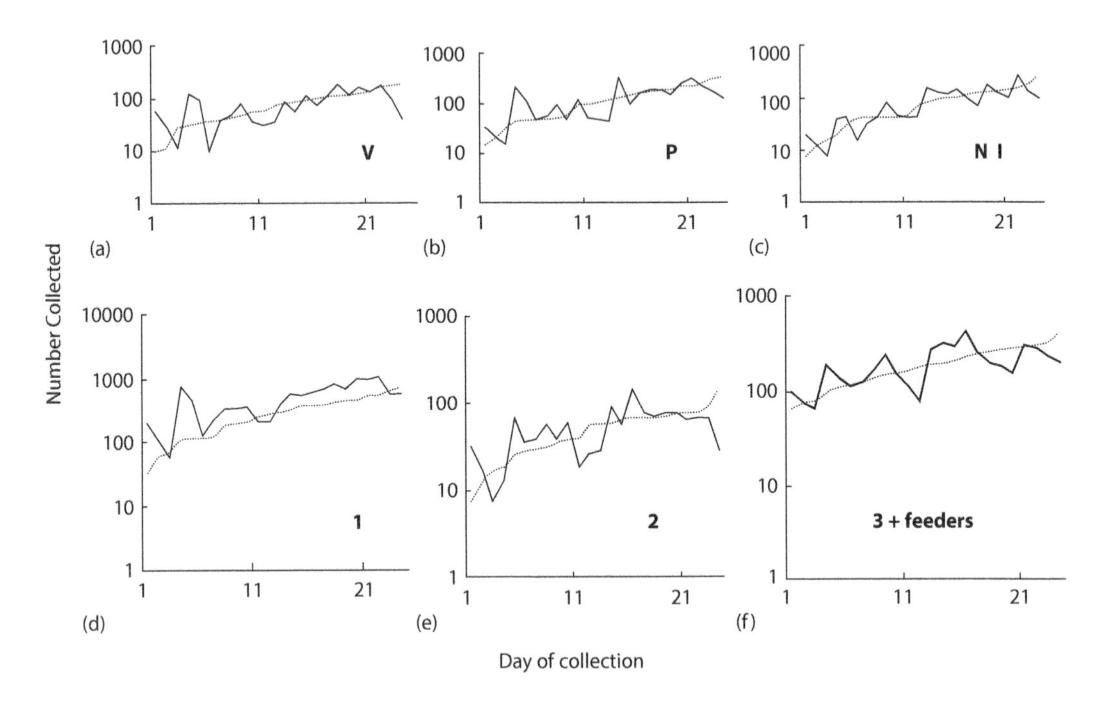

Figure 16.55 Estimated numbers of *Anopheles gambiae* s.l. (largely *A. coluzzii*) by gonotrophic age collected in the Furvela tent trap, Okyereko, during June and July 2010. (a) Virginal mosquitoes; (b) mosquitoes with a mating plug, unfed; (c) nulliparous mosquitoes with ovarioles at Stage I; (d–f) mosquitoes feeding for (d) the first, (e) the second and (f) a subsequent time. (From Charlwood, J.D. et al., *Med. Vet. Ent.*, 26, 263–270, 2012.)

TANZANIA

According to the World Health Organization, 69% of the estimated 663 million malaria cases averted during the past 15 years were attributed to the use of LLINs, 10% to IRS and the remaining 21% to artemisinin combination therapy (ACT) (WHO, 2015a). The development of resistance by the principal vectors of malaria to the pyrethroid insecticides used on mosquito nets is potentially a very serious problem for malaria control. Nowhere is this of greater concern than in the areas around Lake Victoria in Tanzania. The *kdr* gene is 'fixed' in *A. gambiae*, the principal vector in the region, *A. gambiae*. The mosquitoes also show extremely high levels of metabolic resistance to pyrethroids, DDT and bendiocarb. Although new control techniques may eventually involve non-insecticidal approaches, for the moment treatments of nets in such a way that the resistance is minimised may extend the useful life of nets. One such method is to incorporate a synergist along with the insecticide. Piperonyl butoxide (PBO) is a chemical that acts by inhibiting enzymes involved in the natural defence mechanisms of insects, which results in the pyrethroid not being detoxified in the insect and the net remaining potent against mosquitoes despite resistance. Alternative strategies include combining LLINs with IRS using different classes of insecticides.

A four-arm randomised control trial with 24 clusters was undertaken in Muleba District (1° 45′ S 31° 40′ E), covering an area of 944 km^2 to the north of Lake Victoria in which nets with PBO were compared with standard nets with and without IRS (with microencapsulated pirimiphos-methyl). The area is situated between 1100 and 1600 m above sea level with many steep-sided hills. People are generally subsistence farmers growing, in particular, bananas. People tend to live in isolated houses surrounded by their banana plants (Figure 16.56).

There were altogether 29,000 households and an estimated population of 135,000 inhabitants in the study area. Light traps (indoors) and tent traps (outdoors) were used to monitor the mosquito populations and cross-sectional parasitological surveys of malaria prevalence in children 6 months old to 14 years of age were undertaken to estimate malaria transmission.

Prior to the intervention, a total of 29,401 and 2668 mosquitoes were collected indoors in 770 light-trap collections and outdoors in 120 tent-trap collections, respectively. Mean vector density per collection night was 24.5 indoor and 12.7 outdoors. The great majority (95.5%) of the vectors were *A. gambiae* s.l., with the reminder belonging to the *A. funestus* group. Among the 969 *A. gambiae* s.l. tested by PCR for species identification, 4.5% ($n = 44$) were *A. arabiensis*, 2.2% were not amplified and the remaining 93.5% were *A. gambiae*. There was no evidence of a difference in *A. gambiae* s.l. species composition between indoor and outdoor collections (X^2: 3.5, $p = 0.18$).

Mosquitoes from the tent trap were, however, younger than those from indoor collections (either light traps or window traps) Table 16.13.

Two-thirds (599/866) of parous *An. gambiae* had well-defined dilatations (sac stage c or d), indicating that there had been a delay between oviposition and returning to feed. This contrasts with

Figure 16.56 Two views of the Muleba study area. Note the houses, many with tin roofs, surrounded by banana plantations.

Table 16.13 Age Structure of the *A. funestus* and *A. gambiae* from Different Trap Types, Muleba, Tanzania

		Total	Nulliparous				Parous (sac Stage)				Total	Parous
		Dissected	Virgin	Plug	N1	NII	a	b	c	d	Parous	Rate
A. funestus	Light trap	8	0	2	0	2	2	0	2	0	4	0.50
	Tent trap	40	6	3	11	8	8	9	8	3	19	0.48
	Window trap	17	2	0	2	3	4	1	0	6	10	0.59
A. gambiae	Light trap	137	28	19	3	18	4	13	10	42	69	0.50
	Tent trap	1067	324	208	78	137	60	97	134	200	409	0.38
	Window trap	451	190	70	63	76	39	54	60	153	265	0.59

the *A. funestus* in which 24 of the 43 dissected had dilatations (Chi square = 11.75, $p = 0.0007$) and may indicate that the two species were using different breeding sites.

Of the 289 *A. funestus* tested by PCR were *A. funestus*, 15.6% *A. leesoni*, 1.0% *A. parensis* and 1.7% were not amplified. Of 4,311 vectors tested for sporozoites by ELISA, 203 (4.7%) were positive.

There was a considerable variation between clusters with the numbers of vectors collected, ranging from 0 to 930 (Figure 16.57b and c). The variation followed the 80/20 rule (Figure 16.58). Factors that affected the number of mosquitoes in a house included altitude, house construction, owning animals and environmental conditions on the night of collection. As might be expected, given their different larval habitats there was no association between the presence of *A. funestus* and *A. gambiae* in a house.

In Kyamyorwa B, a village where IRS had been implemented, parous *A. gambiae* were eliminated almost immediately after spraying (Table 16.14). The parous rate went from 56% (448 of 798 dissected) to 29% [121 of 414 dissected – parous insects all being 1- or 2-parous. Collected mosquitoes represented the declining output from the (still unknown) breeding site. Very small numbers of *A. gambiae* continued to be collected in tent traps. Almost all of these insects were newly emerged, pre-gravid virgins.

Eleven months after the intervention the species ratio of the *A. gambiae* complex had changed. *Anopheles arabiensis* was now the dominant species. Thus, only 26 (19.3%) of the 135 *A. gambiae* s.l. identified to species and collected in light traps during January and February 2016 were

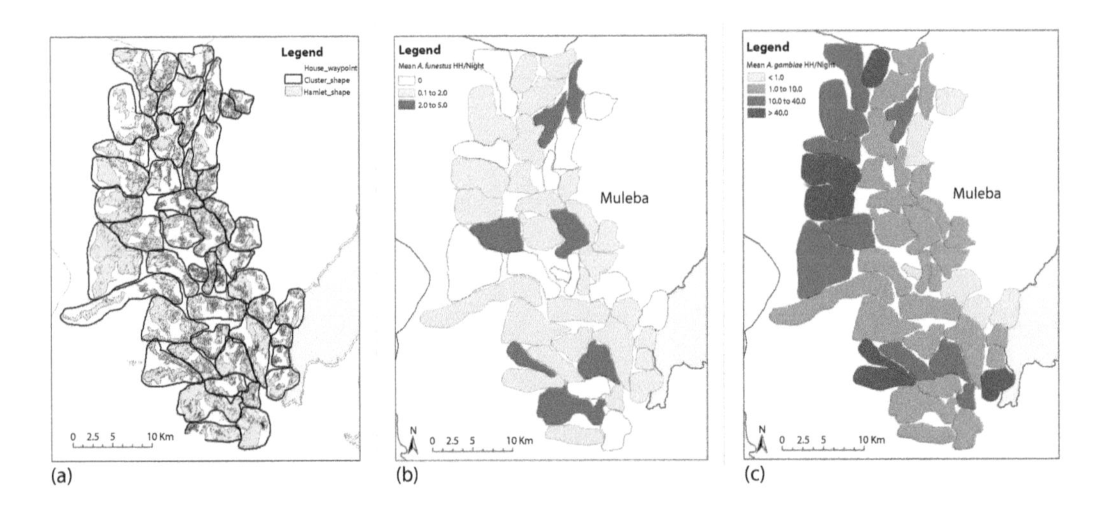

Figure 16.57 (a) A map of the study area bordering Lake Victoria, Tanzania, (b) density of *A. funestus* and (c) *A. gambiae* from light-trap collections made prior to the intervention.

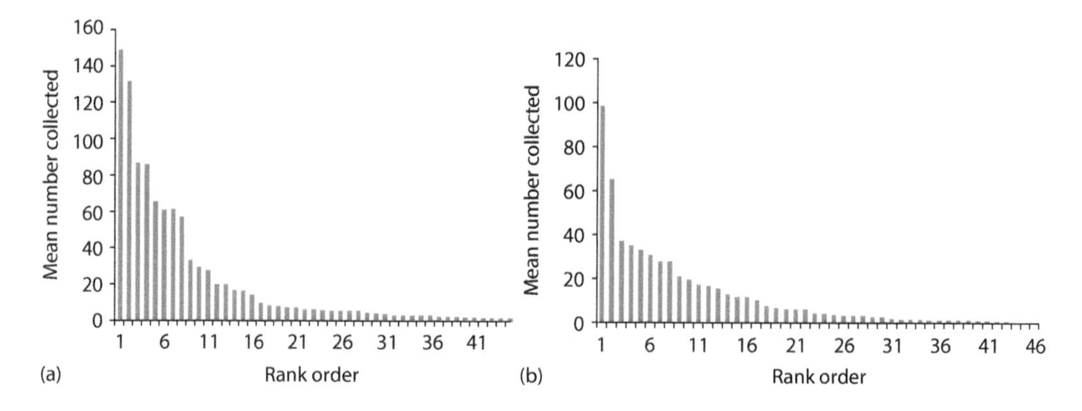

Figure 16.58 Mean number of *A. gambiae* collected by cluster prior to the intervention: (a) light trap; (b) tent trap.

Table 16.14 Parous Rates of *A. gambiae* by Collection Method Before and After the Intervention

	Collection	Total	Parous	Parous Rate
Before	Light trap	107	66	0.62
After	Light trap	24	2	0.08
Before	Tent trap	365	159	0.44
After	Tent trap	335	105	0.31
Before	Window	323	157	0.49
After	Window	55	14	0.25

A. gambiae. A similar proportion was found in the 48 insects identified from Furvela tent-trap collections. The sporozoite rate also declined. Only 44 (1.4%) of the 3200 mosquitoes tested were positive for sporozoites in the year following the intervention. This effect lasted in the study arms that had received PBO-treated nets or IRS but not in the arm that received standard nets (Protopopoff et al., 2018).

During the project a number of small-scale entomological studies were also undertaken in Kyamyorwa.

One of the less well-researched components of the malaria triangle (Figure 16.59) that can have a major impact on transmission is human behaviour. With the control of indoor transmission through the use of LLINs, IRS or housing improvements, possible transmission outdoors becomes more important. It becomes increasingly necessary to know how long people spend outdoors before going inside (where they are protected by the standard interventions) and what they might be doing (Monroe et al., 2019).

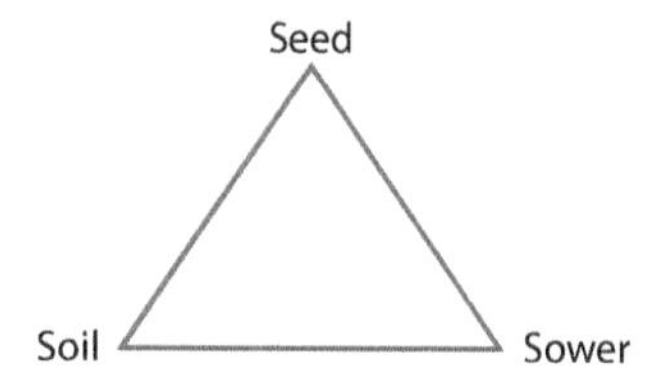

Figure 16.59 The malaria triangle – the seed (parasite), sower (mosquito) and soil (human host). Break any one of the links and malaria is eliminated. (After Gabaldon, A., *Am. J. Trop. Med. Hyg.*, 27, 653–658, 1978.)

Figure 16.60 Map of the walk taken by the census taker counting number of residents outside their houses in the early evening.

In Kyamyorwa in order to determine how much exposure might take place before people go to bed, a villager left his house every hour after sunset, walked a specified route (with a handheld GPS unit so that his track could be viewed) (Figure 16.60) and counted the number of people outside of their respective houses.

The route encompassed 26 houses with 193 residents. The villager reported to a field assistant sleeping in a tent trap at the end point of his route. The track on the GPS was checked, the number of people counted, noted down and the counter reset. This enabled us to determine the time at which people went indoors. When all residents were inside, counting stopped. At this time, the cages on the tent traps were changed and the mosquitoes in a window trap collected. By 21:00 the majority of people had retired inside their houses (Figure 16.61a). At this time, the cages on the tent traps were changed and the mosquitoes in a window trap collected. In collections undertaken prior to the spraying, only 9.7% of the 2117 *A. gambiae* collected were caught in the first three-hours of the night (i.e. before people went to bed) (Figure 16.61b). This was similar to the situation described from an earlier study elsewhere in Tanzania (Killeen et al., 2006). Hence, outdoor exposure was only a fraction of the exposure indoors and malaria was effectively controlled in the village.

One of the potential problems that makes outdoor assessment and control of potential vectors so important is that a small, persistent, population of an effective vector (such as *A. gambiae* or *A. funestus*) may rebound once the insecticidal effect has worn off, or once resistance develops in the mosquito. Alternatives to insecticide-based interventions are required. As the saying goes, 'if your only tool is a hammer then you tend to see every problem as a nail'. We need to invent some chisels, screwdrivers or saws to complete the job!

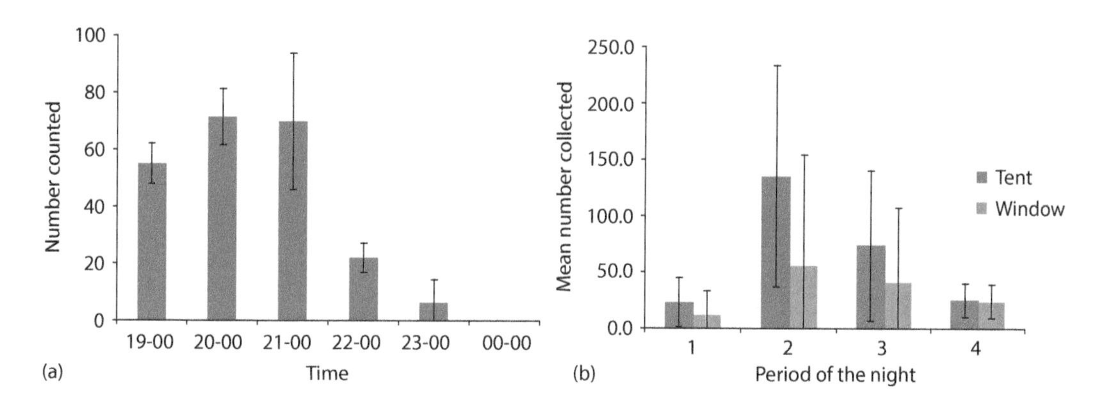

Figure 16.61 (a) Number of residents from 26 houses outside their houses by hour of the night, (b) mean number of *A. gambiae* collected from tent and window traps, Kyamyorwa, Tanzania.

CAMBODIA

According to the WHO, a reduction of 139 million malaria cases during the past 15 years was attributed to artemisinin combination therapy (ACT) (WHO, 2015a). These gains are now seriously under threat due to the development of resistance to artemisinin. Resistance was first described from the area around Pailin in Cambodia. This was the place where resistance to both chloroquine and subsequently to sulphadoxine-pyramethamine (S-P) also originated. Parasites resistant to these drugs spread from S.E. Asia to Africa (Figure 16.62) where they were responsible for millions of deaths.

The fear today is that the same pattern will be repeated with artemisinin. Artemisinin resistance is not only spreading, as did chloroquine, but it is also emerging independently, potentially dooming efforts to build a firewall around the region to contain it.

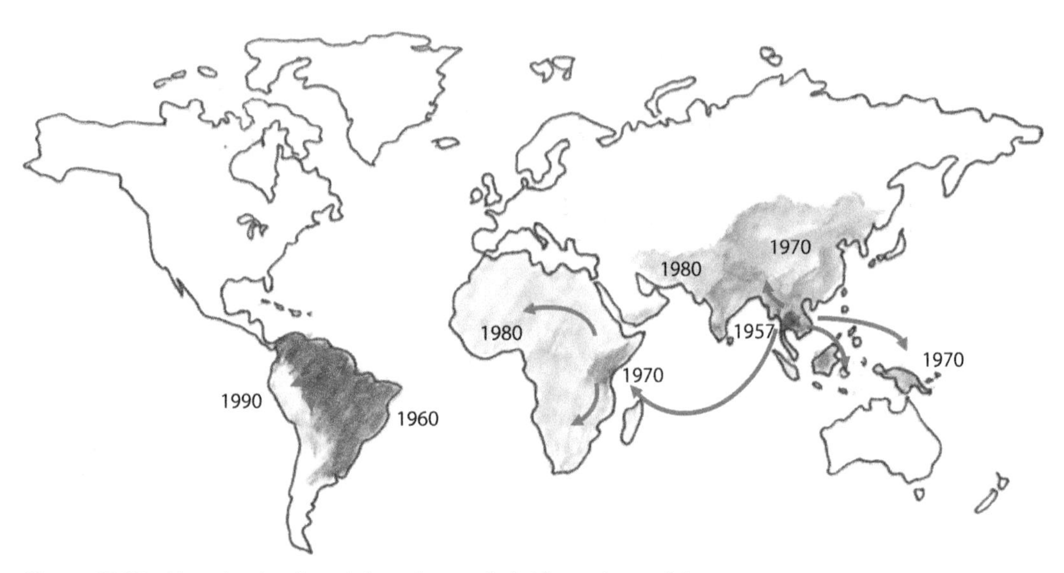

Figure 16.62 Map showing the origin and spread of chloroquine resistance.

Figure 16.63 Pursat, 2014. Previously, this was a forest; the haunt of monkeys and *Anopheles dirus.*

Cambodia lost around 1.59 million hectares of tree cover between 2001 and 2014, and just 3% remains covered in primary forest (Shimada et al., 2014) (Figure 16.63). In Cambodia it is mostly young men who get the disease when they go into the forest for a variety of activities, especially logging.

When in the forest at night, people do not use mosquito nets and so are exposed to the bites of mosquitoes. Therefore, the current challenge is to protect people before they go to bed and while they are outdoors. One way of doing this might be to use spatial repellents to prevent mosquitoes from biting. The development of formulations with high vapour action at ambient temperatures has led to the elaboration of devices that work without requiring the application of heat. Such devices have the additional advantage that they will work for an extended period of time (weeks instead of hours), which makes them an attractive alternative for control of all disease vectors. The effectiveness of one such spatial repellent, metofluthrin, was investigated in four locations in Cambodia, including Pailin, each with a different predominant mosquito species (Figure 16.64).

In addition, a number of fundamental questions regarding the ecology and behaviour of the mosquitoes were posed. In particular, we attempted to determine if post-oviposition behaviour was environmentally determined or was intrinsic to the particular species and, if it was environmentally determined, what was responsible for the differences in behaviour. The factors that make *A. dirus* such a potent vector were also investigated.

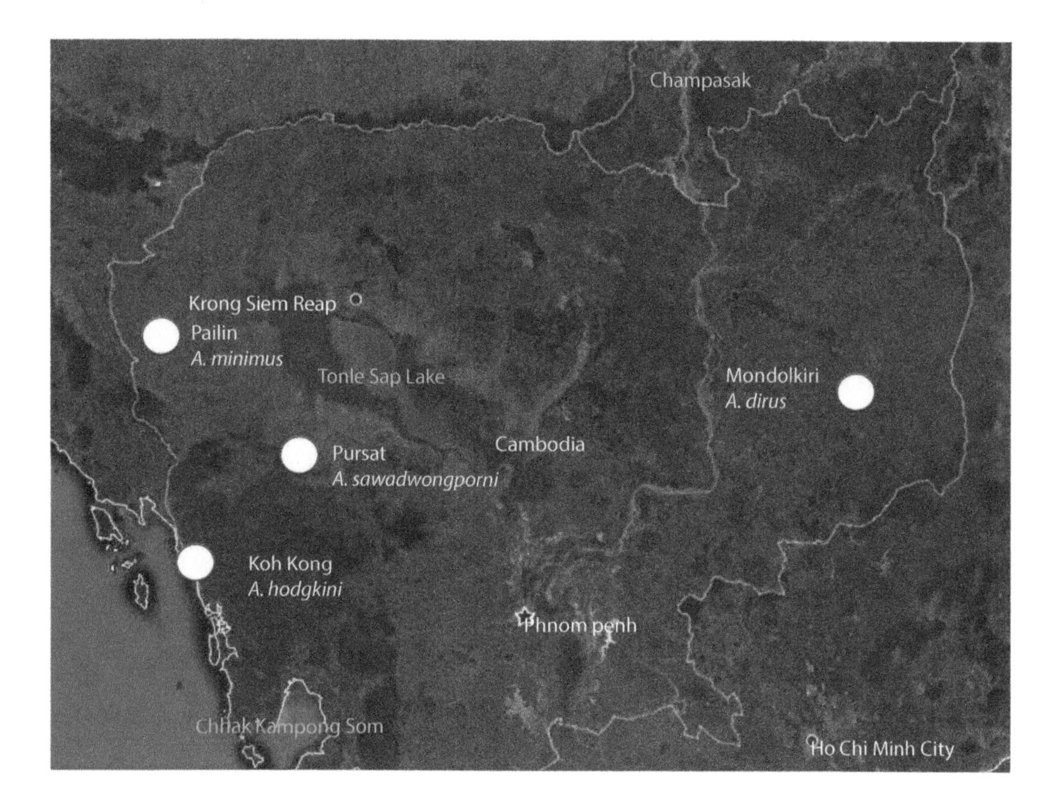

Figure 16.64 Satellite map of Cambodia with the four study locations and the predominant *Anopheles* present.

One of the difficulties associated with working on the *Anopheles* fauna in Southeast Asia is the large diversity of species encountered and the large number of species complexes that occur there. In addition to the principal vector, *A. dirus*, Cambodia is home to a number of secondary, or incidental, malaria vectors, including *Anopheles minimus*, *Anopheles maculatus*, *Anopheles barbirostris* and *Anopheles sinensis*, many of which bite in the early part of the night (Durnez et al., 2013). During the project, a total of 15 species or species groups of Anopheles were morphologically identified from the four study locations.

The other difficulty these days is actually finding a place where *A. dirus* is sufficiently common to be able to assess its behaviour. *Anopheles dirus* was previously common in Pailin and Pursat but was rare or absent in the present study (only 27 specimens being caught in 2121 hours of landing collection). The rapid and almost complete deforestation that has taken place around Pailin and Pursat, and which is occurring elsewhere in Cambodia at an unprecedented rate, is eliminating *A. dirus* and so is reducing transmission. In the present project the only site with sufficient *A. dirus* for ecological studies was the village of Ou Chra in Mondolkiri province (Figure 16.64).

In Pailin, a single emanator was used the numbers of mosquitoes in landing collections were reduced by approximately half and when four emanators were used by two-thirds (Charlwood et al., 2012). Equivalent reductions were seen among species collected in Pursat. In Koh Kong, however, the emanators had no demonstrable effect on landing rates. In Pailin, a single emanator positioned 1 m from the entrance of a tent trap or a light trap hung underneath a house also reduced the numbers of anophelines collected.

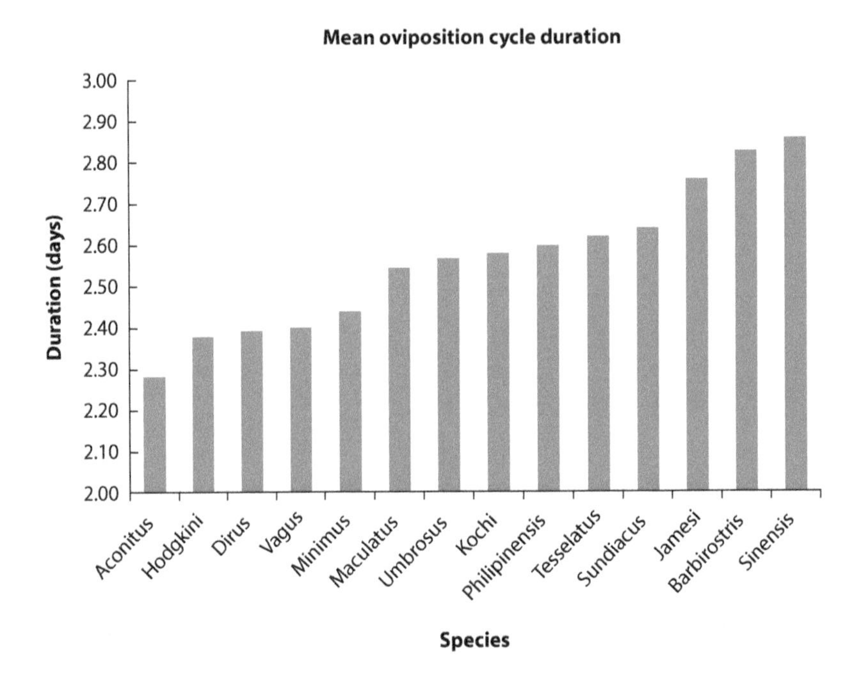

Figure 16.65 Estimated duration of the oviposition cycle among anophelines dissected from the four study areas, Cambodia.

The mosquitoes differed in their ecology in a number of ways. In particular, sac rates, and hence estimated oviposition cycle duration, differed between species collected (Figure 16.65).

They were generally similar between locations in the species dissected in sufficient numbers for adequate comparisons. However, there was an indication that, apart from species-specific factors, the length of the oviposition cycle might vary with humidity. In Ou Chra, as rainfall increased at the start of the wet season (from week 10 to week 25) so did the number of *A. dirus* (correlation between numbers and rainfall = 0.852). Southeast Asian *Anopheles* are physiologically adapted to humid conditions, having wider spiracles than African ones (Schapira and Boutsika, 2012). They may invest in large numbers of eggs per batch to the detriment of longevity and thereby vectorial capacity. This is likely to reduce their ability to survive in hot dry environments, typically associated with gonotrophically discordant species. Indeed, the small number of gravid insects collected during the study implies that, for all species, including the least common ones, gonotrophic discordance was not the survival strategy adopted during the long, hot dry season. Moreover, species such as *A. maculatus* s.l. and *A. aconitus*, previously reported as becoming discordant, did not show evidence of discordance. Thus, for all locations and all species, and at all times, gonotrophic concordance appeared to be the rule.

Moon phase appeared to affect the cycle in *A. minimus* s.l.; it was shorter when moonlight was present prior to or at sunset, compared to nights when no moonlight was present at this time. Moonlight did not appear to affect the duration of the cycle in the forest-dwelling *A. dirus* even though, in their review of factors affecting the biology of this species, Obsomer and colleagues (2013) considered that moonlight was an important determinant of cycle duration. It is possible that in the forest environment, where most ovipostion of *A. dirus* occurs, illumination is of less consequence than other factors in oviposition site and host location.

Densities of many species were very low. Given that the environment was rapidly changing in the study locations, it is possible that we were witnessing the gradual demise of a number of species, including *A. dirus*, rather than rare species maintaining themselves by a set of effective survival strategies.

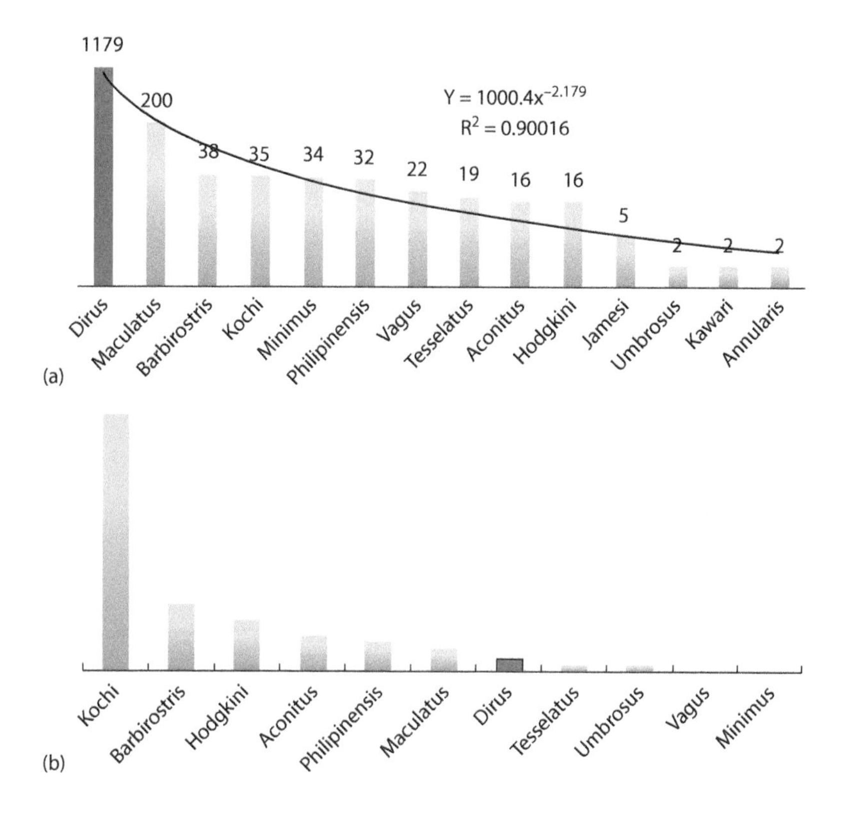

Figure 16.66 Species of anophelines collected from Ou Chra in rank order (on a log scale). (a) CDC light traps inside houses, (b) CDC light traps over a pigsty. (From Charlwood, J.D. et al., *Malaria World J.*, 8, 11, 2017a.)

The host choice of the different species collected also varied. *Anopheles dirus* was strongly anthropophagic. It accounted for 74% of all anophelines collected inside houses in Ou Chra but was only the ninth most common species collected from light traps hung close to pigs (Figure 16.66). Many houses in Cambodia are built on stilts. This would be sufficient to reduce the number of many species entering houses. Numbers of *A. dirus*, however, decreased only slightly with increasing house height, possibly reflecting its habit of feeding on monkeys in trees when they were available.

The species collected close to pigs in Ou Chra were also the ones collected close to pigs in Pursat (Table 16.15), indicating that these mosquitoes are unlikely to be vectors.

Table 16.15 Mean Number of *Anopheles* Species Collected in Light Traps Hung Adjacent to Different Hosts Outdoors, Veal Veng, Pursat, Cambodia

	Host		
Species	**Two Humans**	**Two Cows**	**Three (Little) Pigs**
A. aconitus	0.42	0.24	0.45
A. minimus s.l.	0.62	0.59	0.72
A. maculatus s.l.	5.58	19.72	32.10
A. kochi	4.27	19.59	58.90
A. jamesi	0.23	0.62	1.00
A. barbirostris	0.00	0.17	0.72

Source: Charlwood, J.D. et al., *Malar. J.*, 15, 356, 2016b.

In Ou Chra the greatest numbers of *A. dirus* were caught in the houses closest to the remaining forest (Figure 16.67). Deforestation eliminates *A. dirus* and reduces malaria transmission. Indeed, the presence of members of the *A. dirus* complex is an indication of a healthy forest environment.

It should surely be possible to control malaria without destroying the forest.

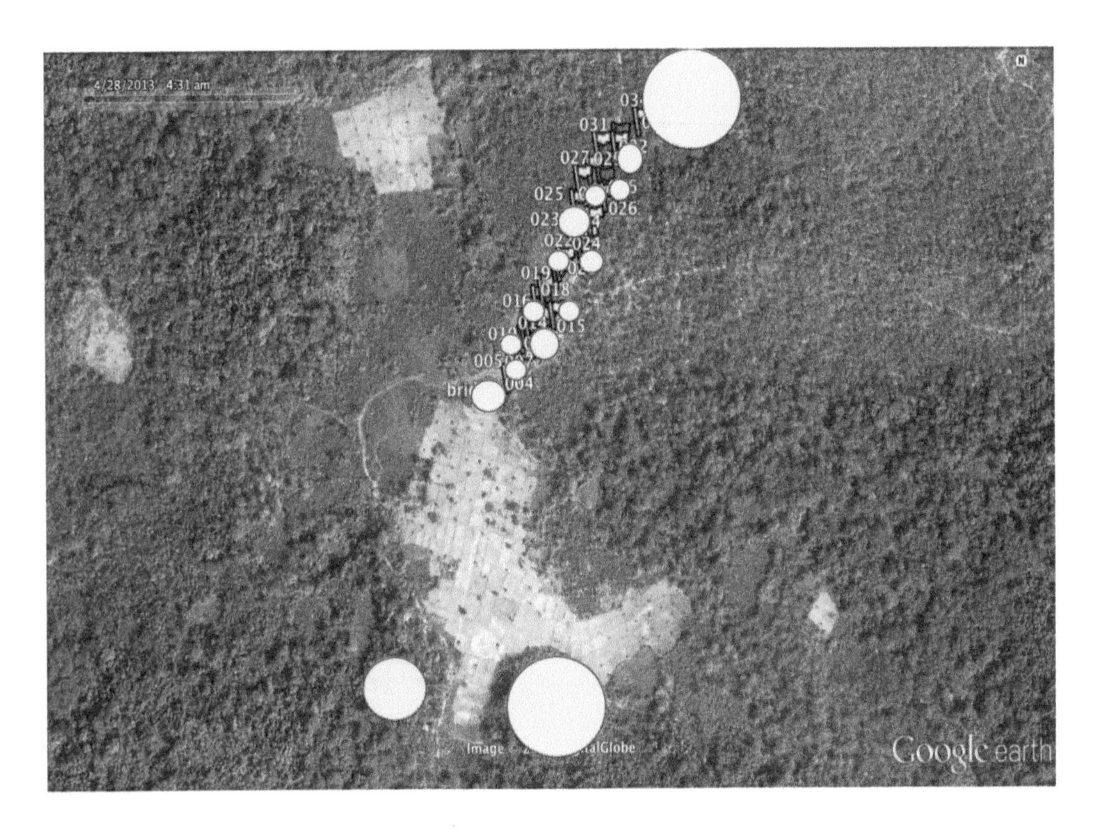

Figure 16.67 Map of Ou Chra. The relative density of *A. dirus* from light traps run inside houses is given according to the diameter of the pie charts. Note the much higher densities in houses closest to the forest.

CHAPTER 17

Some Useful Websites

- **1000 Genomes Project**: http://www.malariagen.net/projects/ag1000g – The *Anopheles gambiae* 1000 Project site provides a foundation for detailed investigation of mosquito genome variation and evolution.
- **MR4**: https://www.beiresources.org/About/MR4.aspx – The purpose of the MR4 is to provide a centralized resource for research reagents to the scientific community that can be used as reference standards or to generate new renewable reagents.
- **WHO**: https://www.who.int/ – The website for the World Health Organization.
- **WHO**: https://www.who.int/malaria/publications/world-malaria-report-2016/report/en/ – The World Malaria Report is the WHO's flagship malaria publication, released each year in December. It assesses global and regional malaria trends, highlights progress towards global targets, and describes opportunities and challenges in controlling and eliminating the disease.
- **Epidemic forecasting**: https://epidemia.sdstate.edu/ – is a highly developed prediction tool to actively support malaria early warning in epidemic-prone regions of the Ethiopian highlands.
- **MARA**: https://www.who.int/tdr/news/2014/mapping-mal-risk/en/ – The Mapping Malaria Risk in Africa (MARA) project was set up as a Pan-African enterprise, not owned by any specific organisation but coordinated by South Africa's Medical Research Council.
- **CDC**: https://www.cdc.gov/malaria/tools_for_tomorrow/research_resources.html – Malaria World.
- **CDC**: https://malariaworld.org/ – A scientific and social network for malaria professionals.
- **CDC**: https://mosquitoweb.ihmt.unl.pt – will help people identify mosquitoes from photographs. The site will eventually be in a position to build up a picture of occurrence of different species. It is primarily designed to identify mosquitoes from Europe but that may change.

GLOSSARY

acquired immunity: immunity acquired over time in people residing in malaria endemic areas through continued exposure to malaria parasites. Although full immunity is not obtained, and low-level parasite infections may still occur, it does generally protect against severe malaria.

ACT: Artemisinin Combination Therapy, an anti-malaria drug treatment, usually as a single pill, combining Artemisinin with a second drug such as lumefantrin.

active case detection: proactive screening of a defined portion of the population.

administrative feasibility: the possibility of creating a national administrative infrastructure that can carry out a malaria elimination programme with a strong, long-term, governmental commitment and a conducive legal environment for elimination.

aestivation: a means by which mosquitoes survive the long, hot dry season found in some parts of Africa. Such mosquitoes tend to survive by resting and taking occasional small blood meals. They may survive for exceptional periods of time and are individually epidemiologically dangerous.

allopatric: species, or populations, that do not share the same environment.

annual blood examination rate (ABER): the number of blood slides examined for malaria parasites as a proportion of the total population in areas at risk of transmission.

annual parasite index (API): a measure of the number of confirmed malaria cases per thousand people per year in a defined geographical area.

***Anopheles*:** the genus of mosquitoes that is responsible for malaria transmission.

artemisinin: wormwood derivative that was among the first drugs used to treat malaria in China, now used worldwide as part of combination therapies. It is rapidly cleared from the host compared to other drugs, which should limit the development of resistance. Resistance has now been shown to exist in a number of S.E. Asian sites.

autochthonous (indigenous, local): transmission acquired locally in an area where malaria regularly occurs.

basic reproductive number (R_0): the number of potentially infected humans that would arise from a single infected human, or the number of potentially infected mosquitoes that would arise from a single infected mosquito, after one complete generation of the parasite. It measures maximum potential transmission, so it describes populations with no immunity and no malaria control.

benign tertian malaria: type of malaria caused by *Plasmodium vivax* whose erythrocytic cycle has a 48-hour periodicity.

bilirubin: toxic by-product of haemoglobin released in large quantities during the rupture of erythrocytes in malaria.

blackwater fever: severe manifestation of infection by *P. falciparum* in which the kidneys are damaged by massive release of haemoglobin from ruptured erythrocytes. Formerly associated with the consumption of quinine.

capture-recapture: equivalent to Mark-Release-Recapture. Animals are collected and given a specific mark using a variety of techniques and released back into the environment and then, with any luck, recaptured at a later date provides information on survival and dispersal of animals.

case, imported: a case whose origin can be traced to a known malarious area outside the area in which it was diagnosed.
case, indigenous: a malaria case likely to have occurred through local transmission.
catholic feeder: a mosquito that feeds on a wide choice of hosts including animals and humans.
cerebral malaria: severe manifestation of infection by *P. falciparum* after parasites enter the central nervous system; characterised by progressive headache, very high fever, psychosis, convulsions, coma and death within hours.
chloroquine: inexpensive drug used to treat malaria to which many strains of *P. falciparum* have developed resistance.
circadian: about a day. A circadian rhythm is an activity that continues in darkness with a near 24-hour periodicity. Antennal erection in *Anopheles* males is one example.
circumsporozoite protein: component of *Plasmodium* sporozoites that contain multiple sets of repeated amino acid sequences; produces vigorous antibody production and protective immunity in some species of mice.
cladism: organisms are grouped according to recency of common ancestry; groups are joined on the grounds that they possess the same derived, highly evolved, characters, and not on primitive characters.
classification: is the process of delimiting, ordering and ranking of taxa within a hierarchical series of groups (species, genera, families, etc.).
controlled reproductive number (R_C): the same as the basic reproductive number (R_O) but considers all of the malaria control measures that have been put into place to slow transmission. It is also a measure of potential for outbreaks.
cost-benefit: ratio of costs to benefits, considering the financial value of an intervention.
cost-effectiveness: ratio of the net cost divided by the number of disability adjusted life years (DALY) averted, or some other metric of morbidity or mortality averted.
distal: at the far end – away from the body.
ecological niche: the *n* variables corresponding to all of the ecological factors relevant for the species, an *n*-dimensional hyper-volume can be defined in the environmental hyperspace between the limiting values permitting a species to survive and reproduce.
entomological inoculation rate (EIR): the product of the human biting rate and the sporozoite rate usually defined as the expected number of infectious bites per person per year.
elimination: the interruption of local mosquito-borne malaria transmission in a defined geographical area, creating a zero incidence of locally contracted cases.
endemic: applies to a malarious area when a sustained measurable incidence of cases and mosquito-borne transmission occur over a succession of years.
endophagic: a mosquito that feeds inside houses.
endophilic: associated with a mosquito resting inside houses.
El Niño Southern Oscillation (ENSO): a warming of the ocean surface, or above-average sea surface temperatures (SST), in the central and eastern tropical Pacific Ocean. Over Indonesia, rainfall tends to become reduced while rainfall increases over the tropical Pacific Ocean.
epidemic: occurrence of many cases of infection that substantially exceeds the expected number in a given place and time period.
epidemiology: is the study of the patterns and the determinants of the spatial and temporal distributions of a disease.
eradication: the permanent reduction to zero of the worldwide incidence of malaria infection. Intervention measures are no longer needed once eradication has been achieved.
erythrocytic cycle: time from the infection of the red blood cell to its rupture.
eurygamous: a mosquito that mates in large volumes of space (such as a swarm). Most anophelines are eurygamous and so need to adapt to laboratory conditions if they are to be colonised.

exophagic: a mosquito that feeds outside of houses.
exophilic: a mosquito that rests outside of houses.
extrinsic cycle: the time that the malaria parasite requires inside the mosquito to develop into sporozoites.
financial feasibility: the ability to establish and sustain the necessary funding to achieve and maintain elimination on a long-term and reliable basis.
focus (foci): a defined and circumscribed locality situated in a current or former malarious area that contains the continuous or intermittent epidemiological factors necessary for malaria transmission.
force of infection: rate per year at which susceptible individuals become infected by malaria.
fundamental niche: is the entire set of conditions under which an animal (population, species) can survive and reproduce itself.
gametocyte carrier: person who has malaria gametocytes in his or her peripheral blood, making the person a potential source of infection. All such people in an area constitute the infectious reservoir.
gametocyte: the sexual stage of malaria parasites, present in the host red blood cells, that are infective to the anopheline vector mosquito.
genome: refers to the entire compliment of DNA contained in a haploid cell of a species or individual.
GIGO: Garbage In, Garbage Out. If your data is rubbish, then so will your conclusions be rubbish.
gonotrophic concordance: one blood meal gives rise to one batch of eggs in a regular fashion.
gonotrophic cycle: the interval between one blood meal and the next (assuming gonotrophic concordance) or one oviposition and the next.
gonotrophic disassociation: the taking of several blood meals during a single gonotrophic cycle – usually associated with aeastivating mosquitoes in areas where there is a long, hot dry season.
holoendemic: permanent intense transmission with a high parasite rate among infants and a well-developed acquired immunity in older children and adults.
horizontal programme: an effort to provide the population with access to all health services and interventions through an integrated health delivery system.
horizontal transmission: describes the passing of a parasite from one individual host to another at one 'period' of time (e.g., malaria transmitted by the bite of a mosquito to a new host).
human biting rate: the number of mosquito bites, per person, per discreet time period, often a year.
hyperendemic: an area with high transmission that is frequently seasonal with parasite rates of parasite rates of 50–70% and considerable morbitity in young children.
hypoendemic: an area with little malaria incidence and a parasite rate of less than 10% in children aged 2–9 years.
identification: is a precise term describing the allocation of an unknown specimen to a predefined group.
importation risk: (also known as vulnerability). The probability of malaria reintroduction based on an area's proximity to other malarious areas and the movement of infected humans or infected *Anopheles* mosquitoes.
incidence: is a measure of the number of new cases of disease occurring over a specified period.
incubation period: defined as the time from infection to development of symptomatic disease.
infected: a mosquito that is developing malaria parasites but does not have sporozoites in its salivary glands and so will not yet transmit it to humans.
infectious: a mosquito with sporozoites in its salivary glands that will therefore transmit malaria to people it subsequently bites.

integrated vector management (IVM): a rational decision-making process for the optimal use of resources for vector control. The WHO highlighted the five major components of an IVM strategy: (i) advocacy, social mobilisation and legislation; (ii) collaboration within the health sector and with other sectors; (iii) an integrated approach; (iv) evidence-based decision-making; and (v) capacity building.

intertropical convergence zone (ITCZ): also known as the 'doldrums', the area under the zenith of the Sun as it passes between the meridians of the Tropic of Cancer and Capricorn. The hotter air rises at this point, drawing in colder air that produces the trade winds. As the Sun passes from north to south in the Southern Hemisphere, it produces the short rainy season (in October) and in the reverse direction the long rains (in April/May). In the Northern Hemisphere, the effect is reversed so that the short rains occur in May and the long rains in August.

internal rate of return (IRR): the percentage rate of interest that represents the economic return on an investment in malaria elimination; it is calculated from the incremental annual costs of an elimination strategy over the baseline costs of a strategy of sustained control over time.

intrinsic rate of increase: the number of births minus the number of deaths in a population; in other words, the reproduction rate less the death rate.

kdr: knockdown resistance – this is usually mediated by target site resistance.

locus: is a specific location in a genome.

macrogamete: large female gamete of *Plasmodium* species, similar to an egg.

malaria: disease characterised by alternating cycles of debilitating fever and chills, anaemia, and more severe manifestations such as cerebral malaria and renal failure during blackwater fever.

malariogenic potential: combination of a region's outbreak risk and importation risk.

male and female gametocyte: stage of malaria parasite that produces either multiple microgametes or a single macrogamete.

malignant tertian malaria: severe type of malaria caused by *Plasmodium falciparum* whose erythrocytic cycle has a 48-hour periodicity.

mark-release-recapture: experimental technique of marking insects and releasing them and subsequently using recaptured marked insects to infer a number of population parameters from the data.

mass drug administration (MDA): presumptive treatment of a defined population with a therapeutic dose of an antimalarial drug or drugs.

merozoites: parasites released into the host bloodstream when a hepatic or erthrocytic schizont bursts, initiating a new cycle of development within the red blood cells.

mesoendemic: an area of intermediate malaria incidence and a parasite rate of up to 50% in children aged 2–9 years.

metapopulation: a group of populations that are separated by space but consist of the same species. These spatially separated populations interact as individual members move from one population to another. The populations may suffer local extinction and subsequent re-invasion.

microgametes: small 'male' gametes of *Plasmodium* species; similar to sperm.

mild tertian malaria: rare type of malaria caused by *Plasmodium ovale* whose erythrocytic cycle has a 48-hour periodicity.

multivoltine: a species that has multiple generations a year. Tropical mosquitoes are multivoltine.

nomenclature: is the aspect of biology that deals with the names applied to organisms, not with how organisms are classified. It helps ensure that names are applied consistently to the same biological entity. Without scientific nomenclature, biology would effectively cease to be international and would quickly become mired in confusion.

nulliparous: a female mosquito that has never laid eggs.

oocyst: stage of *Plasmodium* species' life cycle that follows encystment of the ookinete on the outer wall of the mosquito's stomach; undergoes many rounds of reproduction to produce multiple sporozoites within it.

ookinete: motile zygote of *Plasmodium* species that is produced by fertilisation of a macrogamete by a microgamete.

operational feasibility: the ability to establish and sustain the systems and capacity to effectively implement all the activities needed to achieve and maintain elimination.

outbreak: a case or number of cases of locally transmitted infection greater than would be expected at a particular time and place.

outbreak risk: (also known as receptivity) a measure of the potential of an area or focus to allow transmission to occur, or once elimination has been achieved, the propensity for reintroduced malaria to give rise to malaria outbreaks.

oviposition cycle: see gonotrophic cycle.

parasite rate: (PR) prevalence of asexual blood-stage parasites.

parasitemia: percentage of malaria-infected red blood cells.

parous: a mosquito that has laid eggs at least once.

passive case detection: detection of malaria cases among patients who on their own initiative went to a health post to get treatment, usually for a febrile disease.

Plasmodium falciparum: species of parasite causing the deadliest form of malaria in humans; also has the highest rate of drug resistance.

Plasmodium malariae: species of human malaria parasite that causes quartan malaria.

Plasmodium ovale: least common of the parasites that cause malaria only in humans.

Plasmodium vivax: species of parasite most often responsible for malaria in the temperate zones.

positive predictive value (PV +): the probability that infection is truly present, given a positive diagnostic test result.

pre-elimination phase: malaria control programme reorientation during the period between sustained control and elimination, in which emphasis on surveillance, reporting and information systems increases.

pre-gravid feed: a blood meal in a mosquito that does not result in the development of eggs.

prevalence: is a measure of the total number of existing cases of a disease or condition in a specific population at a particular time.

private sector: all health facilities outside of the government's health system, and all potential malaria contributors that are outside government.

proximal: the nearest end – close to the body

quartan malaria: type of malaria caused by *Plasmodium malariae* whose erythrocytic cycle has a 72-hour periodicity.

quinine: one of the earlier drugs used to treat malaria, to which many strains of *Plasmodium* have developed resistance.

realised niche: is the set of conditions actually used by given animal (population, species), after interactions with other species (predation and especially competition) have been taken into account.

reintroduction risk: the risk following elimination that endemic malaria will be reestablished once surveillance shows a reduction to zero of all locally acquired cases (i.e., not including imported cases), when malaria can be reintroduced to the local environment.

resistance: the development of an ability in a strain of some organism to tolerate doses of a toxicant that would prove lethal to a majority of individuals in a normal population of the same species. Resistance is a genetically inherited characteristic whose frequency increases in the vector or parasite population as a direct result of the selective effects of the insecticide or drugs.

ring form: early stage of merozoite in *Plasmodium*-infected erythrocytes; resembles a signet ring.
schizonts: stage of *Plasmodium* species life cycle in which the infected cell is filled with multiple merozoites produced by asexual reproduction.
sensitivity: (of a test) the percentage of true positives correctly identified by diagnostic test results.
serology: the diagnostic identification of immunoglobulins/antibodies in the serum.
slide positivity rate: (SPR) the proportion of blood slides found positive among all slides examined.
spatial analysis: a general ability to manipulate spatial data (e.g., maps) into different forms and extract additional meaning (e.g., high-risk areas) as a result.
species: a group of living organisms consisting of similar individuals capable of exchanging genes or interbreeding that are reproductively isolated from other such groups. The species is the principal natural taxonomic unit, ranking below a genus and denoted by a Latin binomial, e.g., *Anopheles gambiae.*
speciation: the evolutionary process by which species multiply.
specificity: (of a test) The percentage of true negatives correctly identified by a diagnostic test.
sporozoite: stage of the life cycle in *Plasmodium* species that infects the mammalian host via a mosquito bite.
sporozoite rate: the proportion of mosquitoes with sporozoites in their salivary glands.
superinfection: a new infection occurring in a patient having a preexisting infection.
synanthropic: associated with humans.
stenogamous: mosquito that will mate in small volumes of space.
surveillance: is an organised system of collecting data. It is any procedure or group of procedures that collects the estimates of vector population density we need to predict, prevent or control vector-borne disease.
sustained control: period during which malaria control measures are stabilised and universal coverage is maintained by continued strengthening of health systems.
sympatric: species that share the same environment. Thus, *A. gambiae* and *A. arabiensis* are sympatric over much of their distribution.
synanthropic: likes being around people and houses; e.g., houseflies are synanthropic.
taxonomy: is the theory and practice of describing, naming and classifying organisms. It is the arrangement of similar entities (objects) in a hierarchical series of nested classes.
technical feasibility: the probability that malaria transmission can be reduced to zero in a given area, and that zero transmission can be maintained in that area once elimination has been achieved, using currently available control tools.
topology: is the establishment of spatial relationships among points, lines and polygons.
transmission foci: areas in which malaria transmission is concentrated.
transmission, stable: constant, year-round malaria transmission that is relatively insensitive to environmental changes.
transmission, unstable: malaria transmission with marked fluctuations in intensity due to changing environmental conditions.
univoltine: a population that has one generation per year. Many temperate species of mosquito are univoltine overwintering as eggs or larvae and emerging as adults in the spring.
vectorial capacity: = [Number of vectors feeding on the host per unit time] × [Probability the vector survives the Extrinsic Incubation Period] × [Number of blood meals on people after EIP].
vectorial capacity: is the average number of potentially infective bites that will ultimately be delivered by all the vectors feeding on a single host in one day.
vectorial capacity: the expected number of infectious bites that will arise from all the mosquitoes that bite a single person in one day.

vertical programme: a nonintegrated (e.g., stand-alone) health programme, often aimed at a single disease, group of diseases or target population.

vertical transmission: from one vector generation or stage of metamorphosis also occurs (for example, viruses such as dengue or yellow fever can pass from the female to her eggs).

zoophagic: a mosquito that feeds on animals.

zoophilic: likes animals.

REFERENCES

Adamou A, Dao A, Timbine S, Kassogué Y, Yaro AS, Diallo M, Traoré SF, Huestis DL & Lehmann T. 2011. The contribution of aestivating mosquitoes to the persistence of *Anopheles gambiae* in the Sahel. *Malar Journal* 10: 151.

Amerasinghe FP & Indrajithe NG. 1994. Post irrigation breeding patterns of surface water mosquitoes in the Mahaweli Project, Sri Lanka, and comparisons with preceding developmental phases. *Journal of Medical Entomology* 31: 516–523.

Antonio-Nkondjio C & Simard F. 2013. Highlights on *Anopheles nili* and *Anopheles moucheti*, Chapter 8 in *Anopheles Mosquitoes: New Insights into Malaria Vectors* Edited by Sylvie Manguin, 828 pages, InTech, July 24, 2013 under CC BY 3.0 license.

Babiker HA, Charlwood JD, Smith T & Walliker D. 1995. Gene flow and cross-mating in *Plasmodium falciparum* in households in a Tanzanian village. *Parasitology* 111: 433–442.

Belkin JN 1962. *Mosquitoes of the South Pacific*, Vol 1. University of California Press, Berkley, CA.

Benedict MQ, Charlwood JD, Harrington LC, Lounibos LP, Reisen WR & Tabachnick WJ. 2018. Guidance for evaluating the safety of experimental releases of mosquitoes, emphasizing Mark-Release-Recapture techniques. *Vector-Borne and Zoonotic Diseases* 18: 39–48.

Billingsley PF, Charlwood JD & Knols BGJ. 2003. Rapid assessment of malaria risk using entomological techniques: Taking an epidemiological snapshot Chapter 6 in *Environmental Change and Malaria Risk: Global and Local Implications* Takken W, Martens WJM & Bogers RJ (Eds.) Wageningen, the Netherlands.

Birley MH & Charlwood JD. 1987. Sporozoite rate and malaria prevalence. *Parasitology Today* 3: 231–232.

Birley MH & Rajagopalan PK 1981. Estimation of the survival and biting rate of *Culex quinquefasciatus* (Diptera: Culicidae). *Journal of Medical Entomology* 18: 181–186.

Birley MH & Charlwood JD. 1989. The effect of moonlight and other factors on the ovipositon cycle of malaria vectors in Madang, Papua New Guinea. *Annual of Tropical Medicine and Parasitology* 8: 415–422.

Black WC IV & Kondratieff BC 2005. Evolution of arthropod disease vectors. Chapter 2 in *Biology of Disease Vectors*, 2nd edition, Marquardt WC (Ed.). Elsevier, Amsterdam, the Netherlands.

Bøgh C, Clarke SE, Walraven GE & Lindsay SW. 2002. Zooprophylaxis, artefact or reality? A paired-cohort study of the effect of passive zooprophylaxis on malaria in The Gambia. *Transactions of the Royal Society of Tropical Medicine and Hygiene* 96: 593–596.

Bouma M & Rowland M. 1995. Failure of passive zooprophylaxis: Cattle ownership in Pakistan is associated with a higher prevalence of malaria. *Transactions of the Royal Society of Tropical Medicine and Hygiene* 89: 351–353.

Brady OJ, Godfray HCJ, Tatem AJ, Gething PW, Cohen JM, McKenzie FE, Perkins TA et al. 2016. Vectorial capacity and vector control: reconsidering sensitivity to parameters for malaria elimination. *Transactions of the Royal Society for Tropical Medicine and Hygiene* 110: 107–117.

Briët OJT, Huho BJ, Gimnig JE, Bayoh N, Seyoum A, Sikaala CH, Govella N et al. 2015. Applications and limitations of Centers for Disease Control and Prevention miniature light traps for measuring biting densities of African malaria vector populations: A pooled-analysis of 13 comparisons with human landing catches. *Malaria Journal* 14: 247.

Brunhes J, Le Goff G & Geoffroy B. 1997. Anophèles Afro-Tropicaux. 1.- Descriptions d'espècies nouvelles et changements de status taxonomiques (Diptera: Culicidae). *Annales de la Société Entomologique* 33: 173–183.

Butail S, Manoukis N, Diallo M, Ribeiro JM, Lehmann T & Paley DA. 2012. Reconstructing the flight kinematic of swarming and mating in wild mosquitoes. *Journal of the Royal Society Interface* 9: 2624–2638.

Caputo B, Nwakanma D, Caputo FP, Jawara M, Oriero EC, Hamidadia M, Dia L et al. 2014. Prominent intraspecific genetic divergence within *Anopheles gambiae* sibling species triggered by habitat discontinuities across a riverine landscape. *Molecular Ecology* 23: 4574–4589.

Charlwood JD & Alecrim WD. 1989. Capture-recapture experiments with *Anopheles darlingi* in Rondonia, Brazil. *Annals of Tropical Medicine and Parasitology* 83: 601–603.

Charlwood JD & Braganca M. 2012a. The effect of rainstorms on adult *Anopheles funestus* behavior and survival. *Journal of Vector Ecology* 37: 1–5.

Charlwood JD & Bragança M. 2012b. Some like it cool: The effect of temperature on the size of Anopheles funestus from southern Mozambique. *Journal of Medical Entomology* 49: 1154–1158.

Charlwood JD & Bryan JH. 1987. A capture-recapture experiment with the filariasis vector *Anopheles punctulatus* in Papua New Guinea. *Annals of Tropical Medicine and Parasitology* 81: 429–436.

Charlwood JD & Dagoro H. 1987. Impregnated bed nets for the control of filariasis transmitted by *Anopheles punctulatus* in rural Papua New Guinea. *PNG Medical Journal* 30: 199–202.

Charlwood JD & Edoh D. 1996. Larval distribution of members of the *Anopheles gambiae* Giles complex (Diptera, Culicidae), identified by the PCR reaction, in the environs of Ifakara, *Tanzania Journal of Medical Entomology* 33: 202–204.

Charlwood JD & Galgal K. 1985. Observations on the biology behaviour of *Armigeres milnensis* Lee (Diptera, Culicidade) in Papua New Guinea. *Australian Journal of Entomology* 24: 313–320.

Charlwood JD & Graves PM. 1987. The effect of Permethrin-impregnated bednets on a population of *Anopheles farauti* in coastal Papua New Guinea. *Medical and Veterinary Entomology* 1: 319–327.

Charlwood JD & Hayes J. 1978. Variações geográficas no ciclo de picada do *Anopheles darlingi* Root no Brasil. *Acta Amazonica* 8: 601–603.

Charlwood JD & Jones MDR. 1979. Mating behaviour in the mosquito, *Anopheles gambiae* s.l. 1. Close range and contact behaviour. *Physiological Entomology* 4: 111–120.

Charlwood JD & Jones MDR. 1980. Mating in the mosquito, *Anopheles gambiae* s.l. II Swarming behaviour. *Physiological Entomology* 5: 315–320.

Charlwood JD & Kampango A. 2012. A novel experimental hut for the study of entrance and exit behaviour of endophilic malaria vectors. *Malaria World Journal* 3: 3.

Charlwood JD & Paralupi ND. 1978. O uso de caixas excito-repelentes com *Anopheles darlingi* Root, *A. nuneztovari* Gabaldon e *Culex pipiens quinquefasciatus* Say obtidos em áreas perto de Manaus. *Acta Amazonica* 8: 605–611.

Charlwood JD & Tomás EVE. 2011. Do developing malaria parasites manipulate their mosquito host? – Evidence from infected *Anopheles funestus* (Giles) from Mozambique. *Transactions of the Royal Society of Tropical Medicine and Hygiene* 105: 352–354.

Charlwood JD & Wilkes TJ. 1979. Studies on the age composition of *Anopheles darlingi* Root from Brazil. *Bulletin of Entomological Research* 67: 337–342.

Charlwood JD & Wilkes TJ. 1981. Observations on the biting activity of *Anopheles triannulatus bachmanni* from the Mato Grosso, Brazil. *Acta Amazonica* 11: 411–413.

Charlwood JD, Alcântara J, Pinto J, Sousa C, Rompão H, Gil V & de Rosário VE. 2005. Do bednets reduce malaria transmission by exophagic mosquitoes? *Transactions of the Royal Society of Tropical Medicine and Hygiene* 99: 901–904.

Charlwood JD, Alecrim WD, Fe´N, Mangabeira J & Martins J. 1995a. A field trial with Lambda-cyhalothrin (ICON) for the intradomiciliary control of malaria transmitted by *Anopheles darlingi* Root in Rondonia, Brazil. *Acta Tropica* 60: 3–13.

Charlwood JD, Billingsley PF & Hoc TQ. 1995b. Mosquito-mediated attraction of female European but not African mosquitoes to hosts. *Annals of Tropical Medicine and Parasitology* 89: 327–329.

Charlwood JD, Billingsley PF, Takken W, Lyimo EOK, Smith T & Meuwissen JMET. 1997. Survival and infection probabilities of anthropophagic Anophelines from an area of high prevalence of *Plasmodium falciparum* in humans. *Bulletin of Entomological Research* 87: 445–453.

Charlwood JD, Birley MH & Graves PM. 1986a. Capture-recapture studies of females of the *Anopheles punctulatus* group of mosquitoes (Diptera, Culicidae) from Papua New Guinea. *Bulletin of Entomological Research* 76: 211–227.

Charlwood JD, Birley MH, Dagaro H, Paru R & Holmes PR. 1985a. Assessing survival rates of *Anopheles farauti* (Diptera, Culicidae) from Papua New Guinea. *Journal of Animal Ecology* 54: 1003–1016.

Charlwood JD, Cuamba N, Tomás EVE & Briët OJT. 2013a. Living on the edge: A longitudinal study of *Anopheles funestus* in an isolated area of Mozambique. *Malaria Journal* 12: 208.

Charlwood JD, Dagaro H & Paru R. 1985b. Blood-feeding and resting behaviour in the *Anopheles punctulatus* Donitz complex (Diptera, Culicidae) from coastal Papua New Guinea. *Bulletin of Entomological Research* 75: 463–475.

Charlwood JD, Graves PM & Alpers M. 1986b. The ecology of the *Anopheles punctulatus* group of mosquitoes from Papua New Guinea, A review of recent work. *PNG Medical Journal* 29: 19–27.

Charlwood JD, Graves PM & de C. Marshall TF. 1988. Evidence for a 'memorised' home range in *Anopheles farauti* females from Papua New Guinea. *Medical and Veterinary Entomology* 2: 101–108.

Charlwood JD, Hall T, Nenhep S, Rippon E, Branca-Lopes A, Steen K, Arca B & Drakeley C. 2017a. Spatial repellents and malaria transmission in an endemic area of Cambodia with high mosquito net usage. *Malaria World Journal* 8: 11.

Charlwood JD, Kihonda J, Sama S, Billingsley PF, Hadji H, Verhave JP, Lyimo E, Luttikhuizen PC & Smith T. 1995c. The rise and fall of *Anopheles arabiensis* (Diptera, Culicidae) in a Tanzanian village. *Bulletin of Entomological Research* 85: 37–44.

Charlwood JD, Macia GA, Manhaca M, Sousa B, Cuamba N & Bragança M. 2013b. Population dynamics and spatial structure of human-biting mosquitoes, inside and outside of houses, in the Chockwe irrigation scheme, southern Mozambique. *Geospatial Health* 7: 309–320.

Charlwood JD, Mendis C, Thompson R, Begtrup K, Cuamba N, Dgedge M, Gamage-Mendis A, Hunt RH, Sinden RE & Høgh B. 1998a. Cordon-sanitaire or laissez-faire, differential dispersal of young and old females of the malaria vector *Anopheles funestus* in southern Mozambique. *African Entomology* 6: 1–6.

Charlwood JD, Nenhep S, Protopopoff N, Sovannaroth S, Morgan JC & Hemingway J. 2016a. The effect of the spatial repellent metofluthrin on landing rates of outdoor biting anophelines in Cambodia, S.E. Asia. *Medical & Veterinary Entomology*. doi:10.1111/mve.12168.

Charlwood JD, Nenhep S, Sovannaroth S, Morgan JC, Hemingway J, Chitnis N & Briët OJT. 2016b. 'Nature or nurture': Survival rate, oviposition interval, and possible gonotrophic discordance among South East Asian anophelines. *Malaria Journal* 15: 356.

Charlwood JD, Paru R & Dagaro H. 1984. Raised platforms reduce mosquito bites. *Transactions of the Royal Society of Tropical Medicine & Hygiene* 78: 141–142.

Charlwood JD, Paru R & Dagaro H. A new light bed-net trap to sample anopheline vectors of malaria in Papua New Guinea. 1986c. *Journal of Vector Ecology* 11: 281–283.

Charlwood JD, Pinto J, Ferrara P, Sousa CA, Ferreira C, Gil V & de Rosario VE. 2003a. Raised houses reduce mosquito bites. *Malaria Journal* 2: 45.

Charlwood JD, Pinto J, Sousa CA, Ferreira C & do Rosário VE. 2002b. The swarming and mating behaviour of *Anopheles gambiae* (Diptera: Culicidae) from São Tomé Island. *Journal of Vector Ecology* 27: 178–183.

Charlwood JD, Pinto J, Sousa CA, Ferreira C & do Rosario VE. 2002a. Male size does not affect mating success (of *Anopheles gambiae* from São Tomé). *Medical and Veterinary Entomology* 16: 1–3.

Charlwood JD, Pinto J, Sousa CA, Ferreira C, Gil V & de Rosario V. 2003b. Mating does not affect the biting behaviour of *Anopheles gambiae* from the islands of São Tomé and Príncipe, West Africa. *Annals of Tropical Medicine and Parasitology* 97: 751–756.

Charlwood JD, Pinto J, Sousa CA, Ferreira C, Petrarca V & do Rosario VE. 2003c. A mate or a meal'—Pre-gravid behaviour of female *Anopheles gambiae* from the islands of São Tomé and Príncipe, West Africa. *Malaria Journal* 2: 7.

Charlwood JD, Qassim M, Elsnur EI, Donnelly M, Petrarca V, Billingsley PF, Pinto J & Smith T. 2001b. The impact of indoor residual spraying with malathion on malaria in refugee camps in Eastern Sudan. *Acta Tropica* 80: 1–8.

Charlwood JD, Rowland M, Protopopoff N & LeClair C. 2017b. The Furvela tent-trap Mk 1.1 for the collection of, outdoor biting mosquitoes. *Peer J* 3848.

Charlwood JD, Smith T, Kihonda J, Billingsley PF & Takken W. 1995d. Density independent feeding success of malaria vectors in Tanzania. *Bulletin of Entomological Research* 85: 29–35.

Charlwood JD, Smith T, Lyimo E, Kitua AY, Masanja H, Booth M, Alonso PL & Tanner M. 1998b. Incidence of *Plasmodium falciparum* infections in relation to exposure to sporozoite infected anophelines. *American Journal of Tropical Medicine and Hygiene* 59: 243–251.

Charlwood JD, Thompson R & Madsen H. 2003d. Swarming and mating in *Anopheles funestus* from southern Mozambique. *Malaria Journal* 2: 3.

Charlwood JD, Tomás EVE, Andegiorgish A, Mihreteab S & LeClair CE. 2018. 'We like it wet': A comparison between dissection techniques for the assessment of parity in *Anopheles arabiensis* and determination of sac stage in mosquitoes alive or dead on collection. *Peer J* 5155.

Charlwood JD, Tomás EVE, Bragança M, Cuamba N, Alifrangis M & Stanton M. 2015a. Malaria prevalence and incidence in an isolated, meso-endemic area of Mozambique. *Peer J* 1370.

Charlwood JD, Tomás EVE, Egyir-Yawson A, Kampango A & Pitts RJ. 2012. Feeding frequency and survival of *Anopheles gambiae* from a rice growing area of Ghana. *Medical & Veterinary Entomology* 26: 263–270.

Charlwood JD, Tomás EVE, Kelly-Hope L & Briët OJT. 2014. Evidence of an 'invitation' effect in feeding sylvatic *Stegomyia albopicta* from Cambodia. *Parasites & Vectors* 7: 324.

Charlwood JD, Tomás EVE, Salgueiro P, Egyir-Yawson A, Pitts RJ & Pinto J. 2011. Studies on the behaviour of peridomestic and endophagic M form *Anopheles gambiae* from a rice growing area of Ghana. *Bulletin of Entomological Research* 101: 533–539.

Charlwood JD, Vij R & Billingsley PFB. 1999. Dry season refugia of malaria-transmitting mosquitoes in a dry savannah zone of East Africa. *American Journal of Tropical Medicine and Hygiene* 62: 726–732.

Charlwood JD. 1974. Infra-red TV for watching mosquito behaviour in the 'dark'. *Transactions of the Royal Society of Tropical Medicine and Hygiene* 68: 264.

Charlwood JD. 1980. Observations on the bionomics of *Anopheles darlingi* Root (Diptera, Culicidae) from Brazil. *Bulletin of Entomological Research* 70: 685–692.

Charlwood JD. 1987. Repellent 'soap' for malaria vector control in Papua New Guinea. *PNG Medical Journal* 30: 99–103.

Charlwood JD. 1994. The control of *Culex quinquefasciatus* breeding in septic tanks using expanded polystyrene beads in southern Tanzania. *Transactions of the Royal Society of Tropical Medicine and Hygiene* 88: 380.

Charlwood JD. 1996. Biological variation in *Anopheles darlingi* Root. *Memorias do Instituto Oswaldo Cruz* 91: 391–398.

Charlwood JD. 2001. Zooprophylaxsis – are we in Plato's cave? *Trends in Parasitology* 17: 517.

Charlwood JD. 2003b. May the force be with you: measuring mosquito fitness in the field. Chapter 5, in *Proceedings of a Workshop on Transgenic Mosquitoes* (Edited by Scott T and Takken W). Wageningen, the Netherlands.

Charlwood JD. 2008. Women's groups and the marketing of health interventions a Tanzanian experience. *PNG Medical Journal* 51: 102–104.

Charlwood JD. 2011. Studies on the biology of male *Anopheles gambiae* Giles and *Anopheles funestus* Giles from southern Mozambique. *Journal of Vector Ecology* 36: 382–394.

Charlwood JD. 2017. Some like it hot: a differential response to changing temperatures by the malaria vectors *Anopheles funestus* and *An. gambiae* s.l. *Peer J* 5: 3099.

Cheng L, Abraham J, Hausfather Z & Trenberth KE. 2019. How fast are the oceans warming? *Science* 363: 6423.

Clarke SE, Bogh C, Brown RC, Walraven GE, Thomas CJ & Lindsay SW. 2002. Risk of malaria attacks in Gambian children is greater away from malaria vector breeding sites. *Transactions of the Royal Society of Tropical Medicine and Hygiene* 96: 499–506.

Clarkson CS, Weetman, Essandoh J, Alexander E, Yawson AE, Maslen G, Manske M et al. 2014. Adaptive introgression between *Anopheles* sibling species eliminates a major genomic island but not reproductive isolation. *Nature Communications* 5: 4248.

Clements AN & Paterson GD. 1981. The analysis of mortality and survival rates in wild populations of mosquitoes. *Journal of Applied Ecology* 18: 373–399.

Coetzee M, Hunt RH, Wilkerson R, Della Torre, A, Coulibaly MB & Besansky NJ. 2013. *Anopheles coluzzii* and *Anopheles amharicus*, new members of the *Anopheles gambiae* complex. *Zootaxa* 3619: 246–274.

Coluzzi M, Sabatini A & Petraca V. 1974. Polimorfismi cromosomici loro signilicato adattivo nel complesso gambiae (genre Anopheles). *Parassitologia* 16: 107–109.

Coluzzi M, Petrarca V & Di Deco MA (1985) Chromosomal inversion intergradation and incipient speciation in *Anopheles gambiae. Bollettino di Zoologia* 52: 45–63.

Conn JE, Martha L, Quiñones ML & Póvoa MM. 2013. Phylogeography, vectors and transmission in Latin America, anopheles mosquitoes, Chapter 2 in *Anopheles Mosquitoes: New Insights into Malaria Vectors Edited by Sylvie Manguin*, 828 pages, Publisher: InTech, published July 24, 2013, under CC BY 3.0 license.

Cooper RD, Waterson DGE, Frances SP, Beebe NW, Pluess B & Sweeney AW. 2009. Malaria vectors of Papua New Guinea. *International Journal for Parasitology* 39: 1495–1501.

Corbel V & N'Guessan R. 2013. Distribution, mechanisms, impact and management of insecticide resistance in malaria vectors: A pragmatic review. From *Anopheles* to spatial surveillance: A roadmap through a multidisciplinary challenge. Chapter 19 in *Anopheles Mosquitoes: New Insights into Malaria Vectors* Edited by Sylvie Manguin, 828 pages, Publisher: InTech, published July 24, 2013, under CC BY 3.0 license.

Costantini C, Ayala D, Guelbeogo WM, Pombi M, Some CY, Bassole IHN, Ose K et al. 2009. Living at the edge: Biogeographic patterns of habitat segregation conform to speciation by niche expansion in *Anopheles gambiae*. *BMC Ecology* 9: 16.

Costantini C, Sagnon NF, Ilboudo-Sanogo E, Coluzzi M & Boccolini D. 1999. Chromosomal and bionomic heterogeneities suggest incipient speciation in *Anopheles funestus* from Burkina Faso. *Parassitologia* 41: 595–611.

Crane G, Gibson D, Verral J, Barker-Hudson P, Barker-Hudson BET, Charlwood JD & Heywood P. 1985. Malaria and tropical splenomegaly syndrome in the Anga of Morobe Province. *PNG Medical Journal* 28: 27–34.

Cribb BW & Merritt DJ. 2013. Chemoreception Chapter 24 in *The Insects Structure and Function*, Chapman RF, Simpson SJ & Douglas AE (Eds.), Cambridge University Press, Cambridge.

Curtis CF & Lines JD 1985. Impregnated fabrics against malaria mosquitoes. *Parasitology Today* 1: 147.

Darlington CD. 1969. *The Evolution of Man and Society*. George Allen and Unwin, London, UK, 753 pp.

Darriet F, Robert V, Tho Vien N & Carnevale P. 1984. Evaluation of the efficacy of permethrin impregnated intact and perforated mosquito nets against vectors of malaria. WHO/VBC/84.899.

Darsie RF Jr & Ward RA. 1981. Identification and geographical distribution of the mosquitoes of North America north of Mexico. *Mosquito Systematics* 1(suppl): 1–313.

De Castro MC, Monte-Mór RL, Sawyer DO & Singer BH. 2006. Malaria risk on the Amazon frontier. *Proceedings of the National Academy of Sciences of the United States of America* 103: 2452–2457. doi:10.1073/pnas.0510576103.

de Souza D, Kelly-Hope L, Lawson B, Wilson M & Boakye. 2010. Environmental factors associated with the distribution of *Anopheles gambiae* s.s in Ghana; An important vector of lymphatic filariasis and malaria. *PLoS One* 10: 1371

Delphi 234 Africa Precipitation Map-sr.svg, CC BY-SA 4.0, https://commons.wikimedia.org/w/index.php?curid=35954834.

Detinova TS. 1962. Age-grouping methods in Diptera of medical importance with special reference to some vectors of malaria. Monograph series WHO no 47, 216pp.

Devenport M & Jacobs-Lorena M. 2005. The peritrophic matrix of haemotophagous insects. Chapter 22 in *Biology of Disease Vectors* (2nd edition), Marquardt, W.C. (Ed.). Elsevier, Amsterdam, the Netherlands.

Dia I, Guelbeogo MW & Ayala D. 2013. Advances and perspectives in the study of the malaria mosquito *Anopheles funestus*. Chapter 7 in *Anopheles Mosquitoes: New Insights into Malaria Vectors* Edited by Sylvie Manguin, 828 pages, Publisher: InTech, published July 24, 2013 under CC BY 3.0 license.

Dobens AC & Dobens LL. 2013. FijiWings: An open source toolkit for semiautomated morphometric analysis of insect wings. *G3: Genes, Genomes, Genetics* 3(8):1443–1449. doi:10.1534/g3.113.006676.

Drake JM & Beier JC. 2014. Ecological niche and potential distribution of *Anopheles arabiensis* in Africa in 2050. *Malaria Journal* 13: 213

Drame PM, Poinsignon A, Marie A, Noukpo H, Doucoure S, Cornelie S & Remoue F. 2013. New salivary biomarkers of human exposure to malaria vector bites Chapter 23 in *Anopheles mosquitoes - New insights into malaria vectors* Edited by Sylvie Manguin, 828 pages, Publisher: InTech, published July 24, 2013, under CC BY 3.0 license.

Durnez L & Coosemans M. 2013. Residual transmission of malaria: An old issue for new ideas. Chapter 21 in *Anopheles Mosquitoes: New Insights into Malaria Vectors* Edited by Sylvie Manguin, 828 pages, Publisher: InTech, published July 24, 2013, under CC BY 3.0 license.

Durnez L, Mao S, Denis L, Roelants P, Sochantha T & Coosemans M. 2013. Outdoor malaria transmission in forested villages of Cambodia. *Malaria Journal* 12: 329. doi:10.1186/1475-2875-12-329.

Foster WA & Walker ED. 2002. Mosquitoes (Culicidae). Chapter 12 in *Medical and Veterinary Entomology*, Mullen GR & Durden LA (Eds.). Elsevier, Amsterdam, the Netherlands.

Foster WA. 1995. Mosquito sugar feeding and reproductive energetics. *Annual Review of Entomology* 40: 443–474.

Gabaldon A. 1978. What can and cannot be achieved with conventional anti-malaria measures. *American Journal of Tropical Medicine and Hygiene* 27: 653–658.

Garcia LS. 2007. *Diagnostic Medical Parasitology*. ASM Press, Washington, DC.

Garrett-Jones C & Grab B. 1964. The assessment of insecticidal impact on the malaria mosquito's vectorial capacity, from data on the proportion of parous females. *Bulletin of the World Health Organization* 31: 71–86.

Garrett-Jones C. 1964. The human blood index of malaria vectors in relation to epidemiological assessment. *Bulletin of the World Health Organization* 30: 241–261.

Gibson G & Torr SJ. 1999. Visual and olfactory responses of hematophagous Diptera to host stimuli. *Medical and Veterinary Entomology* 13: 2–23.

Gillett JD. 1971. *Mosquitos*. Weidenfeld & Nicolson, London, UK.

Gillies MT & Coetzee M. 1987. A supplement to the Anophelinae of Africa South of the Sahara (Ethiopian Zoogeographical Region). *Publication of the South African Institute for Medical Research* 55: 1–143.

Gillies MT & de Meillon B. 1968. Anophelinae of Africa south of the Sahara (Ethiopian Zoogeographical Region), 2nd Edition. Johannesburg, *Publication of the South African Institute for Medical Research* 54.

Gillies MT & Wilkes TJ. 1963. Observations on nulliparous and parous rates in a population of *Anopheles funestus* in East Africa. *Annals of Tropical Medicine and Parasitology* 57: 204–213.

Gillies MT & Wilkes TJ. 1965. A study of the age-composition of populations of *Anopheles gambiae* Giles and *A. funestus* Giles in North-Eastern Tanzania. *Bulletin of Entomological Research* 56: 237–262.

Gillies MT. 1961. Studies on the dispersion and survival of *Anopheles gambiae* Giles in East Africa, by means of marking and release experiments. *Bulletin of Entomological Research* 52: 99–127.

Gillies MT. 1989. Anopheline mosquitos: Vector behaviour and bionomics. Chapter 16 in *Malaria: Principles and Practice of Malariology*. Wernsdorfer WH & McGregor I (Eds.).

Gillies MT, Hamon J, Davidson G, deMeillon B & Mattingly PF. 1961. A practical guide for malaria entomologists in the African region of WHO. WHO, Geneva, Switzerland.

Graves PM, Brabin BJ, Charlwood JD, Burkot TR, Cattani JA, Ginny M, Paino J, Gibson FD, Alpers MP. 1987 Reduction in incidence and prevalence of *Plasmodium falciparum* in under 5 year old children by Permethrin impregnation of mosquito nets. *Bulletin WHO* 65: 869–877.

Greenwood BM, Bojang K, Whitty CJM & Targett GAT. 2005. Malaria. *The Lancet* 365: 1487–1498.

Gubler DJ 1998. Dengue and dengue haemorrhagic fever. *Clinical Microbioogical Reviews* 11: 480–496.

Gutsevich A, Monchadskii S & Shtakel'berg AA. 1974. Fauna of the U.S.S.R. Diptera Mosquitoes Family Culicidae. Leningrad, Akadamie Nauk. USSR Zool. N. Ser. 100, 384: 1–408.

Hadis M, Lulu M, Makonnen Y & Asfaw T. 1997. Host choice by indoor-resting *Anopheles arabiensis* in Ethiopia. *Transactions of the Royal Society of Tropical Medicine and Hygiene* 91: 376–378.

Haffer J. 1974. Avian speciation in tropical South America, with a systematic study of the toucans (Rhamphastidae) and jacmars (Galbulidae). *Publications of the Nuttal Orithological Club* 14: 370.

Hagmann R, Charlwood JD, Gil V, do Rosario VE & Smith T. 2003. Malaria and its possible control on the island of Príncipe. *Malaria Journal* 2: 9.

Hamon J, Sales S, Adam J-P & Grenier P. 1964. Age physiologique et cycle d'aggressivité chez *Anopheles gambiae* Giles et *A. funestus* Giles dans la region de Bobo-Dioulasso (Haute-Volta). *Bulletin de la Société Entomologique de France* 69: 110–121.

Hanemaaijer MJ, Collier TC, Chang A, Shott CC, Houston PD, Schmidt H, Main BJ, Cornel AJ Lee Y & Lanzaro GC. 2018. The fate of genes that cross species boundaries after a major hybridization event in a natural mosquito population. *Molecular Ecology* 27: 4978–4990.

Harbach RE 2013. The Phylogeny and Classification of *Anopheles*. Chapter 1 in *Anopheles Mosquitoes: New Insights into Malaria Vectors*. Edited by Sylvie Manguin, 828 pages, Publisher: InTech, published July 24, 2013, under CC BY 3.0 license.

Hemingway J & Ranson H. 2000. Insecticide resistance in insect vectors of human disease. *Annual Review of Entomology* 45: 371–391.

Hemingway J. 2005. Biological control of mosquitoes. Chapter 43 in *Biology of Disease Vectors*, 2nd edition, Marquardt WC (Ed.). Elsevier, Amsterdam, the Netherlands.

Herodotus. 1966. *The Histories*. Edited by: A de Sélincourt. Penguin Books, London, UK.

Hoc TQ & Charlwood JD. 1990. Age determination of *Aedes cantans* using the ovarian oil injection technique. *Medical and Veterinary Entomology* 4: 227–233.

Holmes PR & Birley MH. 1987. An improved method for survival rate analysis from time series of hematophagous dipteran populations. *Journal of Animal Ecology* 56: 427–440.

Homan T, Hiscox A, Mweresa CK, Masiga D, Mukabana WR, Oria P, Maire N et al. 2016. The effect of mass mosquito trapping on malaria transmission and disease burden (SolarMal): A stepped-wedge cluster-randomised trial. *The Lancet* 6736: 30445–30447.

Huijben S & Paaijmans KP. 2018. Putting evolution in elimination: Winning our ongoing battle with evolving malaria mosquitoes and parasites. *Evolutionary Applications* 11: 415–430.

Ijumba JN & Lindsay SW. 2001. Impact of irrigation on malaria in Africa: paddies paradox. *Medical and Veterinary Entomology* 5: 1–11.

Jones MDR. 1979. Possible use of chemicals to change the programming of behaviour in adult mosquitoes in insect neurobiology and pesticide action (Neurotox 1979). *London Society of Chemical industry*. 253–260.

Kampango A, Cuamba N & Charlwood JD. 2011. Does moonlight influence the biting behaviour of *Anopheles funestus* (Diptera: Culicidae)? *Medical and Veterinary Entomology* 25: 240–246.

Kashin P. 1966. Electronic recording of the mosquito bite. *Journal of Insect Physiology* 12: 281–284.

Kasstan B, Hampshire K, Guest C, Logan J, Pinder M, Williams K & Lindsay S. 2019. Sniff and tell: The feasibility of using bio-detection dogs as a mobile diagnostic intervention for asymptomatic malaria in sub-Saharan Africa. *Journal of Biosocial Science* 1–8. doi:10.1017/S0021932018000408.

Kawaguch I, Sasaki A & Mogi M. 2004. Combining zooprophylaxis and insecticide spraying: A malaria-control strategy limiting the development of insecticide resistance in vector mosquitoes. *Proceedings of Biological Sciences* 271, 301–309.

Killeen GF, Kihonda J, Lyimo E, Oketch FR, Kotas ME, Mathenge E et al. 2006. Quantifying behavioural interactions between humans and mosquitoes: evaluating the protective efficacy of insecticidal nets against malaria transmission in rural Tanzania. *BMC Infectious Dis*eases 6: 161.

Killeen GF. 2016. Mass trapping of malaria vector mosquitoes. *The Lancet*. doi:10.1016/S0140-6736(16)30674-2.

Kirby MJ, Green C, Milligan PM, Sisimandis C, Jasseh M, Conway DJ & Lindsay SW. 2008. Risk factors for house-entry by malaria vectors in a rural town and satellite villages in The Gambia. *Malaria Journal* 7: 2.

Kirby MJ. 2013. House screening. Chapter 7 in *Biological and Environmental Control of Disease Vectors*. Cameron MM & Lorenz LM (Eds.). CAB International, Wallingford, UK.

Kiware SS, Chitnis N, Tatarsky A, Wu S, Castellanos HMS, Gosling R, Smith D & Marshall JM. 2017. Attacking the mosquito on multiple fronts: Insights from the Vector Control Optimization Model (VCOM) for malaria elimination *PLoS One* 12(12): e0187680. doi:10.1371/journal.pone.0187680.

Kleinschmidt I, Bradley J, Knox TB, Mnzava, AP, Kafy HT, Mbogo C, Ismail BA et al. 2018. Implications of insecticide resistance for malaria vector control with long-lasting insecticidal nets: A WHO-coordinated, prospective, international, observational cohort study. *The Lancet*. doi:10.1016/S1473-3099(18)30172-5.

Klowden M. 2007. *Physiological Systems in Insects*. Elsevier, Amsterdam, the Netherlands.

Klowden MJ & Zwiebel LJ. 2005. Vector olfaction and behaviour. Chapter 20 in *Biology of Disease Vectors*, 2nd edition. Marquardt WC (Ed.). Elsevier, Amsterdam, the Netherlands.

Knudsen J & Von Seidlein L. 2014. *Healthy Homes in Tropical Zones: Improving Rural Housing in Asia and Africa*. Edition Axel Menges, Berlin, Germany.

Koella JC. 1991. On the use of mathematical models of malaria transmission. *Acta Tropica* 49: 1–25.

Krebs CJ. 1999. *Ecological Methodology*, 2nd edition. Benjamin/Cummings, San Francisco, CA.

Kristan M, Lines J, Nuwa A, Ntege C, Meek SR & Abeku TA. 2016. Exposure to deltamethrin affects development of *Plasmodium falciparum* inside wild pyrethroid resistant *Anopheles gambiae* s.s. mosquitoes in Uganda. *Parasites and Vectors* 9: 100.

Lane RP & Crosskey RW (Eds.). 1993. *Medical Insects and Arachnids*. The Natural History Museum. Springer, Amsterdam, the Netherlands.

Lanzaro GC & Lee Y. 2013. Speciation in *Anopheles gambiae* — The distribution of malaria vectors in Africa genetic polymorphism and patterns of reproductive isolation among natural populations. Chapter 6 in *Anopheles Mosquitoes: New Insights into Malaria Vectors*. Edited by Sylvie Manguin, 828 pages, Publisher: InTech, published July 24, 2013, under CC BY 3.0 license.

Lee P-W, Liu C-T, Rampao HS, Rosario VE & Shaio M-F. 2010. Pre-elimination of malaria on the island of Príncipe. *Malaria Journal* 9: 26.

Lehmann T, Dao A, Yaro AS, Adamou A, Kassogue Y, Diallo M, Sékou T & Coscaron-Arias C. 2010. Aestivation of the African malaria mosquito, *Anopheles gambiae* in the Sahel. *American Journal of Tropical Medicine and Hygiene* 83: 601–606.

Loiseau C, Melo M, Yoosook L, Hanemaaijer MJ, Lanzaro GC & Cornel AJ. 2018. High endemism of mosquitoes on São Tomé and Príncipe Islands: Evaluating the general dynamic model in a worldwide island comparison. *Insect Conservation and Diversity*. doi:10.1111/icad.12308.

Lopes J, Arias JR & Charlwood JD. 1985. Estudo ecologico de Culicidae (Diptera) silvestres criados em pequenos recipientes de agua em mata e emcapoeira no municipio de Manaus Am. *Ciencia e Cultura* 37: 1299–1311.

Mackenzie JS, Gubler DJ & Petersen LR. 2004. Emerging flaviviruses: The spread and resurgence of Japanese encephalitis, West Nile and dengue viruses. *Nature Medicine* 10 (12 Suppl): S98–S109.

Mahande A, Mosha F, Mahande J & Kweka E. 2007. Feeding and resting behaviour of malaria vector *Anopheles arabiensis* with reference to zooprophylaxsis. *Malaria Journal*, 6, 100.

Manoukis NC, Diabate A, Abdoulaye A, Diallo M, Dao A, Yaro AS, Ribeiro MC & Lehmann T. 2009. Structure and dynamics of male swarms of *Anopheles gambiae*. *Journal of Medical Entomology* 46: 227–235.

Marshall JC, Pinto J, Charlwood JD, Gentile G, Santolamazza F, Simard F, della Torre A, Donnelly MJ & Caccone A. 2008. Exploring the origin and degree of genetic isolation of *Anopheles gambiae* from the islands of São Tomé and Príncipe, potential sites for testing transgenic-based vector control. *Evolutionary Applications* 1: 631–644.

Marshall JF. 1938. *The British Mosquitoes*. British Museum (Natural History), London, UK.

Matheson R. 1944. *Handbook of the Mosquitoes of North America*, 2nd ed. Comstock, Ithaca, NY.

Mattingly PF. 1965. Intercurrent resting, a neglected aspect of mosquito behaviour. Cahiers ORSTOM. *Entomologie Medicate*, 34: 187.

Mayagaya VS, Michel K, Benedict MQ, Killeen GF, Wirtz RA Ferguson HM & Dowell FE. 2009. Non-destructive determination of age and species of *Anopheles gambiae* s.l. using Near-infrared spectroscopy. *American Journal of Tropical Medicine and Hygiene* 81(4): 622–630.

Michel AP, Ingrasci MJ, Schemerhorn BJ, Kern M, Le Goff G, Coetzee M, Elissa N et al. 2005. Rangewide population genetic structure of the African malaria vector *Anopheles funestus*. *Molecular Ecology* 14: 4235–4248.

Molyneux DH. 1997. Patterns of change in vector-borne diseases. *Annals of Tropical Medicine and Parasitology* 91: 827–839.

Molyneux, D. 2003. Common themes in changing vector-borne disease scenarios. *Transactions of the Royal Society of Tropical Medicine* 97: 129–132.

Monroe A, Moore S, Koenker H, Lynch M & Ricotta E. 2019. Measuring and characterizing night time human behaviour as it relates to residual malaria transmission in sub-Saharan Africa: A review of the published literature. *Malaria Journal*, 18: 6.

Morgan JC, Irving H, Okedi LM, Steven A & Wondji CS. 2010. Pyrethroid Resistance in an *Anopheles funestus* population from Uganda. *PLoS One* 5(7):e11872.

Müller DA, Charlwood JD, Felger I, Ferreira C, do Rosario VE & Smith T. 2001. Prospective risk of morbidity in relation to multiplicity of infection with *Plasmodium falciparum* in São Tomé. *Acta Tropica* 78: 155–162.

Nijhout HF & Sheffield HG. 1979. Antennal hair erection in male mosquitoes: A new mechanical effector in insects. *Science* 206: 595–596.

Obsomer V, Titeux N, Vancustem C, Duveiller G, Pekel J-F, Connor S, Ceccato P & Coosemans M. 2013. From *Anopheles* to spatial surveillance: A roadmap through a multidisciplinary challenge. Chapter 14 in *Anopheles Mosquitoes: New Insights into Malaria Vectors*. Edited by Sylvie Manguin, 828 pages, Publisher: InTech, published July 24, 2013 under CC BY 3.0 license.

Okech BA, Gouagna LC, Beier JC, Yan G & Githure JI. 2007. Larval habitats of *Anopheles gambiae* s.s. (Diptera: Culicidae) influences vector competence to *Plasmodium falciparum* parasites. *Malaria Journal* 6: 50.

Omer SM & Cloudsley-Thompson JL. 1970. Survival of *Anopheles gambiae* Giles through a 9-month dry season in Sudan. *Bulletin of the World Health Organization* 42: 319–330.

Parker JEA, Angarita Jaimes NC, Gleave K, Mashaun F, Abe M, Martine J, Towers CE, Towers D & McCall PJ. 2017. Host-seeking activity of a Tanzanian population of *Anopheles arabiensis* at an insecticide treated bed net. *Malaria Journal* 16: 270.

Paterson HE. 1985. The recognition concept of species. In *Species and Speciation*, Vrba ES (Ed.), Transvaal Museum Monograph. No. 4, pp. 21–29. Pretoria, South Africa.

Patz JA, Daszak P, Tabor GM, Aguirre AA, Pearl M, Epstein J, Wolfe ND et al. 2004. Unhealthy landscapes: Policy recommendations on land use change and infectious disease emergence. *Environmental Health Perspectives* 112(10):1092–1098. doi:10.1289/ehp.6877.

Pinto J, Donnelly MJ, Sousa CA, Gil V, Ferreira C, Elissa N, do Rosário VE & Charlwood JD. 2003. An island within an island: Genetic structure of *Anopheles gambiae* populations on São Tomé. *Heredity* 91: 407–414.

Pinto J, Sousa CA, Gil V, Ferreira C, Gonçalves L, Lopes D, Petraraca V, Charlwood JD & do Rosario V E. 2000a. Malaria in São Tomé and Principe: parasite prevalences and vector densities. *Acta Tropica* 76: 185–193.

Protopopoff N, Mosha JF, Lukole E, Charlwood JD, Wright A, Mwalimu CD, Manjurano A. 2018. Effectiveness of a long-lasting piperonyl butoxide-treated insecticidal net and indoor residual spray interventions, separately and together, against malaria transmitted by pyrethroid-resistant mosquitoes: A cluster, randomised controlled, two-by-two factorial design trial. *Lancet* 6736: 30427–30436.

Reiner Jr RC, Perkins TA, Barker CM, Niu T, Fernando Chaves L, Ellis AM, George DB et al. 2013. A systematic review of mathematical models of mosquito-borne pathogen transmission: 1970–2010. *Journal of the Royal Society Interface* 10: 20120921.

Rohlf F & Slice D. 1990. Extensions of the Procrustes method for the optimal superimposition of landmarks. *Systematic Zoology* 39(1): 40–59. doi:10.2307/2992207.

Rowland M & Noste F. 2001. Malaria epidemiology and control in refugee camps and complex emergencies. *Annals of Tropical Medicine and Parasitology* 95: 741–754.

Sabatinelli G, Rossi P & Bellii A. 1986. Etude sur la dispersion d'*Anopheles gambiae* s.l. dans une zone urbaine a Ouagadougou (Burkina Faso). *Parassitologia* 28: 33–39.

Saul A. 1987. Estimation of survival rates and population size from mark-recapture experiments of bait-caught haematophagous insects. *Bulletin of Entomological Research* 77: 589–602.

Saul A. 2003. Zooprophylaxis or zoopotentiation: The outcome of introducing animals on vector transmission is highly dependent on the mosquito mortality while searching. *Malaria Journal* 2: 32.

Schapira A & Boutsika K. Malaria ecotypes and stratification. *Advances in Parasitology* 78: 97–167. doi:10.1016/B978-0-12-394303-3.00001-3.

Schreck CE & Self LS 1985. Bed nets that kill mosquitoes. *World Health Forum* 6: 342–344.

Service MW.1976. *Mosquito Ecology*. London: Applied Science.

Service MW. 1993. Mosquitoes (Culicidae). Chapter 5 in *Medical Insects and Arachnids*. Lane RP & Crosskey RW (Eds.). Springer Science+Business Media, Dordrecht, the Netherlands.

Seyoum A, Balcha F, Balkew M, Ali A & Gebre-Michael T. 2002. Impact of cattle keeping on human biting rates of *Anopheles* mosquitoes and malaria transmission around Ziway, Ethiopia. *East African Medical Journal* 79: 485–490.

Shimada M, Itoh T, Motooka T, Watanabe M, Shiraishi T, Thapa R et al 2014. New global forest/non-forest maps from ALOS PALSAR data (2007–2010). *Remote Sensing Environment* 155: 13–31. doi:10.1016/j.rse.2014.04.014.

Silver JB. 2008. *Mosquito Ecology: Field Sampling Methods*, 3rd edition. Springer, Dordrecht, the Netherlands.

Sinka, ME, Bangs, MJ, Manguin, S, Chareonviriyaphap, T, Patil, AP, Temperley, WH, Gething PW et al. 2011. The dominant *Anopheles* vectors of human malaria in the Asia-Pacific region: Occurrence data, distribution maps and bionomic précis. *Parasites and Vectors* 4: 89.

Smallegange RC, Schmied WH, van Roey KJ, Verhulst NO, Spitzen J, Mukabana WR & Takken W. 2010. Sugar-fermenting yeast as an organic source of carbon dioxide to attract the malaria mosquito *Anopheles gambiae*. *Malaria Journal* 9: 292.

Smith DL, Battle KE, Hay SI, Barker CM, Scott TW & McKenzie FE. 2012. Ross, Macdonald, and a theory for the dynamics and control of mosquito-transmitted pathogens. *PLoS Pathogens* 8(4): e1002588.

Smith TA, Charlwood JD, Kitua AY, Masanja H, Mwankusye S, Alonso PL & Tanner M. 1998. Relationships of malaria morbidity with exposure to *Plasmodium falciparum* in young children in a highly endemic area. *American Journal of Tropical Medicine and Hygiene* 59: 252–257.

Smith TA, Charlwood JD, Takken W, Tanner M & Spiegelhalter M. 1995. Mapping the densities of malaria vectors within a single village. *Acta Tropica* 59: 1–18.

Smith TA, Chitnis N, Penny M & Tanner M. 2017. Malaria modelling in the era of eradication. *Cold Spring Harbor Perspectives in Medicine*. doi:10.1101/cshperpect.a025460.

Smith, DL, Smith TA & Hay SI. 2009. Measuring malaria for elimination. Chapter 7 in *Shrinking the Malaria Map: A Prospectus on Malaria Elimination*, Feachem RGA, Phillips AA & Targett GA (Eds.). The Global Health Group, Global Health Sciences, University of California, San Francisco, CA.

Sota T & Mogi M. 1989. Effectiveness of zooprophylaxis in malaria control: A theoretical inquiry, with a model for mosquito populations with two bloodmeal hosts. *Medical and Veterinary Entomology* 3: 337–345.

Sousa CA, Pinto J, Almeida PG, Ferreira C, do Rosario VE & Charlwood JD. 2001. Dogs as a favoured host of *Anopheles gambiae* sensu stricto (Diptera, Culicidae) of São Tomé, West Africa. *Journal of Medical Entomology* 38: 122–125.

Southwood TRE. 1977. Habitat, the templet for ecological strategies? *Journal of Animal Ecology* 46: 337–365.

Southwood, TRE. 1978. *Ecological Methods: With Particular Reference to the Study of Insect Populations.* Chapman & Hall, London, UK.

Steffen W, Rockström J, Richardson K, Lenton TM, Folke C, Liverman D, Summerhayes CP et al. 2018 Trajectories of the Earth system in the anthropocene. *Proceedings of the National Academy of Sciences* 115: 8252–8259.

Strode C, Donegan S, Garner P, Enayati AA & Hemingway J. 2014. The impact of pyrethroid resistance on the efficacy of insecticide-treated bed nets against African anopheline mosquitoes: Systematic review and meta-analysis. *PLoS Medicine* 2014; 11: e1001619.

Takken W, Charlwood JD, Billingsley PF & Gort G. 1998. Dispersal and survival of *Anopheles funestus* and *A. gambiae* s.l. (Diptera, Culicidae) during the rainy season in southeast Tanzania. *Bulletin of Entomological Research* 88: 561–566.

Tangena J-AA, Thammavong P, Hiscox A, Lindsay SW & Brey PT. 2015. The human-baited double net trap: An alternative to human landing catches for collecting outdoor biting mosquitoes in Lao PDR. *PLoS One* 10(9): e0138735. doi:10.1371/ journal.pone.0138735.

Taylor B. 1975. Changes in the feeding behaviour of a malaria vector, *Anopheles farauti* Lav., following use of DDT as a residual spray in houses in the British Solomon Islands Protectorate. *Transactions of the Royal Society of London* 127: 277–292.

Teklehaimanot HD, Teklehaimanot A, Kiszewski A, Sacrramento HR & Sachs JD. 2009. Malaria in São Tomé and Príncipe: On the brink of elimination after three years of effective antimalarial measures. *American Journal of Tropical Medicine & Hygiene* 80: 133–140.

The *Anopheles gambiae* 1000 Genomes Consortium. 2017. Genetic diversity of the African malaria vector *Anopheles gambiae*. *Nature* 552: 96–100. doi:10.1038/nature24995.

Tirados I, Gibson G, Young S & Torr SJ. 2011. Are herders protected by their herds? An experimental analysis of zooprophylaxis against the malaria vector *Anopheles arabiensis*. *Malaria Journal* 10: 68.

Tonnang HEZ, Kangalawe RYM & Yanda PZ. 2010. Predicting and mapping malaria under climate change scenarios: the potential redistribution of malaria vectors in Africa. *Malaria Journal* 9: 111.

Trape JF, Tall A, Diagne N, Ndiath O, Ly AB, Faye J, Dieye-Ba F et al. 2011. Malaria morbidity and pyrethroid resistance after the introduction of insecticide-treated bednets and artemisinin-based combination therapies: A longitudinal study. *Lancet Infectious Diseases* 11.

Vazquez-Prokopec G. 2009. A new, cost-effective, battery-powered aspirator for adult mosquito collections. *Journal of Medical Entomology* 46: 1256–1259.

Vicente JL, Clarkson CS, Caputo B, Gomes B, Pombi M, Sousa CA, Antao T et al. 2017. Massive introgression drives species radiation at the range limit of *Anopheles gambiae*. *Nature Scientific Reports* 7: 46451.

Vicente JL, Sousa CA, Alten B, Caglar SS et al. 2011. Genetic and phenotypic variation of the malaria vector *Anopheles atroparvus* in southern Europe. *Malaria Journal* 10: 5.

Walsh J, Molyneux D & Birley M. 1993. Deforestation: Effects on vector-borne disease. *Parasitology* 106(S1): S55–S75. doi:10.1017/S0031182000086121.

White NJ. 2003. Malaria. Chapter 73 in *Manson's Tropical Diseases*, 22nd edition, Cook GC & Zumla AI (Eds.). Saunders Elsevier, Amsterdam, the Netherlands.

WHO. 1975. Manual on practical entomology in malaria Vol 1 + 2. Offset publication 13 World Health Organization, Geneva, Switzerland.

WHO. 1982. Manual on environmental management for mosquito control with special emphasis on mosquito vectors. WHO offset publication No. 66. WHO, Geneva, Switzerland.

WHO. 1991. Basic malaria microscopy (part I and II). World Health Organization, Geneva, Switzerland.

WHO. 2013a. Test procedures for insecticide resistance monitoring in malaria vector mosquitoes. World Health Organization, Geneva, Swizerland.

WHO. 2013b Malaria entomology and vector control. Guide for participants. WHO, Geneva, Switzerland.

WHO. 2015a. Achieving the malaria MDG target: Reversing the incidence of malaria 2000–2015. World Health Organization, Geneva, Switzerland.

WHO. 2015b. Global Technical strategy for Malaria 2016–2030. WHO, Geneva, Switzerland.

WHO. 2017. Global vector control response 2017–2030. WHO, Geneva, Switzerland.

WHO COP24. 2019. Special report health and climate change. WHO, Geneva, Switzerland.

Wilkes TJ, Matola YG & Charlwood JD. 1996. *Anopheles rivulorum*, a vector of malaria in Africa. *Medical and Veterinary Entomology* 10: 108–110.

Wilson AL, Chen-Hussey V, Logan JG & Lindsay SW. 2014. Are topical insect repellents effective against malaria in endemic populations? A systematic review and meta-analysis. *Malaria Journal* 13: 446.

Woolhouse ME, Dye C, Etard JF, Smith T, Charlwood JD, Garnett GP, Hagan P et al. 1997. Heterogeneities in the transmission of infectious agents, implications for the design of control programs. *Proceedings of the National Academy of Sciences* 94: 338–342.

Zippin C. 1956. An evaluation of the removal method of estimating animal populations. *Biometrics* 12: 163–189.

Zippin C. 1958. The removal method of population estimation. *Journal of Wildlife Management* 22: 82–90.

Bibliography

Babiker HA, Randford-Cartwright LC, Currie D, Charlwood JD, Billingsley PF, Teuscher T & Walliker D. 1994. Random mating in a natural population of the malaria parasite *Plasmodium falciparum*. *Parasitology* 109: 413–421.

Becker N, Petric D, Zgomba M, Boasea C, Dahl C, Madona M & Kaiser A. 2010. *Mosquitoes and Their Control*, 2nd edition. Springer-Verlag, Berlin, Germany.

Beltz LA 2011. *Emerging Infectious Diseases: A Guide to Diseases, Causative Agents, and Surveillance*. John Wiley & Sons, San Francisco, CA.

Caputo B, Santolamazza F, Vicente JL, Nwakanma DC, Jawara M et al. 2011. The 'Far-West' of Anopheles gambiae Molecular Forms. *PLoS One* 6(2): e16415. doi:10.1371/journal.pone.0016415.

Charlwood JD & Jolley DJ. 1984. The coil works (against mosquitoes in Papua New Guinea). *Transactions of the Royal Society of Tropical Medicine & Hygiene* 78: 678.

Charlwood JD, Lopes J & Whalley PC. 1982. Light intensity measurement and the biting behaviour of some sylvatic mosquitoes of the Amazon basin (Diptera, Culicidae). *Acta Amazonica* 12: 61–64.

Charlwood JD, Paru R, Dagaro H & Lagog M. 1986d. The influence of moonlight and gonotrophic age on the biting activity of *Anopheles farauti* (Diptera, Culicidae) from Papua New Guinea. *Journal of Medical Entomology* 23: 132–135.

Charlwood JD, Tomás EVE, Cuamba N & Pinto J. 2015b. Analysis of the sporozoite ELISA for estimating infection rates in Mozambican anophelines. *Medical and Veterinary Entomology* 29: 10–16.

Charlwood JD. 1976. Mating in mosquitoes. PhD thesis. University of Sussex, Brighton, UK.

Charlwood JD. 1979a. Estudos sobre a biologia e habitos de *Culex quinquefasciatus* say em Manaus Amazonas Brasil. *Acta Amazonica* 9: 271–278.

Charlwood JD. 1979b. Observações sobre o comportamento de acasalamento de *Culex quinquefasciatus* Say (Diptera, Culicidae). *Acta Amazonica* 9: 463–470.

Charlwood JD. 1984a. Factors affecting the assessment of man-biting rates of malaria vectors. In *Proceeding of a Conference to Honour* R.H. Black pp. 143–152.

Charlwood JD. 1984b. Which way now for malaria control? *PNG Medical Journal* 27: 159–162.

Charlwood JD. 1985. The influence of larval habitat on the ecology and behaviour of females of the *punctulatus* group of *Anopheles* mosquitoes in Mosquito Ecology, *Proceedings of a Workshop*, Lounibos LP, Rey JR & Frank JH (Eds.). pp. 399–407.

Charlwood JD. 1986a. A differential response by *Anophelines* and *Culicines* to bed-nets. *Transactions of the Royal Society of Tropical Medicine & Hygiene* 80: 958–961.

Charlwood JD. 1986b. Survival rate variation of *Anopheles farauti* (Diptera, Culicidae) between neighbouring villages in coastal Papua New Guinea. *Journal of Medical Entomology* 23: 361–365.

Charlwood JD. 1995. Bednets, A cautionary tale. *Parasitology Today* 11: 184–185.

Charlwood JD. 1997. Vectorial capacity, species diversity and population cycles of Anopheline mosquitoes from indoor light-trap collections from a house in southeastern Tanzania. *African Entomology* 5: 93–101.

Charlwood JD. 2003a. Did Herodotus describe the first use of mosquito repellents? *Trends in Parasitology* 19: 555–556.

Charlwood JD. 2004. Roll Back Malaria: A failing global health challenge Developing a market for bed nets and insecticides is problematic. *British Medical Journal* 328: 1378.

Clements AN. 1999. *The Biology of Mosquitoes, Volume 2: Sensory, Reception, and Behaviour, 1st edition*. CABI Publishing Oxford, Wallingford, UK.

Donnelly MJ, Cuamba N, Charlwood JD, Collins FH & Townson H. 1999. Population structure in the malaria vector, *Anopheles arabiensis* Patton, in East Africa. *Heredity* 83: 408–417.

Eldridge BF. 2005. Mosquitoes, the Culicidae. Chapter 9 in *Biology of Disease Vectors*, 2nd edition, Marquardt WC (Ed.). Elsevier, Amsterdam, the Netherlands.

Gordon RM & Lavoipierre MMJ. 1962. *Entomology for Students of Medicine*. Blackwell Science, Oxford, UK.

Govella NJ, Chaki PP, Geissbuehler Y, Kannady K, Okumu FO, Charlwood JD, Anderson RA & Killeen GF. 2009. A new tent trap for sampling exophagic and endophagic members of the *Anopheles gambiae* complex. *Malaria Journal* 14: 8.

Gubler DJ. 2011. Dengue and Dengue Haemorrhagic Fever Chapter 72 *In Tropical Infectious Disease, Principles, Pathogens and Practice*, Guerrant RL, Walker DH & Weller PF (Eds.). Saunders/Elsevier, Philadelphia, PA.

Haji H, Smith T, Charlwood JD & Meuwissen JME Th. 1996a. Absence of relationships between selected host factors and natural infectivity of *Plasmodium falciparum* in an area of high transmission. *Parasitology* 113: 425–431.

Haji H, Smith T, Meuwissen JME Th & Charlwood JD. 1996b. Estimation of the infectious reservoir for *Plasmodium falciparum* in natural vector populations using oocyst size. *Transactions of the Royal Society of Tropical Medicine and Hygiene* 90: 494–497.

Hayes J & Charlwood JD. 1977. *Anopheles darlingi* evita o DDT numa área de malária resistente a drogas. *Acta Amazonica* 7: 289.

Hayes J & Charlwood JD. 1979. Dinâmica estacional de uma população de *Anopheles darlingi*, numa área endémica de malária no Amazonas. *Acta Amazonica* 9: 79–86.

Huber W, Haji H, Charlwood JD, Certa U, Walliker D & Tanner M. 1998. Genetic characterization of the malaria parasite *Plasmodium falciparum* in the transmission from the host to the vector. *Parasitology* 116: 95–101.

Kampango A, Bragança M, de Sousa B & Charlwood JD. 2013. Netting barriers to prevent mosquito entry into houses in southern Mozambique: A pilot study. *Malaria Journal* 12: 99.

Killeen GF, Tami A, Kihonda J, Okumu FO, Kotas ME, Grundmann H, Kasigudi N et al. 2007. Cost-sharing strategies combining targeted public subsidies with private-sector delivery achieve high bednet coverage and reduced malaria transmission in Kilombero Valley, southern Tanzania. *BMC Infectious Diseases* 7: 121.

Kilombero Malaria Project. 1992. The level of anti-sporozoite antibodies in a highly endemic malaria area and its relationship with exposure to mosquitoes. *Transactions of the Royal Society of Tropical Medicine and Hygiene* 86: 499–504.

Knols BGJ, Takken W, Charlwood JD & De Jong R. 1995. Species-specific attraction of *Anopheles* mosquitoes (Diptera, Culicidae) to different humans in South-east Tanzania. *Proceedings of Experimental and Applied Entomology* 6: 201–206.

Lacey LA & Charlwood JD. 1980. On the biting activities of some anthropophilic amazonian Simulidae (Diptera). *Bulletin of Entomological Research* 70: 495–509.

Lahondère C & Lazzari C.R. 2013. Thermal stress and thermoregulation during feeding in mosquitoes. Chapter 16 in *Anopheles Mosquitoes: New Insights into Malaria Vectors*. Edited by Sylvie Manguin, 828 pages, Publisher: InTech, published July 24, 2013, under CC BY 3.0 license.

LeClair C, Cronery J, Kessy E, Tomás EVE, Rowland M, Protopopoff N & Charlwood JD. 2017. 'Repel all borders': Combination mosquito nets enhance collections of endophilic *Anopheles gambiae* and *An. arabiensis* in CDC light-traps. *Malaria Journal* 16: 336.

Lees RS, Knols B, Bellini R, Benedict MQ, Bheecarry A, Bossin HC, Chadee DD et al. 2014. Improving our knowledge of male mosquito biology in relation to genetic control programmes. *Acta Tropica* 132 Suppl: S2–S11.

Lehane, M. 2005. *The Biology of Blood-Sucking Insects*, 2nd edition. Cambridge University Press, Cambridge.

Lindsay SW, Emerson MP & Charlwood JD. 2002. Professor Angelo Celli's forgotten experiments: Reducing malaria by mosquito-proofing homes. *Trends in Parasitology* 18: 510–514.

Liverani M, Charlwood JD, Lawford H & Yeung S. 2017. Field assessment of a novel spatial repellent for malaria control: a feasibility and acceptability study in Mondulkiri, Cambodia. *Malaria Journal* 16: 412.

Mattingly PF. 1969. *The Biology of Mosquito-Borne Disease*. George Allen and Unwin, London, UK.

McGavin GC. 2013. Prologue in *The Insects Structure and Function*, Chapman RF, Simpson SJ & Douglas AE (Eds.). Cambridge University Press, Cambridge.

Mendes C, Félix R, Sousa A, Lamego J, Charlwood JD, do Rosário VE, Pinto J & Silveira H. 2010. Molecular evolution of the three short PGRPs of the malaria vectors *Anopheles gambiae* and *Anopheles arabiensis* in East Africa. *BMC Evolutionary Biology* 10: 9.

Molyneux DH. 2001. Vector-borne infections in the tropics and health policy issues in the twenty-first century. *Transactions of the Royal Society of Tropical Medicine and Hygiene* 95: 233–238.

Moore CG & Frieier JE. 2005. Use of Geographic Information Systems in the study of vector-borne diseases. Chapter 16 in *Biology of Disease Vectors*, 2nd edition, Marquardt WC (Ed.). Elsevier, Amsterdam, the Netherlands.

Muirhead-Thompson RC. 1991. *Trap Responses of Flying Insects: The Influence of Trap Design on Capture Efficiency*. Academic Press, London, UK.

O'Brochta DA, Subramanian RA, Orsetti J, Peckham E, Nolan N, Arensburger P, Atkinson PW & Charlwood JD. 2006. hAT element population genetics in *Anopheles gambiae* s.l. in Mozambique. *Genetica* 127: 185–198.

Obsomer V, Defourny P & Coosemans M. 2007. The *Anopheles dirus* complex: Spatial distribution and environmental drivers. *Malaria Journal* 6: 26.

Pennington JE & Wells MA. 2005. The adult midgut: Structure and function. Chapter 21 in *Biology of Disease Vectors*, 2nd edition, Marquardt WC (Ed.). Elsevier, Amsterdam, the Netherlands.

Pinto J, Donnelly MJ, Sousa CA, Gil V, Ferreira C, Elissa N, do Rosário VE & Charlwood JD. 2002. Genetic structure of *Anopheles gambiae* (Diptera: Culicidea) in São Tomé and Príncipe (West Africa): Implications for malaria control. *Insect Molecular Biology* 11: 2183–2187.

Pinto J, Lynd A, Vicente JL, Santolamazza F, Randle NP, Gentile G, Moreno M et al. 2007. Multiple origins of knockdown resistance mutations in the Afrotropical mosquito vector *Anopheles gambiae. PLoS One* 11: e1243.

Pinto J, Sousa CA, Gil V, Gonçalves D, do Rosario VE & Charlwood JD. 2000b. Mixed-species malaria infections in the human population of São Tomé island, West Africa. *Transactions of the Royal Society of Tropical Medicine and Hygiene* 94: 256–257.

Rafael JA & Charlwood JD. 1980. Idade fisiologica, variaçao sazonal e periodicidade diurna de quatro populaçoes de Tabanidae (Diptera) no campus Universitario, Manaus, Brasil. *Acta Amazonica* 10: 907–927.

Rajan CM & Charlwood JD. 1984. *Plasmodium vivax multinucleatum* in Papua New Guinea? *Transactions of the Royal Society of Tropical Medicine & Hygiene* 78: 422.

Rejmánková E, Grieco J, Achee N & Roberts DR. 2013. Ecology of larval habitats. Chapter 13 in *Anopheles Mosquitoes: New Insights into Malaria Vectors.* Edited by Sylvie Manguin, 828 pages, Publisher: InTech, published July 24, 2013 under CC BY 3.0 license.

Roberts L. 2016. Drug resistance triggers war to wipe out malaria in the Mekong region. *Science* 10.

Russell TL, Lwetoijera DW, Maliti D, Chipwaza B, Kihonda J, Charlwood JD, Smith TA et al. 2010. Impact of promoting longer-lasting insecticide treatment of bed nets upon malaria transmission in a rural Tanzanian setting with pre-existing high coverage of untreated nets. *Malaria Journal* 9: 18.

Salgueiro P, Lopes AS, Mendes C, Charlwood JD, Arez AP, Pinto J & Silveira H. 2016. Molecular evolution and population genetics of a Gram-negative binding protein gene in the malaria vector *Anopheles gambiae (sensu lato). Parasites & Vectors* 9: 515.

Shelley AJ, Pinger RR, Roraes MAP, Charlwood JD & Hayes J. 1979. Vectors of *Onchocerca volvulus* at the river Toototobi, Brazil. *Journal of Helminthology* 53: 41–43.

Smith TA, Charlwood JD, Kihonda J, Mwankusye S, Billingsley PF, Meuwissen J, Lyimo, E, Takken W, Teuscher T & Tanner M. 1993. Absence of seasonal variation in malaria parasitaemia in an area of intense seasonal transmission. *Acta Tropica* 54: 55–72.

Sudomoa M, Ariantia Y, Wahidb I, Safruddinb D, Pedersen EM & Charlwood JD. 2010. Towards eradication: Three years after the tsunami has malaria transmission been eliminated from the island of Simeulue? *Transactions of the Royal Society of Tropical Medicine and Hygiene* 104: 777–781.

Thomsen TT, Madsen LB, Hansson HH, Tomas EVE, Charlwood JD, Bygbjerg IC & Alifrangis M. 2013. Rapid selection of *Plasmodium falciparum* chloroquine resistance transporter gene and multidrug resistance gene-1 haplotypes associated with past chloroquine and present artemether-lumefantrine use in Inhambane District, Southern Mozambique. *American Journal of Tropical Medicine & Hygiene* 88: 536–541.

Wilkes TJ & Charlwood JD. 1979. A rapid gonotrophic cycle in *Chagasia bonneae* from Brazil. *Mosquito News* 39: 137–139.

Index

Note: Page numbers in italic and bold refer to figures and tables, respectively.

A

J

K

L

M

N

O

P

R

S

T

U

V

W

Z